Are Science And Mathematics Socially Constructed?

A Mathematician Encounters Postmodern Interpretations of Science

Are Science And Mathematics Socially Constructed?

A Mathematician Encounters Postmodern Interpretations of Science

Richard C Brown
University of Alabama, USA

NEW JERSEY · LONDON · SINGAPORE · BEIJING · SHANGHAI · HONG KONG · TAIPEI · CHENNAI

Published by
World Scientific Publishing Co. Pte. Ltd.
5 Toh Tuck Link, Singapore 596224
USA office: 27 Warren Street, Suite 401-402, Hackensack, NJ 07601
UK office: 57 Shelton Street, Covent Garden, London WC2H 9HE

British Library Cataloguing-in-Publication Data
A catalogue record for this book is available from the British Library.

ARE SCIENCE AND MATHEMATICS SOCIALLY CONSTRUCTED?
A Mathematician Encounters Postmodern Interpretations of Science

Copyright © 2009 by World Scientific Publishing Co. Pte. Ltd.

All rights reserved. This book, or parts thereof, may not be reproduced in any form or by any means, electronic or mechanical, including photocopying, recording or any information storage and retrieval system now known or to be invented, without written permission from the Publisher.

For photocopying of material in this volume, please pay a copying fee through the Copyright Clearance Center, Inc., 222 Rosewood Drive, Danvers, MA 01923, USA. In this case permission to photocopy is not required from the publisher.

ISBN-13 978-981-283-524-6
ISBN-10 981-283-524-5

Printed in Singapore.

To Phyllis who supported the effort, to Andres, John, David, and Lydia who gave both of us more time, and to Tootsie who was happy to put a paw in to help.

Contents

Preface ix

Acknowledgments xvii

1. Rip van Winkle Awakes 1

2. A Golden Age and its End 19

3. Ingredients in the PIS Bouillabaisse 27

4. A Canary in the Mine 37

5. The Unmasking of Reason 45

6. Thought Styles and Thought Collectives 57

7. The Reluctant Revolutionary 63

8. Anything Goes 87

9. The Sociological Attack 107

10. The Deconstruction of Mathematics 149

11. Epistemic Issues 169

12.	The Fallibility of Conventionalism and Fallibilism	215
13.	Madison 1973	243
14.	Kto Kogo?	275
Bibliography		293
Index		307

Preface

This book is a history, analysis, and criticism of what the author calls "postmodern interpretations of science" (PIS)[1] and the closely related "sociology of scientific knowledge" (SSK). This movement traces its origin to Thomas Kuhn's revolutionary work *The Structure of Scientific Revolutions* (1962), but is more extreme. It is characterized by a belief that science is a socially constructed, "situated," historical product whose theories are generated by contextual factors such as class interest, ideology, or laboratory politics rather than nature. Moreover, instead of being a glorious and progressive achievement of the Western world since the seventeenth century it has been a deeply flawed enterprise which has degraded the environment, oppressed women, minorities, the Third World, and is presently a tool of corporate capitalism and the military industrial complex. These views are often joined to a profound anti-realism. Since nature cannot be interpreted independently of the conceptual structures scientists bring to the task, there is *no* "way the world is" apart from these structures. Scientific theories are not "caused" by nature. Instead, they *constitute* nature. So-called "scientific facts" do not correspond to mind-independent properties of the universe, but merely represent the biases of scientists. And since these biases inevitably reflect class, gender, race, or other sociopolitical factors so does what we call "nature." It follows that vaunted "objectivity" of science is an illusion. Science is an ideology like any other. At best it may have some pragmatic or technological value, but there is no epistemological reason for it to be preferred over "other ways of knowing" such as religion and myth.

Such views, which radically conflict with the positive vision of science dominant in the West from the seventeenth century at least through the

---

[1] We use the same abbreviation in either the singular or plural case.

1950s, have become quite fashionable in the humanities since the 1970s. They are now particularly to be found among faculty in English, Cultural, Women's, and Post-Colonial Studies, as well as in the relatively new discipline of Science, Technology and Society (STS). Not even the History of Science has been immune. For example, Arnold Thackray who in the 1970s became editor of *Isis*, the preeminent American journal in the history of science,[2] thought that the central questions to be decided in the history of science were:

> Can logical and historical distinctions between science and magic be sustained? Is science culture free in any but a trivial sense? Is progress or truth a useful concept around which to organize historical work? ... Is science essentially an ideology or form of oppression?[3]

Examination of several years worth of issues of *Isis* and other journals in the 1970s and 1980s supports the view that such questions (with negative answers to all but the last) were popular areas of investigation.

SSK/PIS has been attacked by Paul Gross and Norman Levitt,[4] Noretta Koertge,[5] Alan Sokal and Jean Briccmont,[6] and others. The resulting conflicts comprised the so-called "Science Wars of the 1990s". These authors view the entire movement as an intellectual freak show and concentrate on demolishing it. They emphasize "extremists," especially feminists, Afrocentrists, deep ecologists, and radical social constructivists, or in the case of Sokal and Bricmont the more exotic and unintelligible representatives of French post-structuralism (such as Derrida, Foucault, and Lacan). While the author basically agrees with this point of view, this book is intended to be more than another broadside in the now nearly exhausted topic of the Science Wars. The authors cited above have done their job well, and it is not necessary to repeat it. Instead, we will try to understand the emergence of a family of attitudes towards science which represents as great (though hopefully not as potent) an intellectual change as did the Enlightenment. What are its intellectual origins? What were its causes? Was it the product of some social or political transformation peculiar to the United States or

---

[2] Equivalent in prestige, say, to the *Transactions of the American Mathematical Society* in Mathematics.
[3] Quoted in Novick (1988), p. 537.
[4] *Higher Superstition* (1994)
[5] *A House Built on Sand* (1998)
[6] *Intellectual Impostures* (1998)

Europe in the 1960s? Is SSK/PIS an early warning of possible political consequences in the same sense as the writings of the *philosophes* of eighteenth century France were?

To properly deal with such questions we would ideally require something equivalent to E. J. Dijsterhuis' magisterial *Mechanization of the World Picture: Pythagoras to Newton* (1961), but devoted to SSK/PIS instead of to the development of the mechanical philosophy in the sixteenth and seventeenth centuries. This is beyond our power; perhaps also the ignobility of the subject is unworthy of such effort. Instead, we will mainly concentrate on offering a series of "snapshots" of people whose doctrines illustrate what we think are the main intellectual stages in the progression to full-blown SSK/PIS—roughly in the period 1925 to 1975. To this end, we will look at the thought of Karl Popper, Karl Mannheim, Ludwik Fleck, Thomas Kuhn, Paul Feyerabend, David Bloor, Steve Shapin, Harry Collins, Bruno Latour, and several others, as well as PIS-like doctrines in mathematics. We will also describe various philosophical contributions to PIS ranging from the Greek sophists to 20th century post-structuralists. Our focus on the most important actors, however, means that detailed consideration of several thinkers who are at least "fellow travelers" of the movement will be omitted. We will have only a few remarks, therefore, concerning French poststructuralism, the American philosophers Richard Rorty and Willard Quine, and the wider phenomenon of postmodernism as it has affected other areas of culture.[7]

But PIS is not only an event in late twentieth century intellectual history, it is also deeply political. Therefore, an equally important goal will be to understand the social and political environment that allowed it to flourish. Ironically, our method here will be similar to that which SSK uses to deconstruct science. To use the language of Karl Mannheim we wish to "unmask" PIS and reveal the hidden "extra-theoretic" goals and needs it serves. Specifically, we will argue that the disturbed political atmosphere of the Vietnam war era was a critical catalyst in the rise of PIS. This is hardly a novel thesis, but the documentation for it is ample and worth presenting.

The motivation for writing the book is autobiographical. From 1958 until 1963 I had been in fairly close contact as a Berkeley history student with Thomas Kuhn.[8] My interests, however, changed from history to mathematics in 1967, and I have had a conventional, reasonably successful research

---

[7]But see Chapter 13.

[8]My generation of graduate students included Paul Forman, J. L. Heilbron, and Karl Hufbauer, who all went on to distinguished careers.

oriented career in this discipline until my retirement from the University of Alabama in 2002. In 1996 the Sokal hoax and the Gross and Levitt book came to my attention, and in the summer of that year I was saddened to learn of Kuhn's death and impressed by lengthly obituaries he received. These events made me aware of the Science Wars. As was the case for most scientists, they had been completely invisible to me as my intellectual interests were mostly limited to my own narrow technical research. Despite my background, I no longer paid any attention to what was going on in the humanities. Therefore it was a shock to learn of the profound and negative change—representing a kind of counter Enlightenment—in the vision of science on the part of many humanistic intellectuals which had developed since my days as a graduate student in history over 30 years earlier. This stimulated my curiosity and made me want to find out what exactly had happened and why.

The effort to write the book has been immense. It required catching up in a year or two on forty years of unread books and articles—clearly a daunting task. Much of it I found to be junk, which to quote an apocryphal mathematical review filled "a much needed gap in the literature"—despite in some cases having been written by holders of endowed chairs at the better universities.[9] Few of the ideas in this literature had the difficulty of even elementary mathematical concepts found in undergraduate textbooks. Most, in fact, if not nonsense, were fairly trivial—their appearance of profundity being due to the obscurity of expression rather than to any inherent worth. But there was a lot of material to sift through! To my amazement I soon discovered that thousands of articles and hundreds of books with a PIS point of view had been written since the 1960s. In the face of this astounding mass of verbiage, I sometimes felt like a ragged urchin—whom one sometimes sees in news magazine photos—picking through an immense pile of garbage, near some third world city, but finding little of value. Yet at the same time, some of the earlier writers who seemed to have inspired PIS had articulated ideas that needed serious attention, regardless of one's opinion of their validity. There is a difference between the work of thinkers such as Mannheim, Fleck, or Kuhn and the tirades of some feminist or Afrocentric assistant professor who may cite them.

---

[9] The quotation has often been attributed to a review of a bad mathematical paper by a famous mathematician, for example Paul Halmos. It does not however ever seem to have appeared in print. For its amusing history see Jackson (1997). Branding something "junk" does not mean that I merely disagree with it. I have read much brilliant high quality scholarship with deeply offensive theses. Rather, "junk" means that the work is unintelligent and would not have earned a B- in a Berkeley seminar c. 1960.

Despite its serious intentions, some—to use fashionable terminology—may find that my book is "transgressive." It is written by an outsider and violates disciplinary boundaries. Its author is neither a sociologist, philosopher, or historian of science and has had only a slight exposure to these disciplines in the 1960s. This means that there will be errors. The author has no doubt that important sources have been missed and others misinterpreted. These shortcomings as well as his general point of view may offend specialists. For such errors and naiveté, the author apologizes and takes full responsibility. Yet, he would be prepared to defend the overall truth of the main theses of the book, and consoles himself with the fact that his errors may not be as bad as those made by postmodernists when they discourse on such things as nonlinearity, Gödel's theorems, Relativity, Chaos Theory, Quantum Mechanics or Topology. These forays into the sciences and mathematics are almost always at least technically wrong and often pure nonsense.[10] More generally, those who attempt to create a "science of science" and who argue that science is "socially constructed" and should not be "epistemologically privileged" over other cultural practices such as literary criticism or shamanism generally commit worse errors than any the author has in his book. I have never ceased to be amazed that those without *any* hands on experience of scientific or mathematical research—or even any concrete knowledge of the details of a technical field can write with absolute self-confidence and authority on the nature of science or mathematics.[11] Such chutzpah exceeds anything in this book!

To sketch the organization of the book: we begin with a semi-autobiographical chapter which explains in greater detail how the book came to be written. This chapter also summarizes the theses of SSK/PIS in the form of fifteen propositions to which we will often refer. Chapter 2 discusses the positive image of science in the first two decades of the Cold War as well as logical positivism, the philosophy supporting that image. We

---

[10] This has been convincingly demonstrated to the satisfaction of anyone who has any understanding of these subjects in the book in *Higher Superstition*. See also Sokal and Bricmont's *Intellectual Impostures*. Especially enjoyable examples of technical incompetence may be found in Bruno Latour's treatment of Relativity (Latour (1988b)) or in N. Katherine Hayes' book *Chaos Bound* (1990).

[11] A historian of science once justified this practice to me by claiming that "it is not necessary to be old to be specialist in geriatrics." I leave the reader to decide if this analogy is apt. There are some honorable exceptions, however. Kuhn himself was a well trained physicist as was his student J. L. Heilbron. In fairness we should also admit Steve Shapin and David Bloor with whose ideas the writer strongly disagrees have also had scientific training. Yet in comparison with the vast army of SSK/PIS advocates they are exceptions that prove the rule.

also describe the heady post-Sputnik scientific climate of the 1960s and its abrupt decline after 1970. Chapter 3 is a survey of some of the philosophical resources exploited by PIS together with the intellectual changes in the early twentieth century that may have prepared the way for it. Chapter 4 focuses on Karl Popper who has been accused of being the hidden father of PIS. We disagree and see him only as a relatively indirect harbinger of the changes to come. Chapter 5 shifts emphasis away from both philosophy and the philosophy of science to sociology where we think the main inspiration for PIS lies and centers on Karl Mannheim whose sociology of knowledge (SK) originated many key concepts which SSK was able to apply to science. There follow chapters on Ludwik Fleck, Thomas Kuhn, and Paul Feyerabend. We see these individuals, together with Mannheim, as constituting the most direct route leading to PIS. Only in Chapter 9 do we attempt to unravel the tangled nest of ideas that characterizes SSK and PIS. We concentrate on Michael Mulkay, Steve Woolgar, Bruno Latour, Harry Collins, and particularly on the creators of the so-called "Edinburgh Strong Program," principally David Bloor and Barry Barnes. Attention is also given to Steve Shapin and Simon Schaffer in the form of an analysis of their provocative book, published in 1985, *Leviathan and the Airpump*.[12] We then give several examples to show how standard PIS attitudes have negatively affected the images of "great scientists," especially Newton and Pasteur. In Chapter 10 we look at the application of PIS to mathematics. This focuses on the work of David Bloor, Sal Restivo, Imre Lakatos, Paul Ernest, Reuben Hersh, and the pronouncements of Ludwig Wittgenstein on the nature of mathematics. Through Chapter 10 we have tried to present the ideas involved in PIS or SSK as straightforwardly and fairly as possible. Although our unfavorable opinion of many of them is probably evident, serious criticism is reserved for the last three chapters. Since PIS is also linked to perennial philosophical issues, we look in detail at some of these in Chapters 11—14. Our attitude is one of semi-agnosticism. We agree with Colin McGinn, that no serious philosophical problem can be "solved." This means that the anti-realism SSK and PIS generally invoke can neither be refuted or proven, as indeed is the case with scientific realism. But although this is so, we personally feel that the consequences of the philosophy behind PIS are more confused and unbelievable than those of scientific realism. More fundamentally, we argue that anti-realism considered as a philosophical doctrine will not support the weight PIS puts upon

---

[12] Recently awarded the Erasmus Prize by the Dutch Government.

it; it is perfectly compatible with a view, e.g., in its Machian or Copenhagen interpretation form, that scientific theories are tools serving to organize experimental data, "save the appearances," and make predictions. Further, we think that the characteristic PIS theses that observations are completely "theory-laden," that carefully done experiments cannot falsify a scientific theory, that science consequently is the product of "political" negotiation, that nature is a social construct, etc., etc., are simply false. Chapter 13 examines the problem of historical causation. Why has SSK/PIS achieved the influence it did when it did? Why did its ideas which had been anticipated for a long time "jell" to become a political force only after the mid-1960s? To answer this question it is necessary, as we have already remarked, to apply the sociology of knowledge to the ideas of PIS or SSK, that is, to explain the social and political functions they serve and why they have emerged. Having done this, we are licensed in true SK fashion to "unmask" or "disintegrate" them, without taking the arguments of their defenders too seriously. In Chapter 14 we review some of the points made earlier and close with a speculative (and deeply Machiavellian) analysis of the possibility of a new outbreak of the Science Wars resulting from of a collision of interests between PIS and institutional science. Such an outbreak will become inevitable if, as we fear is already happening, science bureaucrats, funding agencies, and activists exploit the ideology of PIS for political ends.

After completing most of the composition of our book, we discovered that it covers some of the same ground as John H. Zammito's magisterial *A Nice Derangement of Epistemes: Post-Positivism in the Study of Science from Quine to Latour* (2004). By "post-positivism" Zammito means roughly what we regard as the union of PIS and SSK. Zammito seems to have read virtually every primary and secondary source concerning these movements and offers a post-Quine intellectual history whose thoroughness and detail we cannot match. Yet, there are differences in point of view, emphasis, and content between our work and his. Zammito seems to regard the emergence of post-positivism as purely a matter of change within the philosophy of science. With the exception of a few perceptive remarks[13] he more or less omits the political aspect. Also, he sees the work of Quine as critical, while we view Quine as a secondary figure. Instead, as we have already mentioned we trace the origins of SSK/PIS to Karl Mannheim and analyze his thought in some detail. Zammito does mention Mannheim in

---

[13] For example, Zammito (2004), p. 124.

several places,[14] but for him Mannheim's importance is not central. We also have chapters on Ludwik Fleck whom Zammito omits and mathematics, a subject which Zammito does not consider. Nevertheless, Zammito's book is excellent! We recommend it as a supplement or corrective to our own.

Two final remarks: the University called "Persepolis State" which will appear fairly frequently in this book is *not* the author's own University. It is intended as a Platonic archetype of a standard public university or redbrick found in the US and UK. As such, it is a distillation of several well-known institutions: for instance, Paterno University in College Stadium, Pennsylvania, Volunteer State University in Fort Sanders, Tennessee, Hogarth Polytechnic and the University of Rummidge, which are located respectively in Royal Pavilion and Rummidge, UK, and even a bit of Euporia State University's campus by the bay in Plotinus, California. Likewise, any reference to a Professor, Chairperson, Dean, Provost, or other administrator does not describe any person living or dead, with or without tenure.

---

[14] *Ibid.*, pp. 124, 126, 138–140.

Acknowledgments

The author has benefitted from the support and encouragement of many people in writing this book. He wishes in particular to thank Max Hocutt, James Cook, David Edmunds, Maarten Ultee, and Zac Dover. They and the author did not always agree, but their conversations on many of the issues considered here have been stimulating and helpful. He is also grateful to his beloved wife Phyllis not only for her input and careful proofreading of the manuscript, but for her toleration of a messy house, littered with books and reprints, and of prolonged absences at the computer. Appreciation is also due to the staff of Gorgias Library of the University of Alabama, especially to Barbara Brosier and Pat Causey who were of great assistance in procuring material for my research and who forgave many late returns. Finally, special thanks to his editor Ms. Lai Fun Kwong and his technical advisor Rajesh Babu of World Scientific Publishing Co. for their assistance in preparing the manuscript for publication.

Chapter 1

Rip van Winkle Awakes

In the early summer of 1996 my wife and I were spending a few weeks in Brighton where I was visiting a colleague at the University of Sussex and writing a now nearly forgotten paper on weighted Sobolev embeddings. Each morning it was my habit to gratify the local Pakistani newsagent by purchasing the three major British newspapers, the *Guardian*, *Times*, and *Telegraph*, supplemented if I were in a raffish mood by the *Sun*. I enjoyed savoring the various ideological wars conducted in these newspapers, or to use postmodernist language the "incommensurable discourses" found in them. On a particular June morning a large photograph of Thomas Kuhn was on the front page of each newspaper with the announcement that he had passed away at age 73. Multi-column obituaries followed.

This brought back vivid three decade old memories. Between 1958 and 1963 while a student at Berkeley I had taken both an advanced undergraduate course and a graduate seminar from Kuhn. Although I was an immature somewhat uneven student, Kuhn treated me with consideration and patience. We were not close friends, but I had been to his home, briefly worked as one of his research assistants, and had frequent conversations with him in his office. On one of these occasions in 1963 the telephone rang with an offer from Princeton. Prior to the publication of *The Structure of Scientific Revolutions* in 1962 Kuhn, although respected for his earlier book *The Copernican Revolution* (1957), was a relatively ordinary member of the Berkeley faculty, holding a joint appointment as Associate Professor in the History and Philosophy Departments. He had come from Harvard in 1957 where he had failed to achieve tenure. Probably Michael Polanyi (1891–1976), Stephen Toulmin (1922–), Norwood Russell Hanson (1924–67), and Paul Feyerabend (1924–94)—who after 1958 was also in the

Berkeley Philosophy Department—were then more established than Kuhn.[1] I can recall clearly when *Structure* was published in 1962. Like many of Kuhn's students, I immediately bought a copy which I still own. I think that I understood it imperfectly, but could tell that it had a highly original thesis. I remember arguing about it with other graduate students. It is fair to say that none of us, probably not even Kuhn himself, realized that the book would become the single most influential work in the history of science in the twentieth century.[2] I left Berkeley in 1965, spent a few years of teaching history and philosophy at small institutions, and soon lost contact with Kuhn. By 1967 I had turned to the serious study of mathematics, and had almost forgotten about the history of science. The only evidence that I had ever studied the subject is a book collection in my library that is reasonably up to date as of 1964.

By 1996 I had been a mathematician for nearly 30 years and was pursuing a quite conventional and modestly successful career of teaching and research at a Southern university. The life, in spite of the endless cycle of teaching calculus to a mostly captive audience of uninterested and ill prepared students, was not a bad one. There was, after all, the occasional good student who, however briefly, made teaching seem worthwhile. Telling students about derivatives, integrals or related rate problems was less stimulating than telling them about Robespierre, the Committee of Public Safety, or the September Massacres as I had done decades earlier. On the other hand, although I took my teaching duties seriously, the very difficulty of the subject meant that the average student in my classes could directly experience something resembling the atmosphere of the Republic of Virtue and Terror without having to read about it. The grading was also easier than in the humanities. One did not have to spend endless hours reading mind rotting essays—or in the rare case that they were of some worth trying determine if they had been downloaded from the internet or purchased from a term paper mill.[3] A red pen and the political will to write a large number of zeros in the margin was all that was required. My teaching duties thus allowed me time to concentrate on my research which gave me great satisfaction. The job also entailed many travel opportunities. Because my previous interests had in part been directed to European history, I especially sought out mathematical contacts and travel opportu-

---

[1] Fuller (2000), p. 5.
[2] Karl Popper's *Logic of Scientific Discovery* probably ran second.
[3] Popular ones at Persepolis State include "CheatHouse.com", "School Sucks.com," or "Paperstore.net."

nities in both Eastern and Western Europe. Yet, although I enjoyed looking into the political and cultural history of the places I happened to visit,[4] I was—like most mathematicians and scientists—unconsciously living in a hermetically sealed mental world. My subject was difficult, perhaps more so for me than for younger colleagues, since I had begun graduate work in it at the relatively late age of 28. Teaching, refereeing, committee work, not to mention the occasional *coup d'état* plotted or narrowly averted in the Department, plus thousands of hours spent writing research papers consumed my time and energy. Even though I had started out in the humanities and knew a little intellectual history I was totally unaware of the current Zeitgeist of many humanistic intellectuals. From the newspapers I had seen the word "postmodernism." I knew that it was controversial, had something to do with the humanities, and in some fashion was associated with an occult, possibly disreputable, practice of English faculty known as "deconstruction." Beyond this, it was just a word; I had no more awareness of its meaning than of some obscure technical term found in a chemistry or accounting textbook. In this ignorance I was only slightly better off than most of my mathematical and scientific colleagues. I may have remembered Tom Kuhn as a person quite clearly, while only a few other members of the Department had even heard of him. But none of us were aware of Bruno Latour, the Edinburgh Strong Program, or could define "social constructivism."[5] We all certainly knew about "paradigm shift," but only because it functioned as a cliché in automobile or deodorant advertisements, or as a phrase much employed by political op-ed writers.[6]

In a medium-sized University such as mine, there may be a thousand or more faculty in the various Departments of the College of Arts and Sciences, many of them intelligent and worth knowing. The increasingly difficult academic market in the 1970s in almost all the core subjects necessarily and rapidly increased the quality of those who managed to land (and keep) tenure track jobs. Even at second-tier institutions like mine the new faculty were easily comparable to many (if not to the very best) of the faculty hired at higher status universities in the period from 1940 to the late 1960s.[7]

---

[4]For instance, the origins of the Iron Guard in Jassy Romania, the deadly politics of 1930s Bucharest, or Goering's authorization of Oberwolfach in 1944.

[5]This ignorance was not just a characteristic of science faculty at my university; I also found it among colleagues at Wisconsin or Pennsylvania State.

[6]As in "Bush's Iraq policy represents a paradigm shift ...." An occasional Dean or Provost was also fond of the term to describe his "reforms" or latest Five Year Plan, which in practice meant more meaningless busy work for the faculty.

[7]In the mid-1960s academic jobs were absurdly easy to get, especially in mathematics

But unless one was on some official committee there was neither the time or opportunity to meet them; in my case as I aged other departments, located on the far side of campus from the mathematics building, gradually became too distant to visit on foot, and parking on an ever more crowded campus became impossible. My declining energy, moreover, had to be restricted to research and teaching to the exclusion of socialization. For most research active faculty in the sciences or mathematics, one's friends were mainly colleagues in the same research area. Only a few at most would be at one's department or university. Instead, they might be found in such places as Kent State, Texas A&M, Lima, Prague, Manchester, Plovdiv, Sussex, Zagreb, or Novosibirsk. Mutual contact was restricted to meetings, exchange of preprints, email, and in a few cases sabbatical visits.[8]

Presumably a similar situation must have been true for people in the humanities. But they usually had not the training to understand the *technical* content of science and mathematics, and so were probably even more isolated from the science faculty than the science faculty was from them. This is not intended as a criticism of humanists. An equally profound mutual isolation is often true of scientists with respect to technical research areas other than their own, and is an unavoidable consequence of the incredible specialization and opaqueness of high level research within science. A common joke among mathematicians, for instance, is that two individuals could have exactly the same AMS-MOS research classification number and still be totally unable to understand each other's work. This phenomenon means that the reasons for attending mathematical conferences are usually quite different from their officially stated purpose. One does not participate in order to listen to or absorb the papers being presented by other mathematicians. Instead major (but unacknowledged) goals include securing an invitation as a plenary speaker (a valuable sign of academic worth to administrators), the informal exchange of ideas and gossip with old friends, hiring and networking, the consumption of interesting food and drink in exotic foreign locations,[9] and perhaps of greatest value a brief Cinderella-like escape from students, administrators, and the micropolitics of one's department.

In theory, a scientist or mathematician can, provided he can spare the

---

and science. As a consequence, many of the people hired then were by all objective measures weaker than those struggling to find academic jobs a generation later.

[8] The miracle of email in particular meant that one could conduct joint research or even write a paper with someone one had *never* physically met.

[9] Very valuable in this respect were the wine lockers (situated in the lecture rooms) and afternoon teas at the German mathematical institute Oberwolfach.

time, appreciate an occasional production in history or philosophy if it is written in some approximation to normal literary English. But the same is just not possible for the humanist vis à vis, say, mathematics or organic chemistry except on the most popular and non-technical level. Even so, this is a rare accomplishment for a scientist. In the first place, he hardly ever *has* the time. Secondly, many of the humanities have (aping the sciences) developed an opaque, painful to read, jargon of their own. As a consequence, the average scientist or mathematician is more on the level of the *History Book Club* or *American Heritage* magazine than the *American Historical Review*. In his spare time he might read a novel by Hardy or Conrad, but not Terry Eagleton's or Frank Lentricchia's books on literary theory. Thus it is not easy for him to be aware of important intellectual currents in the humanities; C. P. Snow's "two cultures" are evidently still alive, well, and almost totally failing to communicate with each other. Despite my own background in the humanities I, in particular, did not have the slightest idea of what was going on either in the humanities departments of my university or in the disciplines they studied.

Like almost all scientists and mathematicians, I had a simple and straightforward view of the nature of science. Any perturbation caused by reading *The Structure of Scientific Revolutions* more than thirty years earlier had long since dissipated. It was clear to me that science produced objective, reliable knowledge about a world whose properties exist independently of both the scientist and the ambient social, religious, or political atmosphere in which he or she is immersed. Neutrons are exactly the same for liberal democrats, voters for George W. Bush, Muslim fundamentalists, and North Korean communists. To be sure, on rare occasions science has been contaminated by ideology, for instance, under both communist and fascist regimes; but this has been universally acknowledged to have been fatal to its integrity. Science, furthermore, achieves its goals by reliance on "scientific method." I was aware that philosophers could and did quibble about its details as well as other issues such the ontological status of theoretically posited but unobservable entities; yet as a whole, the method was fairly obvious. Scientists, by careful observation and rational reflection, accumulate evidence and formulate theories in order both to explain known and to predict new phenomena. Often, especially in the physical sciences, this involves the construction of mathematical models which are idealizations of the concrete situation in which the essential features become mathematical abstractions and the less important are ignored.[10] These theories and predictions are then tested by well-designed experiments. If they fail such tests they are rejected. If experiments verify them they are provisionally accepted. While no amount of experimental evidence can ever absolutely verify a theory, the better the verification, the greater in some sense is the probability that the theory is correct.[11] Especially well verified theories may be regarded as yielding "laws of nature." Furthermore, some discrete claims about nature are so well verified, such as e.g., the earth is not flat and is approximately spherical, the blood circulates, or the moon is about 60.263 earth radii from the earth that they become absolute truths or "facts." Both laws of nature and facts represent permanent features of the world which existed before human beings discovered them, just as the North American continent existed before 1492 or the earlier Viking voyages. These features are also independent of the human mind. Mt. Everest or positrons are what they are regardless of our thoughts or belief structure about them. Of possibly greater importance, however, than the exact details of scientific method is the moral *objectivity* of science. This demands that the scientist separate his politics, desires, ego, gender, social position, and all other extraneous features from his work. In particular, theories should never be judged in terms of their ideological consequences. These may indeed be present, but they have nothing to do with scientific adequacy, which can only be tested by observation and experiment. On occasion, this neutrality of the scientist concerning social or political factors will be difficult or even impossible to attain. But an honest attempt must be made; only in this way does the resulting science have any chance of universality.

---

[10]For example, the mathematical model of projectile motion essentially due to Galileo yields the equations

$$H(t) = (v_0 \sin \theta)t - gt^2/2$$
$$D(t) = (v_0 \cos \theta)t,$$

where H, D, θ, v_0, and g stand respectively for the height, horizontal distance traveled, angle of release (with respect to the D axis), velocity of release, and acceleration of gravity. This model ignores air resistance, the mass of the projectile, the attraction of the moon, the inverse square law of gravitational attraction, the curvature of the earth, temperature, and many other factors. Hence, while the equations are easily solved the solution is not very accurate, except for heavy dense projectiles at low velocity and close to the earth's surface. But the point is that they can be modified to obtain a model as accurate as we wish. In this case computer methods are necessary for solution.

[11]Considered logically, however, this inference is invalid. If T denotes a theory and O a predicted observation, we cannot go from $T \Rightarrow O$ to $O \Rightarrow T$, which is an elementary logical fallacy.

By such means science develops, facts accumulate, and our understanding of the natural world and its laws approaches the truth. To say this in a different way, the collective human mind becomes a more and more accurate mirror of nature. I recognized, of course, that scientific beliefs especially at the research frontier can turn out to be mistaken and change when additional evidence becomes available. Yet even if this is true, most (if not all) contemporary science *probably* consists of true and well justified beliefs. And because it at least *attempts* to faithfully reflect reality, scientific knowledge is different from the claims generated by other belief systems. The existence, say, of neutrons has a different epistemological status than the theological claim that the Holy Ghost proceeds from the Father to the Son, dialectical materialism, or the theses of American neoconservatives, Salafist Muslims, African witch-doctors, professors of cultural studies, and so forth. To express this more bluntly, science (and mathematics) represents the best and most reliable knowledge the human race can possess; it is the most rational enterprise of which we are capable. The other areas of human intellectual activity are mostly a confused (though sometimes interesting) mixture of cultural myth, ideology, and value judgement. Their existence is explained by social and historical contingency rather than rational criteria.

It follows as a result of the emphasis put on scientific inquiry in the last three hundred years that the human race in the modern world possesses far more accurate knowledge about nature than in the Middle Ages or in the ancient world. This in turn implies that one of the most obvious and uncontested features of science has been its cumulative progress. With respect to applications, science has made modern technology possible which in turn has greatly improved communication, the ability to travel quickly and (for the most part) safely, human economic possibilities, and (unfortunately) also weaponry. The application of the scientific outlook to medicine has also improved the human condition by eliminating many of the ancient scourges of mankind. With antibiotics or vaccines we can cure TB and no longer die of smallpox, diphtheria, typhus, cholera, polio, and other terrible infectious diseases.[12] Physicians no longer bleed people to death or apply hot irons to the temples of the dying Charles II; instead they perform heart or kidney transplants and greatly prolong the life of cancer patients even when the underlying condition may not be curable. In fact, a personal medical situation, requiring some very advanced (and expensive) modern

---

[12] Of course, much of this progress is being called into question with the spread of antibiotic resistant bacteria. The situation is especially serious for some forms of TB. Yet, ever more potent drugs are being developed.

medical technology to resolve, reinforced my hero worshipping attitude towards medical giants like Pasteur, Lister, Koch, or Cushing. I was very much in the same mental world as Paul de Kruif's *Microbe Hunters* (1926) or Arturo Castigioni's *History of Medicine* (1947). Such judgements as these seemed to me too obvious to be contested—since the Enlightenment they merely represented, so I thought, the common sense of reasonably well educated people at least in the developed world.

I also believed that human moral categories do not often apply to science. The existence of neutrons, for instance, has nothing to do with moral values. Science certainly can be used for good or evil purposes, but it has no inherent moral qualities of its own except a commitment to truth and honesty in the research process. Whether we use nuclear physics to produce bombs or CAT scans is the responsibility of political leaders or society generally, not that of scientists. It is especially not the responsibility of those whose fundamental or "pure" research lies a great distance from technological applications; thus Rutherford or Chadwick should not be blamed for nuclear weapons. Moreover, even if some applied research is morally problematic, this has nothing to do with the validity of the results obtained; these stand or fall on their own scientific merits. For example, we can disapprove of Pentagon funded research seeking to improve the stickiness of napalm, but the conclusions of such research are true or false independently of moral factors. In spite of the fact that science can be and too often has been exploited for evil ends, I continued to believe that it remains one of the intellectual glories of the modern *Western* world. No other civilization in history has come close to producing the scientific miracles that Europe has since the Scientific Revolution of the seventeenth century. This view of science was also tied into a wider set of cultural values. I valued knowledge for its own sake, and thought that in general more of it was a good thing. I believed that truth was an objective property which in many cases we could discover by comparing an assertion about something to what was actually the case; even when this is difficult or impossible to do careful and *objective* study can at least clarify the issues. In short, I was still (in the 1990s!) a child of the Enlightenment and believed in the power of reason to uncover the nature of a common physical and social world. Reason, despite the historical excesses in its application, was far better for humanity I thought than myth and superstition.

I knew that the 1960s had discovered the pleasures of debunking politicians, generals, writers, and philosophers, particularly if they were white and male. In a mild way I—especially when younger—had enjoyed participating in the sport. But such treatment could not apply to medical heros such as Pasteur, or to great scientists like Newton or Einstein. Certainly, not all of them were "nice" people or of sterling moral virtue; I probably would not have enjoyed a social evening with Newton or a political conversation with Pasteur; nor would I have trusted Einstein with any young and attractive female relative.[13] But this situation is not surprising, since these scientists were merely human beings. But no criticisms of this kind can possibly discredit their monumental achievements.

Such were the clichés which described my state of mind around 1995. Alas, like Monsieur Jourdain who discovered that he had been thinking and speaking "prose" all his life, I soon learned that I had been thinking and speaking "modernism" all of mine. But while "prose" was still presumably a viable option for Jourdan, modernism was now definitely passé for anyone with any pretence to sophistication. In my ignorance I had not realized that most, perhaps all, of my attitudes concerning science were viewed with contempt by advanced intellectuals. Locked in my narrow technical world composed of splines, classical inequalities, weighted Sobolev embeddings, or the spectral properties of differential operators, I was unaware that since the 1970s there had been a "postmodern" shift concerning science, mathematics, and reason almost as profound as the shift beginning in the seventeenth century and completed in the Enlightenment, but in the opposite direction.

An awakening from my dogmatic slumbers came in the spring of 1996 when I discovered Alan Sokal's marvelous hoax. Sokal, a physicist at New York University, had placed a completely nonsensical article in the cultural studies journal *Social Text*.[14] The article was written in a ridiculous jargon and was full of mistakes, yet it had been refereed and accepted by a journal that included such intellectual heavyweights as Andrew Ross, Frederic Jameson, and Stanley Aronowitz among its editors. The ensuing howls of rage by victims of the hoax and Sokal's replies made the national press.[15]

---

[13] But this aspect of Einstein although real is often exaggerated. The most reliable account of Einstein's love life may be found in Isaacson (2007).
[14] Deliciously entitled "Transgressing the boundaries: Toward a transformative hermeneutics of quantum gravity." The original article was published in *Social Text*, 46/47 (1996), pp. 217–252. A reprint may be found in Appendix A of Sokal and Bricmont (1998) or in Sokal (2008). One should be a physical scientist or mathematician to fully appreciate the unsavory bouquet of this paper. Perhaps its most remarkable feature is that it was not noticeably sillier than some of the other papers in the same issue of the journal.
[15] Several of the responses to the hoax have been republished by the editors of *Lingua Franca*; see The Sokal Hoax (2000).

There were many letters to the editor and op-ed pieces in major newspapers. One would have had to been a permanent resident of an organic chemistry lab (as some of my colleagues in the Chemistry Department were), a desert island, or even a petri dish to have been unaware of the controversy. Then I saw a splenetic review (I forget by whom) of Gross and Levitt's 1995 book *The Higher Superstition* which piqued my interest, and I decided to read it. The book was a devastating portrait of what the authors conceived to be the ideological zoo of contemporary "science studies." In this way I learned of the existence of the "Science Wars" which were apparently raging—without any awareness on my part—all around me, possibly even in certain departments of my own university. A little later I was led to Sokal and Bricmont's *Intellectual Impostures* (1998), a demolition of postmodernist French intellectuals like Lacan, Irigaray, Baudrillard, et al., and also to the critical essays in *A House Built on Sand*, edited by Noretta Koertge (1998). As a result of this initial experience, I started an accelerated course of reading several of the sources considered by these authors. My first reaction was to compare myself to Rip van Winkle. After a nearly thirty year nap I had awakened to a new and strange intellectual world. I was amazed by it just as Bernal Diaz, a soldier of Cortez's army, was when he first gazed at Tenochtitlan, the fantastic capital of the Aztecs and learned of the sacrificial feeding of the gods with living human hearts.

The writers I examined had attitudes that were polar opposites to mine. Most shared a common stance that science is just another form of socially constructed "discourse" divorced from an independently existing natural world (which itself is a social construction). At best it was one "way of knowing," peculiar to the West, not to be privileged over others, and whose theoretical modes of explanation tell us more about the political agendas or social situation of scientists than about an essentially unknowable "reality".

More precisely, the following is a representative sample of the doctrines I encountered. They are given in a list of fifteen propositions. I have arranged them in a quasi-logical order, going from the philosophical to the stridently political. A few contain striking quotations from prominent writers, but most are abstractions from many sources.

(i) Science claims to "discover" the underlying regularities of nature by observation, but this is an illusion. We can establish such regularities only by inductive argument. However, the validity of induction has been destroyed by the objections of Hume. The regularity of nature is part of the belief system of scientists; it is not a property of the world.

(ii) Furthermore, "observation" itself is a problematic concept. What a scientist observes is always determined wholly or partially by the theoretical and/or ideological apparatus he or she brings to the task. Observation is "theory-laden." We see only what our theories permit us to see. We are not a *tabula rasa* which the world inscribes; rather it is our concepts and theories which "construct" the world.

(iii) More generally, scientists have no access to the world as it is. All we can know are our representations of it—which from (ii) are in great part supplied by us. Hence the claim that nature and its properties exist independently of our representations of it is meaningless, for how could we ever verify this? We can only compare one representation with another, not a representation with the reality it is supposed to represent. A consequence is that the correspondence theory of truth is false.

(iv) Whatever the status of observation, scientific theory is always underdetermined by the evidence. This evidence is always finite since it necessarily consists of finitely many experiments or observations. But a sufficiently complicated theory—and there are potentially infinitely many of these—can explain any finite amount of evidence. Duhem and Quine have also shown that it is impossible for a single crucial experiment to falsify a theory since it is possible to explain away failure by changing one or more of the auxiliary assumptions involved in the experiment.

(v) Since observation is theory-laden and experiments settle nothing (by (ii) and (iv)), scientific claims or theories are not "true" in the sense of corresponding to or describing the underlying facts of nature, but only reflect the consensus of scientists. What we call "science" is a network of "interpretations" hammered out by quasi-political negotiations among scientists, just as Party dogma is hammered out by Party ideologists or theology by the Holy Office. "Nature" may play little or no role in the formulation of these interpretations; at the very least she does not "compel" them. In the words of Bruno Latour's "Third Rule of Method":

> Since the settlement of a controversy is the *cause* of nature's representation, not its consequence, we can never use this consequence to explain how and why a controversy has been settled. [italics in original][16]

Thus, since nature does not exist independently of human representations and our so-called knowledge of it is a purely human product and not a reflection of a mind-independent reality, it follows that scientific theories are "social constructions" rather than "discoveries" about the world. Like other human cultural activities they are "situated" relative to a particular social and historical context.

(vi) The negotiations determining scientific theories may not even be "rational" processes. Different theories are embedded in much larger intellectual structures called "paradigms." Paradigms dictate what problems scientists should

---

[16] Latour (1987), p. 258.

investigate, what questions theories should answer, and and what constitutes valid scientific explanation. Paradigms are mutually "incommensurable", i.e., there are no standards accepted by both sides which can decide between them. As in political disputes scientists talk past each other. Consequently, the decision must rely on rhetoric, persuasion, networking, social pressure, etc.—in a word, "politics."

(vii) What in particular are "scientific facts" or "truths"? In view of (v) and (vi) they are just those interpretations agreed on by the dominant scientific group or victorious paradigm. They are an expression of social power and consensus, not a "mirror" of reality. Consequently, "facts" are not stable. They develop and change with time. One era's facts or truth are a latter era's mistakes. Like other human cultural expressions, they and the theories behind them are relative to their social and historical context.

(viii) Scientific progress considered as an cumulatively more precise understanding of nature is an illusion. Accounts of such "progress" are merely "Whig history"—in other words edifying propaganda produced by those whose conceptions momentarily dominate science.

(ix) External political, social, and other contextual factors also play a critical role in the formation of science. Race, class, gender, political ideology, or the demands of the state or the military determine (often through funding decisions) what problems the scientist will investigate and what methods he will use. More than this, they also influence the *content* of scientific theories. As is implied by (ii)-(vi), the scientist's interpretation of data and the hypotheses he will entertain are functions—perhaps unconsciously—of his ideological baggage, and the ambient ideological climate can determine the social acceptability of his conclusions. The belief, therefore, that science is or should be value free, objective, or that its development depends on "internal" factors only is a myth. Science is as affected by external political and social factors as much as any other cultural "discourse." The same is true even of mathematics. The mathematician's belief in his personal autonomy is a delusion. "The point is that neither logic or mathematics escapes the 'contamination' of the social."[17] To express the mutual entanglement of science with politics more abstractly: "Solutions to the problem of [scientific] knowledge are solutions to the problem of social order."[18]

(x) The nature of these external factors as well as who deserves the credit for "unmasking" them has been well expressed by Alan Sokal:

> ... most recently feminist and poststructuralist critiques have demystified the substantive content of mainstream Western scientific practice, revealing the ideology of domination concealed behind the facade of 'objectivity'. It has thus become increasingly apparent that physical 'reality', no less than social 'reality' is at bottom a social and linguistic construct; that scientific 'knowledge' far

---

[17] Aronowitz (1988), p. 346.
[18] Shapin and Schaffer (1985), p. 332.

from being objective, reflects and encodes the dominant ideologies and power relations of the culture that produced it... [19]

(xi) Not only does science encode ideologies of domination *internal* to the West, but even at its purest it is basically a Western ethnoscience "molded on the twin templates of capitalist greed and imperialist expansion" and thus is a tool for the oppression of the Third World.[20]

(xii) We conclude from (i)–(xi) that science does not give "objective" (whatever this can possibly mean) knowledge about the world. At best such a belief is a product of wishful thinking, and in many cases it is a scam practiced by scientists to enhance the power of the ruling elites or of the West generally, as well as their own social prestige. Thus science should not be "privileged" over "other ways of knowing" to be found in one's own society or in other cultures. For example, it is not to be ranked above intellectual systems such as Chinese traditional medicine, American Indian cosmology, Vedic science, astrology, Zuni rain dances, or African magic. Because of its ties to Western patriarchy, capitalism, and colonialism it may indeed be inferior to these alternative systems.

(xiii) It follows from (v)–(xii) that science is relative to the tradition or frame of reference in which it is embedded. While science or another set of beliefs may be a valid consequence of a particular frame of reference or perspective, the frame itself, is at the ultimate level, not given by nature, but rather is a culturally determined interpretation of nature. Secular humanists, for example, believe in Darwin because they still live in a culture of bourgeois capitalism which supplied the metaphors Darwin exploited and encoded in the *Origin of Species* and think from the perspective of metaphysical naturalism, associated since the Enlightenment with that culture. On the other hand, those embedded in the perspectives of the Bible incline to "Creation Science." There is no "neutral" evidence that can decide between the two, because the evidence is interpreted differently by the two perspectives. Fossils, for example, are either the relics of long extinct species or the relics of animals living before the Flood, a few thousand years ago, or perhaps even a trick introduced by Satan to cause men to lose their souls.[21] Each is a perfectly rational "discourse" relative to the frame of reference supporting it. From this and other examples a profound epistemological relativism follows, with the result that:

> It is no more easy to defend non-context dependent, non-culture dependent beliefs of things or objective scientific truths than be-

---

[19] This is a direct quotation from Sokal's hoax article, quoted in Sokal and Bricmont (1998), p. 200; but since it expresses one of the characteristic attitudes of the postmodern interpretation of science—and, after all, the article was accepted—we felt at liberty to use it.

[20] Nanda (2000), p. 208. Nanda claims to be presenting the views of Sandra Harding.

[21] The writer heard this theory fervently endorsed over lunch by a prominent NSF funded mathematician at another southern university.

liefs in gods and demons.[22]

(xiv) Given the dependence of Western science on contextual—often "oppressive"—factors (see (ix)–(xiii)), it can be changed if we change these factors. A good start will be to purge scientists with the wrong social, political, gender, or ethnic characteristics and to replace them by underrepresented groups having the characteristics we prefer, and also to change the sources of financial support behind science—such as by eliminating military or corporate funding. We can then produce a Jewish science, a Feminist science, an Aryan science, a Hispanic science, an African Mathematics, or a socially responsible "postmodern and liberatory science," etc.[23] Some of these products will be "better" or more "Strongly Objective" than the conventional science existing today.

(xv) On the other hand, science may be so deeply flawed that it would be better to simply junk it or replace it by literary criticism, the utopian politics of youth, or poetry as the preeminent cultural activity of mankind. As Richard Rorty has asserted:

> My utopia, as I have often said, is one in which poets rather than scientists... are thought of as the cutting edge of civilization, and are heros and heroines of culture.[24]

Let us denote the collection of sentiments (or any substantial part thereof) revealed by the above list as a "postmodern interpretation of science" (or as we have abbreviated it "PIS"). PIS has at least four main components. There is first of all a borrowed philosophic anti-realism encoded in statements like (i)–(iii) which independently of any application to science has a long and honorable tradition. Then there are constructivist theses like ((ii), (v)–(vii)) combined with relativism, e.g., (vii), (ix), and, finally and perhaps of most significance, a political attack on the "objectivity" and neutrality of science ((v)–(vii), (ix)–(xiv)). Many of the latter two categories have been the contributions of largely UK sociologists who have since the early 1970s created a subdiscipline which they call the "sociology of scientific knowledge" (or SSK) which argues that not only the institutional structure and culture of science is socially determined, but its very *content* is as well. SSK is a continuation, applied to science, of the

---

[22] Mary Douglas, embedded in a quotation of Yehuda Elkana defining relativism in terms similar to ours, as recorded in Cole (1991), p. 11.

[23] Of course, no one today speaks of an Aryan or Jewish science, although such notions were fashionable in certain countries in c. 1935. I introduce them here only to emphasize that in 1935 they followed from arguments *structurally* similar to those given here. I am indebted once again to Sokal's hoax article for the adjective "liberatory" although it pervades the literature he is satirizing. See Sokal and Bricmont (1998), p. 200.

[24] Rorty (1995), p. 32.

so-called "sociology of knowledge" (henceforth "SK") mainly developed by Karl Mannheim (1892–1947). Whether or not it is a really a variety of PIS in the sense that it shares the postmodern characteristics of the latter is debatable. Certainly, the two are "postmodern" in the trivial chronological sense. Yet, as we shall see below, SSK thinks of itself as a full fledged "science of science" which obtains objective knowledge about science (just as science purports to obtain objective knowledge about nature); this is a thoroughly "modern" rather than a "postmodern" attitude. There are certainly also varieties of PIS which do not explicitly involve sociology such as various strands of French poststructuralism. On the other hand, we have chosen to regard SSK as a member of PIS family both because it supplies much of the intellectual muscle (even if unacknowledged) behind other forms of PIS and because it generally views any science (other than itself) as something to be deconstructed as does PIS.

Perhaps none of those of a PIS temper who are not incarcerated in a mental institution believe in *all* the statements in our list, but they are all individually close to what I have found in the literature. Some who are closest to the postmodern stance in the humanities, that is, those who see "truth," "reason," and "objectivity" as code words for various kinds of domination and who are dissatisfied with the political or ethnic status quo in the sciences emphasize (ix)–(xv). Others are simply philosophic anti-realists, buying into the remaining propositions only mildly. Additionally, some social constructivists allow nature or even "evidence" a role (how much is a matter of ongoing dispute in the PIS community) in the construction of scientific theories and certainly would not regard them as delusions. They and the anti-realists may also value science and would admit that science progresses in the sense that, although it does not converge on "truth," its theories become evermore refined instruments of prediction and control.[25] There is, moreover, evidence that the more extreme ideas on our list are now a bit passé, having had their strongest appeal in the 1970s and 1980s.[26] But common to many residents of the PIS camp is at least a certain degree of relativism together with the belief that critical influences on the development of science include almost any factor other than rational argument or evidence; hence scientific facts are (in some sense) constructed or invented rather than discovered and may have no better warrant than other forms of cultural discourse. There are also endless variations (both weaker and stronger) on all the statements on our list. We have also omit-

---

[25] This, indeed, is a characteristic positivist view.
[26] See Hess (1997a).

ted doctrines such as Bruno Latour's "actor-actant network theory" as we find it impossible to express his ideas coherently, although they are surely part of PIS.[27] We recognize that this analysis has been rough and inexact. The reader may think of other characteristics of PIS which we have omitted or disagree with some we have given. But we are trying to describe an intellectual miasma or fog, not a precisely defined doctrine. Our intention has only been to indicate the "flavor" of one side of the Science Wars. Also, what we have been describing is just one small pond within the much larger sea of postmodernism, a movement which has affected almost every area of culture since the 1970s. There are, for example, doctrines parallel to what we call PIS in art, literature, philosophy, architecture, theater, and in many other fields. When we describe the social changes leading to PIS (Chapter 13) we will briefly describe some of these; but we point out here that one characteristic common to all aspects of postmodernism is a rejection of the Enlightenment belief in the power of reason to accurately represent (and improve) the world. For many postmodernists this is not because of a failure of mental power, but because there *is* no world or reality that exists independently of human culturally determined concepts that purport to describe it. Instead, there are many worlds, each as actual as any other but often mutually incommensurable, which these concepts *construct*. What we call "reason" or "truth" is never absolute, but is relative to the set of concepts we use; these in turn depend on our culture, historical period, etc. No God's eye view is available to us.

What is perhaps the most remarkable feature of the PIS position on science is that it is not just the product of a minor academic sect; rather it has become in some form the "mother's milk"[28] or set of common postulates of a new and substantial discipline, usually called "Science and Technology Studies" or "Science, Technology, and Society" (STS). STS hardly existed in the 1960s, but now (as a casual internet search shows) it is a rapidly growing field represented in some form in more than fifty US universities, many offering Ph.D.s, as well as several abroad, especially in the UK.[29]

---

[27]But see Chapter 11 for an attempt.
[28]Levine (1996), p. 126.
[29]These include Cornell, MIT, Rensselaer, and Virginia Tech for interdisciplinary STS Ph.D. programs and Northwestern, UC Davis, UC San Diego, University of Wisconsin, and the University of Illinois at Urbana Champaign for disciplinary programs. Additionally, several other universities offer STS flavored Ph.D.s in departments such as history, philosophy of science, or sociology. At present STS faculty seems a mixture of historians and philosophers of science, anthropologists, sociologists, and cultural study types. Perhaps as the field continues to mature they will be replaced, as in the earlier case of computer science, by people having degrees in the field.

Moreover, constructivism, attacks on the objectivity, status, or epistemological privileging of science, and a general radicalism[30] have spread beyond departments associated with STS and are now found throughout the humanities; they particularly infest English, Cultural Studies, Comparative Literature, and even as we pointed out above the History of Science.

The literature of PIS is also huge. Thousands of articles and books sympathetic to it have been written since 1970. This is not to say that there are no dissenting voices. A substantial number of scholars working within STS oppose epistemological relativism, and, for example, study the sociological factors that affect the growth or organization of science without critiquing its content on ideological grounds.[31] The most vociferous dissent, however, comes from outside STS in, for example, the books mentioned above. But the most surprising aspect of all this activity is, as we have indicated, its near invisibility to most practicing scientists. There are no nefarious reasons for this; it is just an effect of the specialization and isolation within the modern university.

Had I never studied the history of science, I would have dismissed PIS as just another of the twentieth century's, if not *traison*, at least *folie des clercs*. After all, intellectual history is replete with bizarre doctrines, and for me it was more fun and more rewarding to try to show that the best constant K in the Opial-type inequality

$$\int_0^1 |yy'| \leq K \int_0^1 (y')^2,$$
$$\int_0^1 y = 0.$$

is $1/4$ or to study Sobolev-like embeddings on a domain $\Omega \subset \mathbb{R}^n$ satisfying weak conditions on its boundary (such as having Minkowski dimension $< n$) than to wade through material, much of which seemed on my initial encounter with it as appetizing as the contents of an overfull and long disused septic tank installed by the Civilian Conservation Corps in the 1930s which—while afflicted with flies and an excruciating case of poison oak—I had to clean out during a college summer job in 1957.

Many of the writers considered by Gross and Levitt, Sokal and Bricmont, or Koertge supported their conclusions by historical "case studies"

---

[30] The political flavor that STS can have is proudly announced in the Mission Statement of Rensselaer's program: "faculty and students pursue studies of power, gender, race, colonialism, and the interactions between research and activism." See the internet website: *http://www.sts.rpi.edu/index.php.*

[31] See Hess (1997a) for an informative survey of the many interests of STS.

and claimed inspiration from Thomas Kuhn's ideas, even as they had gone far beyond him. Could this really be true of the kind, scholarly, and slightly shy person I knew? But if it was, the fact that I had been to quote Dean Acheson "present at the creation," and yet was totally ignorant of developments since the early 1960s was embarrassing. I had left the field towards the end of an age in many respects dominated by the values of George Sarton. How and why did the older conventional views of science vanish among humanistic intellectuals, especially those in universities, to be replaced by PIS, encoded in the rapidly metastasizing discipline of STS? By what weird historical transformation did Steve Shapin, Bruno Latour, or Harry Collins supplant Alexandre Koyré, C. Coulton Gillespie, or I. B. Cohen in the academic study of the nature and history of science? How did we get from *From the Closed World to the Infinite Universe* to *Leviathan and the Air Pump* or from *Franklin and Newton* to *Pandora's Hope*? This radical change had happened, moreover, in little more than a generation—during a period when science was developing at an exponential rate and had achieved many of its most spectacular successes—and in a modern increasingly secular society which had been more committed to the development and exploitation of science than any other in history. This book is an attempt to find some answers to these questions.

Chapter 2

A Golden Age and its End

To appreciate what a tremendous change PIS has brought in the intellectual climate it is useful to recall that for around two decades after World War II many, perhaps most, humanistic intellectuals in the US seem to have had a healthy respect for (and perhaps also a salutary fear of) science and mathematics. In the context of the period, this was a natural attitude. After all, had not scientific advances in the form of radar, the atomic bomb, sonar, proximity fuses, operations research, etc., been a decisive factor in winning the war and saving democracy from the threats of fascism and Japanese militarism? And during the Cold War were not these same advances, especially in the form of nuclear weapons, protecting the Free World from the menace of totalitarian communism? By 1960 some intellectuals had also been intimidated by C. P. Snow's fashionable 1959 Rede lecture *The Two Cultures and the Scientific Revolution* which asserted that ignorance of the Second Law of Thermodynamics was equivalent to ignorance of Shakespeare and ignorance of the concepts of mass and acceleration was the equivalent of illiteracy in one's native language. Science according to Snow was more rigorous and at a "higher conceptual level" than literary argument. Scientists, he said, "have the future in their bones" while literary intellectuals are "natural Luddites" who wish "the future did not exist" and "have never tried, wanted, or been able to understand the industrial revolution, much less accept it." Scientists, moreover, while sensitive to the tragic aspects of human existence want to improve the human social condition while reactionary literary intellectuals such as Yeats, Pound or Wyndham Lewis have brought "Auschwitz that much nearer." Certainly, in comparison to the tremendous progress made by science in the twentieth century, achievements in the humanities seemed paltry indeed. Quite a few of the literati were pessimistic. The initially politically radical poet and literary critic Max Eastman, for instance, wondered if science might soon answer every human problem and that literature would have "no place in such a world." In 1970 Eugene Ionesco seconded this opinion speculating that, in comparison with the "enormous progress that scientists were making, art may have reached a "dead-end."[1] Michael Yudkin, himself a research chemist with humanistic inclinations, warned in an article critical of Snow:

> ... of a real danger that the problem of the 'two cultures' may gradually cease to exist. There will be no building of bridges across the gap, no appearance of modern Leonardos, no migration of scientists to literature. Instead there will be the atrophy of the traditional culture, and its gradual annexation by the scientific—annexation not of territory but of men. It may not be long before a single culture remains.[2]

This period was still a pre-egalitarian age which had not yet shed a Puritan belief in the moral value of discipline and suffering. Also there still survived a few sheds of aristocratic values among the old-money WASP upper classes which yielded a lingering respect for intellectual elitism generally and science in particular. Such attitudes were reinforced by the tens of thousands of emigrants, largely Jewish, who fled the persecutions of the Third Reich for the United States. Many were highly educated and had been professionals. Probably a majority were on the Left and a significant minority were even Marxist (unlike the WASP establishment they encountered), but Marxism viewed itself as a science, and indeed as the heir and final perfection of the Enlightenment. These political attitudes were combined with a general respect for learning. Then too, it was widely agreed (in a period entirely innocent of postcolonial guilt) that the cultivation of science since the seventeenth century had defined modernity and had propelled the West to a position of knowledge, power, and wealth far ahead of any other civilization past or present. All these psychological elements helped to enhance the prestige of science.

Probably the most significant factor, however, after 1945 favoring both the reputation of and opportunities in the sciences was the Cold War. After the shock of Sputnik in 1957 the US seemed to the general public to be "losing" a scientific and mathematical race with the Soviet Union, the out-

---

[1] For the above quotations see Snow (1959), Chapters I and II and Barash (2005) which contains additional material about Snow.
[2] Leavis (1963), p. 64.

come of which would determine the fate of good things like Free Enterprise, Democracy, and the Free World in the competition with a ruthless totalitarianism. This perception caused a rapid decade-long (c.1958–1969) influx of Federal money into higher education mainly in technical areas, but with a generous trickle-down effect in the humanities. Faculty in the scientific disciplines, especially high-energy physics and mathematics, soon became among the most important and well-paid on campus. Grants and jobs were easily obtained and tenure was often quickly earned.[3] Pure mathematics in particular enjoyed an inordinately high prestige and funding often from Pentagon related sources, while the applied branches were considered little better than engineering and the territory of the second rate.[4] Research papers were full of "arrow chasing" diagrams, words like "epimorphisms," and emphasized topics like Category or Homology Theory. Books were written with titles like *Mathematics Made Difficult*.[5] Formalism was the official mathematical doctrine and deduction from axioms was the only kosher way to present mathematics.[6] A topologist could be paid to deliver lectures on the Separation Axioms $T1$–$T4$ to a bemused audience of Lockeed missile engineers.[7] The Army Ordnance Office could fund work on model theory, by mathematical logicians like Alfred Tarski.[8] It was almost believed that the mere touch of a pure mathematician could solve any technical problem, just as a touch of a medieval king could cure the scrofula. Educationally this was the age of the "New Math," an abstract curriculum invented by the

---

[3] By 1965 a talented Ph.D. in mathematics had merely to call up a university where he wanted to work to be almost certain of obtaining a job offer. The following anecdotes are typical: In 1965 the writer used to hike with a graduate student in Algebra at Berkeley who had made so much money consulting with Bay Area defense contractors that he could afford his own private plane. A few years later he became acquainted with a prominent analyst, a full professor in the New York State University system, who was bitterly resentful that he had to wait a year for tenure, while others in his class received it with their first job!

[4] In recent years the situation has reversed. Pure mathematicians like algebraic topologists, logicians, or number theorists are a drug on the market and some very talented people are now imprisoned in minor branch campuses or unemployed. At major research institutions, on the other hand, the demand is increasingly for applied types and experts in computer modeling of, say, the churning of stomach acids, since they can get grants from the government and corporations.

[5] This book by Carl Linderholm was actually published in 1971 well after the climate I am describing began to fade. However, its mathematical values mirrored the ultra-purity of the mid-60s.

[6] The arid but now passé works of Bourbaki (actually a group of Paris mandarins headed by Jean Dieudonné) were characteristic of the period.

[7] An experience of a late colleague of the writer.

[8] Personal recollection.

School Mathematics Study Group, consisting of top mathematicians, and which was enthusiastically inflicted on the high schools. Tenth graders, for example, were taught the Hilbert-Pasch axioms of plane geometry instead of the old-fashioned accessible but logically flawed Euclidean approach. There was also much hysteria over "functions", "relations," and "sets."[9] Similar programs, whose content was transmitted to the unwashed by NSF funded Summer Institutes, existed for the other sciences. These imperious educational demands were reinforced by the discourse of politicians who asserted that all students had a patriotic duty to either study some hard science, engineering, or mathematics or at least to "appreciate" their role in the defense of democracy. Even leftist intellectuals tended to have a soft spot for technical subjects even if they condemned their military applications for, as we have mentioned, many believed it possible to complete the work of the Enlightenment in rationalizing society by "scientific socialism" or other forms of scientific planning. In short, this period, lasting less than fifteen years, was a golden age for professionals in the hard sciences or mathematics which was without precedent (even during the forced mobilization of science in World War II) and is unlikely ever to be repeated.[10]

True, there were dissenting voices, but these were an assortment of socially peripheral groups like beatniks or romantically inclined literary types and artists, the heirs of William Blake, Goethe, or followers of continental philosophers like Heidegger (thoroughly discredited, however, by his cooperation with the Nazis). The Cambridge literary critic F. R. Leavis was certainly no enthusiast and viciously attacked Snow's worship of science. Snow, he said, was completely ignorant of history and literature; his Rede Lecture "exhibits an utter lack of intellectual distinction and an embarrassing vulgarity of style," and "not only is he not a genius; he is intellectually as

---

[9] In practice, however, the New Math resembled the Coptic liturgy. It was not understood by many of its teachers just as the liturgy written in the Coptic language (a derivative of ancient Egyptian, but written in the Greek alphabet) was sometimes not understood by its priests. Both were simply chanted to the audience.

[10] It was the same in the Soviet Union at the time. "We got Black Sea vacations, good apartments, imported TV sets that didn't explode, and short listed for a Moscovitch car. Institutes grew up like mushrooms in the forest ..." the writer was told by mathematicians, now earning less money than the janitors at their Institutes, at an Analysis meeting he attended in Moscow in 1995. To bring this desirable Cold War state of affairs back, they and he agreed, over far too much fruit flavored vodka, to support xenophobic and belligerent politicians in their two countries, perhaps some rabid hyper-nationalist in Russia and Jesse Helms or Strom Thurmond in the US. Time will tell if our joking wishes will be fulfilled. But given the current (2008) quality of leadership in both countries, the omens are encouraging.

undistinguished as it is possible to be."[11] Some anti-intellectual types of the Eisenhower era certainly regarded scientists as impractical longhairs, perhaps over-inclined to radical politics, but even they valued the technology science was able to produce—especially the weaponry. If English professors were viewed as effeminate wimps by the large anti-intellectual component of the public, so also scientists were considered nerds or quite possibly insane. The "mad scientist" had a spectacular career in well-attended B-rated horror movies of the period (often starring Vincent Price). Although every one admitted that the A-bomb was scary and the fruit of science, neither this fact, the opinions of anti-scientific intellectuals, or negative popular portrayals of scientists could disturb the general political and social consensus concerning the value of science.

One of the most fundamental and universally agreed-upon characteristics of science in this era was that it enjoyed a different epistemological status than the humanities. Since the late nineteenth century humanists had become increasingly aware of the uncertainty and historical contingency in their own work. What nontrivial judgements in history, political science, philosophy, and the like could be said to be unambiguously *true*? What orthodoxy in these fields could endure the eroding effects of historical and political change? Their essence appeared to be perpetual revisionism. Each new generation of scholars was driven by the need to supplant their academic elders, an urge which accelerated in the late 1960s and early 70s, at first because of the political radicalism fueled by Vietnam war and—as time went by—by the desperate need to capture jobs in an increasingly poverty stricken and saturated academic market. The dictum of Nietzsche "...facts [are] precisely what there is not, only interpretations."[12] seems the only solid fact about the humanities (but if so, the statement must be an immediate self-contradiction). On the other hand, those residing in the swamps of humanistic logomachy might look with envy upon science and mathematics. *These* subjects seemed to consist of an ever increasing amount of real *knowledge*. The theorems of Euclid, Archimedes, or Gauss, Newton's inverse square law of gravitation, or the theory behind nuclear weapons were as true today as they were in the age of the dinosaurs. Even the average journal article by a $45,000 a year contemporary Assistant Professor of Mathematics at Persepolis State University contained theorems, however

---

[11] Leavis (1963), pp. 28ff. Leavis' argument is almost entirely *ad hominem* and illustrates his poisoned *ressentiment* and envy more than any intellectual defects of Snow. See also Barash (2005).
[12] Nietzsche (1968), §481.

shallow, which were *eternally* true. Such judgements were reinforced by the still potent (although it had begun a terminal decline) philosophy of logical positivism as represented in the US, for instance by Rudolf Carnap (1891–1970), Hans Reichenbach (1891–1953), or Carl Hempel (1905–1997). There were, of course, many differences, sometimes vociferous, among logical positivists concerning technical issues in the philosophy of science such as the nature of induction and probability; but as a whole they tried to give a rigorous foundation to the standard view of scientific method held by this writer prior to his exposure to PIS in 1995. Scientific knowledge, they believed, is built on empirical foundations, and its theories can be tested by experiments whose results can be expressed in a neutral observation language. But the most famous thesis of the positivists was the "Verifiability Criterion of Meaning" which asserted that a statement had meaning if and only if it was logically or empirically verifiable. An immediate consequence was that most statements in the humanities (especially metaphysics) did not even enjoy the privilege of being "false"; instead they were simply meaningless. Such attitudes must have bred feelings of insecurity and resentment in humanists, which were further enhanced by a certain patronizing contempt on the part of hard scientists, mathematicians, and engineers (especially obvious on university committees)—not to mention the salary differential between the humanities and the sciences.[13]

The post-Sputnik golden age of science we have described began to come to an end after 1970. In part this was inevitable and a consequence of mathematical necessity. Jobs, not only in the sciences, but also in the humanities had been extraordinarily plentiful during the previous decade because of the constant expansion of the universities. The number of students multiplied by at least fivefold between 1950 and 1970. Graduate enrollment and the production of Ph.D.s—many of whom went on to produce more of their kind in newly created graduate programs—continually increased. This expansion in turn was fueled by ever increasing Federal and State funding for higher education. None of these processes which amounted to educational versions of a Ponzi scheme could continue indefinitely.[14] There was a finite upper limit to the number of universities and graduate programs. Also

---

[13] This persists to the present day. But while chemists, physicists, and other hard scientists tend to be paid more than faculty in, say English or Philosophy, applied scientists such as engineers and the like are usually paid much more than either the humanities or the "pure" hard sciences. Some nonscientific fields, however, such as Marketing, Accounting, or Law are paid more than anyone else.

[14] A Dean once told the writer in a job interview: "I saw the boom in the stock market in 1928. Get in while the going is good. This can't last forever."

it was just not true (as was almost assumed) that every household would require its resident scientist or mathematician along with the TV set and refrigerator. And since the bulk of the newly hired faculty was young and would not retire for decades the number of faculty openings would abruptly decline when the saturation level was reached. Nor could funding increase or even stay level in inflation adjusted dollars; the enormous expense of the Vietnam War, the competing demands of various entitlement programs, and the need to build ever more prisons saw to that. The result was that by the early 1970s it seemed that the door had almost been slammed shut on the possibility of university employment.[15]

Not only did job prospects and research funding begin to fail after 1970, but the intellectual climate concerning science began to change. There developed an increasingly widespread and vociferous counterattack by humanists on the prestige and epistemological privileges of science. PIS is the main form this counterattack has taken. Many of the old confidently accepted theses on the nature of science have been upended and replaced by some variation on the PIS-like propositions (i)–(xv). To account for the change we face two distinct problems: what are the intellectual origins of PIS and what political or historical factors helped cause it to bloom when it did? We will defer consideration of the second question until later,[16] and just observe here that a profound change in the Zeitgeist always involves more than purely intellectual causes. It may build upon previous intellectual currents, but it does not come about like an advance in science where, for example, a new theory is proposed to correct a gap in an existing theory; nor is it like a discovery in mathematics where a new theorem may change the outlook of an entire field. Instead, it usually parallels a wider social transformation. PIS comprises a group of quasi-political ideologies and is not just (as is sometimes claimed) a natural response to some perceived technical failure of the positivist project to justify science. We should expect therefore its genesis to be linked with non-intellectual

---

[15] Although the collapse of academic employment opportunities severely affected scientists and mathematicians, they at least could usually get some kind of job related to their skills. There were still jobs at minor branch campuses, community colleges, or industry. The situation was much worse in the humanities and it came earlier. Only a minority could find tenure track jobs at all. A typical career involved an endless sequence, at best, of one year jobs or adjunct positions. When the writer resigned in 1970 from teaching history and philosophy at a Pennsylvania community college to pursue mathematics, there were several dozen applications to replace him from Harvard, Berkeley, Yale, or other elite university Ph.D.s.

[16] Chapter 13.

changes in society just as is the case with other (temporarily) successful ideologies, say, for example, bourgeois liberalism after the French Revolution or European fascism after World War I. But, at the same time, PIS *does* have intellectual origins, just as bourgeois liberalism developed from the secular humanism and scientific optimism which emerged between the seventeenth century and the Enlightenment. In the next few chapters we will look at these origins, beginning with the contributions of philosophy from the classical period until the twentieth century.

Chapter 3

Ingredients in the PIS Bouillabaisse

To give a complete analysis of the intellectual origins of PIS we should have to probe deeply into the history of philosophy. Postmodernists are opportunists and use weapons wherever they can find them. Many of their arguments contain elements that are borrowed from the vast necropolis of past philosophic doctrines. However, the task of exhibiting them is made more difficult because the ideas influencing PIS are seldom clearly stated; they are often presented in a fragmented form—sometimes as if they are the cloudy memories of a long past Philosophy 101 class. It would be inaccurate and unfair to view all of PIS as the intellectual equivalent of the contents of the author's 1957 septic tank; but one can reconstruct the bouillabaisse of ideas PIS has consumed (with some idea of their relative importance) from the residue present in its texts, just as one can reconstruct the diet of the early patrons of the actual septic tank from analysis of the seeds, fruit pits, rinds, and bone fragments found in its contents. Less offensively, we could simply seek the recipe of the steaming intellectual mess set before us. But however we think of it, let us don protective garments and get on with the task.

The oldest ingredient in the bouillabaisse certainly comes from the Greek sophists, especially those hostile to the *Demos*. There is a pungent aroma of thinkers such as Callicles or Thrasymachus of Chalcedon who argued that truth is conventional and relative to social groups, and that ethical doctrines simply reflect the will of the stronger. This doctrine is a close parallel to a typical PIS thesis that scientific theories and "fact" serve the interests of social and economic elites in modern capitalist societies. Indeed, one can think of not only the Science Wars but modern Culture Wars in general as a continuation of the ancient quarrel which has periodically erupted over the centuries between the heirs of the sophists and those disposed to some form of Platonism. There is one difference, however, between the ancient sophists and modern representatives of PIS: for the sophists it was a good and natural thing that justice should be the slave of power, but for PIS it is a criticism. But these ideas are not the only contributions of the Greeks to PIS. We also find pinches of the skepticism of Gorgias of Leontini, Pyrron, or Sextus Empiricus, all of whom deny the possibility of knowledge. Gorgias' famous but now lost book *On the Nonexistent*, for instance, is said to have maintained: (1) There is nothing; (2) If there is anything we cannot know it; (3) If there is anything and we can know it we can't communicate this knowledge. Had it survived, his book might have been favorably reviewed by Stanley Fish and serialized in a journal like *Social Text*. From the the medieval world there is the strong seasoning provided by the nominalism of William of Ockham, yielding the claim that universals (common nouns) do not denote anything actually existing in the world. There are no "natural kinds." Universals merely mirror the way *we* choose to organize or divide up a world consisting only of individuals with our concepts. Had we chosen differently, the world would look very different to us. Proceeding to the seventeenth and eighteenth centuries, we detect the presence of Descartes' systematic doubt and a cup or two of David Hume's critique of induction well mixed with his general skepticism. Berkeley certainly accounts for the idealist flavor of proposition (iii) in Chapter 1; but Kant is also significant for propositions (ii) and (iii). His argument that the mind constitutes the fundamental properties and structure of what we call reality, and that we cannot know "things in themselves" makes knowledge essentially subjective and makes impossible a correspondence theory of truth. A true *a postiori* statement does not correspond to some state of an unknowable reality; rather it is a testimony of consistency among phenomena whose synthetic *a priori* properties are imposed by our minds. To derive theses like (ii) and (iii) it remains only to remove the universality of Kant's thesis—that action of the mind on reality is common to the whole human race—and make it *particular* to the minds of differing groups of scientists. *Their* minds construct the theories that constitute reality in different, possibly contradictory ways. We will see Kantian arguments of this kind repeated over and over again by postmodernist philosophers of science. German historicism after Kant, as represented by Hegel and Marx, is also a basic ingredient linked to relativism and the belief in the social or political determination of science which is characteristic of PIS. Just as Heroditus and the sophists, who presumably read him, were drawn to relativism because of the bewildering, colorful array of incompatible customs, laws, and religions they confronted outside the Greek world, so historically oriented philosophers like Hegel were drawn to a certain relativism by the endless variety and transitory nature of human intellectual systems developing in time. Hegel's system is an attempt to make this relativism respectable by thinking of past doctrines as "moments" in the dialectical evolution of the "World Spirit." Hegel also argued that human reason and action served purposes of which the agent was unconscious and did not intend. Man is ensnared by the "cunning of reason," so that the ends he serves are not his own. Marx enlarged on this theme, turning Hegel upside down. Human thought instead of reflecting the World Spirit is an emanation of class interest and the organization of the means of production, both of which being determining factors of whose influence the thinker may be unaware. Following this line in the case of science gives rise to propositions like (ix) and (x). Of a quite different nature and a little prior to Hegel and Marx would be the romantic hostility to Newtonian science exhibited by William Blake and Goethe. Blake's phase "single vision and Newton's sleep" and Goethe's contempt for Newton's "mechanistic" color theory parallels a dislike of abstract mathematically oriented science found in PIS, especially (as we shall see below) among feminist sympathizers who like to contrast their own closeness and sympathy with nature with the male urge to "torture" and dominate it. It is also a psychological feature of philosophers like Richard Rorty who prefer poetry to the colorless abstractions of science.[1]

Most of the philosophers just mentioned have in common the thesis that man in some sense constructs the world. The world is not "out there" waiting to be discovered; its qualities and structure are somehow contributed by us.[2] Nietzsche carried this idea to its limits. For him the world was a chaotic flux satisfying no pattern at all except that of "eternal return," a concept that negates both the Enlightenment/bourgeois myth of progress and the Christian idea of redemption.[3] All our so-called knowledge in the form of religion, ethics, political doctrine or science are purely human creations and interpretations by which we attempt to impose order on our inherently meaningless experiences. This "order" helps us survive, but its actual existence is an illusion. The connection of such a doctrine with propositions (iii) and (v)–(vii) is obvious. But it is a general anti-realist at-

---

[1] cf. proposition xv.
[2] Note that this position is a clear consequence of nominalism.
[3] Grenz (1996), p. 94. For Grenz and other advocates of postmodernism Nietzsche deserves to be called the "patron saint" of postmodernism.

titude, shared by many philosophers, rather than sharply defined doctrines that is of most importance in the recipe for PIS.

Proceeding to the twentieth century, we find several cups of the sociology of knowledge (itself a casserole of many of the previous ingredients) which, as we will argue below, is the decisive component helping to form SSK in the PIS bouillabaisse. There are also several lesser but nourishing ingredients in contemporary philosophy that the PIS has consumed. On a technical level there was by the 1960s a widely perceived failure of the Vienna Circle and its successors to ground science on a convincing epistemology. Then there are Karl Popper's arguments that we can never verify but only falsify a scientific theory on the basis of experimental evidence, so that science can never consist of certain knowledge. Even the possibility of falsification, however, was contested by the Duhem-Quine thesis and Quine's notion that scientific theories are underdetermined by the evidence (cf. proposition (iv)). On a more sophisticated level, Quine undermined the epistemological claims of science by asserting: (i) Reference is "inscrutable," meaning roughly the nominalistic thesis that the connection between word and object is ambiguous. There is no one-to-one correspondence between a word and the object it references. At a fundamental level the meaning of words referring to objects is determined by properties of the language, not just the object. (ii) Translation of a sentence from a hitherto unknown language to another is "indeterminant" in the sense that it is always possible to offer translations of the same sentence which conflict with each other depending on the hypotheses one must make about the language—which in turn are also indeterminant.[4] Such claims make problematic the pretensions of science to describe the world in terms of features which are independent of the language the scientist employs. On a different plane but of equal significance are various critiques of scientific realism by Richard Rorty, Nelson Goodman, and the later work of Hillary Putnam. These Gorgias-like doctrines, which have gained influence especially in the United States since World War II, all argue either that there is no such thing as reality as it is "in itself" independent of and uncontaminated by human concepts[5] or, if there is, we cannot faithfully represent it. The latter possibility would require a "God's eye view" which is unavailable to us. What we call the world *may* exist independently of us, but it is

---

[4]For a clear discussion and references to (i) and (ii) see Zammito (2004), Chapter 2.
[5]See especially the "internal realism" of Hillary Putnam as expounded, for instance, in Putnam (1981).

given structure and meaning via our conceptual structures; therefore when these structures change the world changes also, the result being that there are many "ways of world making" which can produce different worlds both temporally and simultaneously among different groups. A related consequence of this view, developed especially by Richard Rorty,[6] is that the traditional epistemological project of finding a body of self-evident truths mirroring a mind-independent world that can serve as a secure foundation of knowledge is bankrupt. For Rorty all this means that we should give up the search for truth since it cannot be established by comparison with a mind or culture independent reality and replace it by hermeneutics. Philosophy is a continuing "conversation" aimed at persuasion, not a discipline that can discover truth.

A more pervasive and less technical influence on PIS, which is also a practical consequence of some of the above anti-realist themes, has been pragmatism. For William James and John Dewey what we call "truth" is not a timeless property of a judgement reflecting its agreement with the way the world is, but a social product reflecting possibly incompatible human perspectives. It is a compliment that human beings confer on a idea because it has in James' words "cash value." In the end, true statements are exactly those in which it is advantageous for us to believe. This means that truth is not stable; what may be true for the purposes of one era may not be true for the purposes of another.[7]

Finally, an account of the philosophic ingredients in PIS would not be complete without mention of French poststructuralism represented principally by Michel Foucault and Jean-Francois Lyotard, both of whom provide an exotic spice to the recipe. For Foucault, whose thought seems a refined and subtle mixture of that of Thrasymachus and Nietzsche, knowledge is not "justified true belief" concerning an objective reality but an artifact of institutional power. What counts as knowledge is part of some sanctioned body of "discourse" or "regime of truth" enforced by the power emanating from professional bodies, laws, academic disciplines, schools, or other social agencies authorized to produce it. Furthermore, objects of knowledge *are* just the appropriate discourses imposed on a meaningless flux of experience; consequently they have no independent existence. Disputes over knowledge and truth—even in the sciences—are not rational, disinterested debates but contests of power. Lyotard in his famous book *The Postmodern Condition:*

---

[6] Rorty (1979).
[7] Novick (1988), pp. 152–153.

A Report on Knowledge (1979) which is a *locus classicus* of postmodernist doctrine spoke of the collapse of all the grand "metanarratives" originating in the Enlightenment that purport to be universally valid rational constructions, legitimizing and undergirding modern society. Capitalism, Marxism, Democracy, Progress, The Advance of Science, etc., etc., are myths that are now losing their potency and will be replaced as previous myths have been. Science in particular as the Enlightenment understood it has come to an end. It no longer represents a grand unified effort to understand the universe, but has broken into many narrow, mutually isolated "language games" whose goal is not to find truth or free humanity from superstition but to carry out the concrete technical goals of its corporate or government sponsors; the only criterion of its legitimation is "performativity."

Philosophy, however, is not the only inspiration for PIS, and we suspect that many of its ideas were not born from the imperatives of philosophical argument but were a delayed consequence of a major intellectual shift first making its appearance in other disciplines early in the twentieth century. We find interesting anticipations of the relativistic attitudes characteristic of PIS among historians and anthropologists beginning shortly after 1918. The former were influenced both by the breakdown in the historical consensus after World War I and by revolutionary changes in science that seemed to undermine the intellectual foundations of their discipline. Historians had witnessed (and in many cases participated in) the churning out of historically oriented propaganda during the war and the intractable conflict over war guilt in the 1920s. That this could happen among professional and highly trained scholars seemed to contradict the ideal of historical objectivity.

Prewar historians had modeled their profession on science in order to imitate what they felt was its success. By close examination of the documents, Baconian induction, and the testing of hypotheses according to the canons of scientific method, they felt that it ought to be possible to discover objective historical truth. But the bewildering developments in science from around 1900 through the 1920s such as quantum mechanics, relativity, and non-Aristotelian multi-valued logics not only defied common sense, but dealt a deadly blow to any hope that history could be anchored in traditional empirical science as it had been conceived by nineteenth century positivists. As misunderstood from popularizations—and sometimes the over ambitious pronouncements of scientists themselves—Einstein's relativity or Bohr's quantum mechanics seemed to imply cognitive or even moral relativism.[8]

How could knowledge be "objective," independent of the knower and his community if as Relativity seemed to imply all observations depended on the observer's frame of reference or if the very act of observation according to Heisenberg's Uncertainty Principle changed what was being observed at the subatomic level? These postmodern views fueled not only by the new science, but also by World War I and the chaotic events following it caused historians such as Crane Brinton, Carl Becker, Charles Beard, and others to conclude that historians in particular had no privileged access to bare historical reality or "what really happened" which is independent of the conceptual scheme or perspective he uses to make sense of the facts. How else could one explain that two equally well trained historians looking at the *same* documents could arrive at absolutely contradictory conclusions, for instance, concerning German war guilt? History therefore is necessarily nonobjective, and given the strangeness of its revolutionary "discoveries" perhaps this was true of science too.[9]

The development of cultural anthropology especially in the hands of Franz Boas also led to a cultural and ethical relativism which displaced a previous moral absolutism characteristic of the Victorian era. As in the case of history or philosophy, this transformation was accelerated by the experience of World War I. Prior to 1914 the West was confident in the superiority of its civilization, which in turn provided an intellectual justification for colonialism. Aside from its obvious cruelty and exploitation, colonialism had some positive features. It attracted the best and brightest of generations of young European men to dangerous and primitive places to spread what they considered the benefits of civilization. It also sometimes stimulated a sympathetic study of the cultures they encountered. Much of the resulting scholarship remains valuable today.[10] But not for a moment

---

[8] According to P. W. Bridgeman, for instance:

> We can no longer think of the object of knowledge as separated from the instrument of knowledge. We can no longer think of the object of knowledge as constituting a reality which is revealed to us by the instrument of knowledge ... (*Ibid.*, p. 138).

[9] See *Ibid.*, Chapter 6 for a thorough discussion of the development of historical relativism including the impact of science on historians in both the United States and Germany.

[10] In the universal condemnation of colonialism, it has been almost forgotten that the structure, vocabularies, and mutual relationships of many obscure native languages were carefully analyzed and in many cases given for the first time written form by nineteenth century scholars.

did colonial anthropologists or the administrators who employed them view these cultures in any sense as "equal" to the West. Most appeared to sincerely believe that they were the benign guardians of those whom Kipling called "Lesser breeds without the Law." By 1918, however, the superiority of the West appeared to consist in only the ability to scientifically and efficiently kill millions of people. From this experience cultural relativism blossomed not just as a methodology for anthropologists, but as a moral and intellectual imperative,[11] reflecting not only insight and moral progress but also a failure of nerve. An enduring consequence of World War I was that the West had lost its self-confidence.

This emerging relativism in early twentieth century history and anthropology, as well as some technical doctrines in philosophy ranging from the classical period to the near present must in some sense have prepared the way for PIS. It is not clear, however, what the exact relations between all these ideas and PIS actually are. To borrow again from our septic tank analogy, the residue of many intellectual ingredients are clearly present in PIS—Hume's critique of induction, epistemological relativism, and various forms of philosophical anti-realism being especially obvious. Certainly also, these ingredients have furnished PIS with a potent armory of opportunistic arguments. But whether the original meal consisted of one dish or many cannot be determined. The intellectual residue we find in the PIS tank may consist of side dishes prepared according to a shared inspiration, or appetizers to the main course. In other words, they may not be determining factors behind PIS, but merely independent manifestations of the same cultural forces that produced it and later a convenient source of arguments for a phenomenon that had a different origins. In fact, we will argue below (Chapter 13) that PIS became pervasive for political reasons quite other than a disinterested desire to explore the implications of certain conclusions of philosophers, historians, anthropologists, and sociologists.

We can also think of all these intellectual trends as symptoms of a much broader intellectual change signaling skepticism towards all traditional norms and values in art, literature, philosophy, and science. The early form of this change or what is called "modernism" began towards the end of the nineteenth century and accelerated in the chaos following World War I. It produced not only the ideas in philosophy, history, and anthropology we have described, but profound and very well-known changes in many other areas, especially art, music, and literature. It is then tempting

---

[11] Novick (1988), p. 144.

to argue that "postmodernism," and PIS in particular—in part characterized by a denial of the possibility of objective knowledge transcending the temporal or social circumstances of the knower—is a natural though delayed consequence of modernism; it is a typical example of what sociologists call "reflexivity," and amounts to a kind of auto-immune disease or self-immolation of modernism whose fierce intellectual weapons, having disposed of the pieties of the Victorian era, then turned on itself. An important question then becomes "why the delay?" To use our culinary/septic tank analogy, the ingredients of the PIS bouillabaisse fermented for decades peacefully along with science before the full potency of the dish was realized and could cause the intellectual dyspepsia of the Science Wars. Why did this happen? What historical and social circumstances then caused PIS to appear when it did rather than decades earlier? Why did it even appear at all? After all, many of the key ideas later exploited by PIS were prepared at a time when science enjoyed an unparalleled cultural prestige, and as late as the 1950s had little influence on the interpretation of science. The logical positivist program of the Vienna circle was still dominant. But by the 1970s the tide had turned; a conceptual revolution had occurred and several of the propositions (i)–(xv) had become almost uncontested in the humanities. How and why did this happen? Can we trace the bouillabaisse's development and assign responsibility for its synthesis from the raw ingredients we have described to the ambitions of a particular chef or group of chefs? These will be the questions we will look at in the following chapters.

Chapter 4

A Canary in the Mine

To what intellectual chef then do we owe the beginning of PIS? One answer has been given and relentlessly defended[1] by David Stove: He locates culinary guilt in the philosophy of Karl Popper (1902–1994). For Stove, Popper was the unconscious *fons et origo* of the sickening flavor of the PIS bouillabaisse—that is, of all the fashionable irrationalism that was to come in the philosophy of science. More precisely, in relation to these later currents, Stove compares Popper's role to that of the liberal, westernized Stepan Verkhovensky who played with "advanced ideas" and whose son Peter as a result became a terrorist in Dostoyevsky's novel *The Possessed*.[2] Popper's philosophy, according to Stove, reflected the chaos of post-World War I Vienna, the city where he resided until his emigration to New Zealand in 1937 because of the Anschluss.[3] It was a typical product of the Jazz Age which was characterized by the desire to overturn all established opinion in every field, and, concerning science, was the philosophic embodiment of Cole Porter's "anything goes" and "day's night today," "good's bad today."[4]

What was Popper's original sin? Stove's answer is that Popper accepted Hume's demolition of induction (and hence the first of our PIS propositions). For Hume there is *no* valid reason to accept an inductive argument. The only reason we expect a correlation between two sequences of events to continue in the future is Pavlovian psychological conditioning; we are

---

[1] Stove (1982); Stove (2000).
[2] Stove (2000), p. 16. Dostoyevsky's Peter Verkhovensky was modeled on the anarchist Nechayev.
[3] Popper, born of middle class Jewish parents, was educated at the University of Vienna and later taught in a secondary school.
[4] Stove (1982), p. 5 ff.

like the chicken in the hen house (to use an example of Russell[5]) who on the basis of induction expects to be fed by the farmer each day—until the farmer comes one morning and wrings its neck. Thus, if Hume's thesis is correct we have no grounds to believe that any scientific theory which has been verified any finite number of times (no matter how large) is either true or probable. Consequently, while ordinary scientists (and the logical positivists whom Popper opposed) regarded a theory as having (in some sense) an increasing probability of being true as the number of its confirmed instances or successful predictions increases, Popper denied on the basis of Hume's argument that this ever can be the case. Even worse, he tried to prove by probabilistic reasoning that the probability of such a theory being confirmed in an infinite number of trials is 0.[6] More generally, the probability that a theory is true is inversely related to its explanatory power. A highly probable theory is trivial and says little. Science should therefore seek bold, powerful, and *improbable* theories. Popper concluded that:

> Science is not a system of certain, or well-established statements; nor is it a system which steadily advances to a state of finality. Our science is not knowledge (epistēmē): it can never claim to have attained truth, or even a substitute for it, such as probability.[7]

Furthermore, according to Popper induction plays no part in science, not even in the formulation of the initial conjecture. Popper points out that observation is theory-laden, but not in the sense (as several later philosophers have argued) that "theory" determines *what* we see. Rather observation is always in the service of some prior theory. We *never* look at the world with a totally empty mind, as Bacon prescribed, to see what theories we can establish by "inductions," but only to try to understand previously existing theories.

This rejection of any role for induction led Popper to his famous theory of falsifiability which replaces induction by deduction. Since we can never verify or increase the probability of a theory, however many times it is confirmed, we should instead try to falsify it—which can be done without any appeal to induction. A scientific theory worth taking seriously will always make a deductively justified prediction. Therefore, if we find by experiment that this prediction is false, then we must reject the theory.

---

[5] Russell (1999), p. 63.
[6] Popper (1968), Appendix vii.
[7] *Ibid.*, p. 278.

If it passes this test, we then design a sequence of every more rigorous tests of other predictions in order to try to falsify the theory in the same way by an experimental failure. In this way we can explore the theory's consequences and gain deeper insight into its structure. In the short run the theory may be successful in that it has survived more and more severe attempts at falsification. But our theories are never more than conjectures, and the main business of science is to make ever bolder conjectures followed by attempts at refutation. To Popper, this possibility of falsifying a theory also marks the demarcation between science and non-science. Nonscientific systems like religions, ideologies such as Marxism or psychoanalysis do not have this property. They provide explanations for *all* possibilities. Unlike science they forbid nothing. Every event in the world happens through Divine Providence. Every situation in society has a Marxian interpretation. Every state of the human psyche is grist for a psychoanalytic explanation. Indeed, any attempt to criticize or falsify these systems is forbidden, and can be diagnosed as a result of sin, bourgeois false consciousness, or as an Oedipal rebellion against the authority of the Father.

By eliminating induction either in the formulation of a scientific conjecture or in its confirmation, Popper feels that he has "solved" Hume's problem of induction. But the price has been high: knowledge and truth have been eliminated in so far as they relate to science, preparing the way for a more universal skepticism. Beyond agreeing with Hume, Popper's entire system turned long standing common sense views of science upside down. He had substituted falsifiability for verifiability, deduction for induction, bold conjectures for scientific caution. This Jazz Age deconstruction certainly opened the flood gates for more radical questionings of science and may indeed—like other specimens of the period (e.g., psychoanalysis)—have owed something to the disordered postwar state of Viennese culture in the 1920s. And as Bertrand Russell maintained: "The growth of unreason throughout the nineteenth century and what has passed for the twentieth is a natural sequence to Hume's destruction of empiricism."[8] The philosopher of science Newton-Smith also essentially agrees with Stove, calling Popper an "irrational rationalist." He gives technical arguments designed to show that Popper's rejection of induction leaves "no way in which he can justify the claims that there is growth of scientific knowledge and that science is a rational activity."[9] Popper, he feels, is mistaken in thinking that adoption

---

[8] Quoted in Popper (1972), p. 1.
[9] Newton-Smith (1981), p. 52.

of the falsification principle "will save him from simple skepticism."[10] We would agree then that there is a fair amount of truth in Stove's thesis.

There are many possible criticisms of Popper's system. One of the most obvious is that it fails to accurately describe how scientists actually behave. In a particular case scientists may only rarely agree on what constitutes falsifiability, or if they do and the theory is apparently falsified, they will not abandon it. Suppose, for instance, that an astronomical theory has already passed many attempts to falsify it and has made a large number of successful predictions, but that one experiment goes bad. In that case the astronomer may always explain away the "falsification" by a Duhem-Quine type argument: something is wrong with the experiment; perhaps the procedure or the cosmic ray counter, etc., is defective. Or perhaps, the problem is not with the experimental procedure or the instruments. What if the theory was incorrectly applied, some important variable was ignored, or the interpretation of the data was mathematically defective? All these possibilities (and more) will need to be closely examined before a definite conclusion can be reached. Even if further investigation still cannot explain the result and the astronomer agrees that it is incompatible with his theory, he will put it aside since the rest of the theory is successful and treat it as a problem to be solved later.[11]

For similar reasons, one can argue that falsifiability in contradiction to Popper's assertions fails to demarcate science from non-science. It may work for Marxism and psychoanalysis since these systems can "explain" everything (including criticisms of them), but suppose that our astronomer is confronted by an reputable astrologer who has cast a large number of valid horoscopes. Both he and the astronomer may agree in principle that their respective systems are falsifiable. However, confronted with a blatantly incorrect horoscope the astrologer can argue, as the astronomer did. He can point out the well-known and astrologically justified fact that *very* small errors in the observations behind the horoscope can cause hugely erroneous predictions—such as Jerome Cardan's prediction of long life and a happy reign for Henry VIII's son by Jane Seymour Edward VI,[12] which as he points out is a classic example of the ill-conditioning of a valid science analyzed in every graduate level astrology textbook. In spite of this error Cardan, the astrologer emphasizes, was a great scientist. His ironclad

---

[10] *Ibid.*

[11] This situation is what Kuhn calls an anomaly and is common in the practice of "normal science." See Chapter 7 below.

[12] Edward died at age fifteen after a reign of little more than six years.

proof, for example, that the position of the planets in the year of Christ's birth necessarily implied His crucifixion is still regarded as a masterpiece of astrological reasoning and continues to serve as a Kuhnian exemplar in the field.

Something more, then, is needed than falsification to compare the quality of disparate theories. Even if theories can never attain absolute truth, we feel that in some sense one theory may be closer to this impossible goal than another. If this were not so, the idea of scientific progress would be an illusion. Popper attempts to give meaning to this intuition by his concept of "verisimilitude" which, roughly speaking, is an attempt to compare the quantity of "truth" in each theory. More exactly, Popper said that a theory t_2 has greater verisimilitude or is "closer to the truth" than t_1 if: "(a) the truth-content of t_2, but not the falsity-content of t_2, exceeds that of t_1 and (b) the falsity-content of t_1, but not its truth-content, exceeds that of t_2."[13]

But there is a severe problem with this definition when both theories, e.g., the astronomer's and astrologer's, each make an indefinite, possibly infinite number of true statements. As Popper admits, he can find no available "measure" to determine the relative truth content of two such competing theories. Such a comparison would seem to depend on an inductive argument: one theory may so far have made more true predictions (equivalently withstood more severe tests) than the other; but from Hume's critique of induction we can have no assurance that this will continue in the indefinite future. However, there is a more immediately fatal objection to Popper's idea of verisimilitude; it can be shown his definition implies that any two comparable false theories (i.e., theories such that the set of consequences of one is a subset of the set of consequences of the other, with each theory yielding some false statements) belonging to a quite general class have the *same* verisimilitude. This result means that verisimilitude at least in the form considered by Popper is a "dismal failure."[14]

From the above analysis Popper seems similar indeed to Stepan Verkhovensky. But we also need to recognize that in many other respects Popper is quite conventional and has a very different conceptual framework than PIS. Although science cannot be knowledge, the striving for knowledge and the search for truth are still the strongest motives for scientific

---

[13] Popper (1965), p. 233.
[14] Newton-Smith (1981), p. 184. For verisimilitude, see Popper (1972) and Popper (1965). For criticism, see Newton-Smith (1981), pp. 52–76. The fact that two comparable false theories are equivalent with respect to verisimilitude was shown independently by Tichý (1974) and Miller (1974).

discovery."[15] He opposes instrumentalism as a "new betrayal" in the tradition of Cardinal Bellarmine of the search for scientific truth that began with Galileo.[16] Popper, moreover, emphatically rejects relativism which he sees as the "main philosophical malady of our time"[17] and believes in the correspondence theory of truth as modified by the logician Tarski. He is also a realist and emphatically rejects theses like propositions (ii) and (iii) in Chapter 1. It is not the case that there is a world of appearances only or a world of appearances behind which there is some unknowable ultimate reality. There is no distinction for Popper between the reality of primary and secondary qualities as seventeenth century essentialists thought. As science digs deeper its conjectures reveal successive higher level worlds that explain and stand behind more accessible worlds, beginning with the ordinary world. These are all "layers" of the world and are all equally real.[18] The three main layers which Popper identifies he calls World 1, World 2, and World 3. World 1 consists of the ordinary world of physical objects. World 2 is the world of mental states, and World 3 is the world world of "intelligibles or *ideas in the objective sense*"[19] such as "objective knowledge" and mathematics. All these worlds are real. The mathematical objects in World 3 for instance are just as real as cats and dogs in World 1. Popper admits that World 3 is our creation, but once created it is autonomous; we may have created the sequence of natural numbers, but having done this there are facts about it such as the properties of prime numbers which exist independently of us and may be discovered.[20]

As we have seen, for Popper there is additionally a real epistemological difference between science and non-science. We may find Popper's distinction based on the possibility of falsifying real science unsatisfactory, but at least the distinction is there. As should be obvious, this is not the case for full-blown PIS since PIS regards both science and non-science as merely different ideological systems. Also, because science proceeds by "conjectures and refutations" Popper believes that internal factors alone determine scientific change. He would certainly reject the sociological and political aspects of PIS. For Popper falsification of a scientific conjecture is in principle

---

[15] Popper (1968), p. 278.
[16] Popper (1965), p. 97 ff.
[17] See the Addendum to the 1962 edition of Popper's, *Open Society and its Enemies*.
[18] Popper (1965), p. 115.
[19] Popper (1972), p. 154 [emphasis in original].
[20] *Ibid.*, p. 118. There is an element of Platonism here: Popper credits Plato with the discovery of the reality of World 3, although Plato did not realize that it was a human creation (*Ibid.*, p. 132).

fairly straightforward, although it can be technically demanding. Hence a scientific decision is not the result of a power play or process of negotiation among scientists, but is forced by the results of experiment. Lastly, Popper differed from PIS in having an extremely high regard for science, viewing it as one of the finest creations of the human spirit, and believing (perhaps inconsistently considering his position that scientific theories cannot be knowledge) that it is genuinely progressive: "in science (and only in science) can we say that we have made genuine progress: that we know more than we did before."[21] In all these attitudes Popper was actually closer to the logical positivists who had shunned him in Vienna than he was to what was to follow. Like them he was searching for prescriptive rules that valid science should (but unfortunately sometimes does not) obey. Both he and the positivists rejected the thesis that the nature of science can be characterized by its history. Human nature being what it is, the history of any human activity, including science, is a messy, sometimes irrational business, with many false starts and dead ends. This again is very different from PIS which rejects the prescriptive approach and frequently appeals to history in the form of "case studies" to demonstrate what it regards as the irrationality, contingency, or political contamination of science.

Stove's indictment, then, seems much too strong. Although Popper was an extremely prominent philosopher of science during his life—who received almost every honor it is possible to win including a knighthood—and who has had continuing impact on technical issues in the philosophy of science, he is from the point of view of PIS a rather archaic figure and too much of a realist and rationalist to be a precursor, except concerning induction. He was essentially a bystander in the kitchen responsible for no more than some seasoning in the bouillabaisse. In terms of Stove's analogy, he may share some guilt, but not as the liberal father of the terrorist son; at worst he is a great uncle whose only contamination by liberalism was some sympathy for Alexander II's liberation of the serfs. One might think of him as either the very beginning of what will become a fundamental bend in the road or as a canary in a mine, the peculiarities of whose song being an early warning of the possibility of poisoning by carbon monoxide. Even if it is the case as both Stove and Newton-Smith believe that his system (certainly contrary to his intentions) logically implies that science is irrational, this has had probably little significance for the development of PIS. As we have remarked, a movement of this immense size and significance

---

[21] Popper (1970), p. 57.

is not derived like some mathematical theorem from the observation of a logical gap in the arguments of a predecessor, and in particular cannot just be a logical consequence of Hume's view of induction.[22] We shall have to look elsewhere, perhaps even outside of the philosophy of science, for the real twentieth century ancestry of PIS.

---

[22] Proponents of PIS will, of course, opportunistically use weapons where they find them; so they do praise Hume and accuse scientists and each other when quarreling of "inductivism."

Chapter 5

The Unmasking of Reason

A common characteristic of many of the intellectual movements sketched in Chapter 3, especially as they developed after around 1850, is a progressive narrowing of the scope of human reason together with a destruction of belief in its integrity and autonomy. As Max Horkheimer observed this "end of Reason" followed inexorably from the very triumph of the power of reason since the Enlightenment.[1] In a process of self-immolation similar to that which we have argued was instrumental in the transition from modernism to postmodernism, Reason, via its ruthless critique of all existing institutions, ultimately also undermined itself. This suicidal course, a nearly precise but negative parallelism to the process celebrated by Auguste Comte, went through three stages of which Horkheimer identifies two. The first was already evident in the rationally founded skepticism of Hume and can even be traced back to the sophistic opposition to Plato and Ockham's or Siger of Brabant's destructive criticism of Thomas Aquinas' scholastic rationalism.[2] By the Renaissance Reason had effectively lost its role as an arbiter of religious truth. Also, the problem of the relationship of the Self to the external world raised by Descartes which bedeviled seventeenth century philosophers—and especially the solution to it given by Kant— along with the corrosive effects of Hume's skepicism seemed to destroy the Platonic vision that Reason could be either "the herald of eternal ideas which were only dimly shadowed in the material world" or the belief that it could "discover the immutable forms of reality in which eternal reason was expressed."[3] But if Reason could not give comforting answers to ultimate religious, metaphysical, or epistemological problems, perhaps it still

---

[1] Horkheimer (1941), p. 317.
[2] *Ibid.* However, we feel it more historically accurate to substitute "sophistic" for Horkheimer's "Socratic" and "Ockham" for "Roger Bacon."
[3] *Ibid.* See also Horkheimer (1947).

could by contemplating the ultimate structure of reality, so the thinkers of the Enlightenment thought, construct a rational society and dissolve the archaic structures of the Old Regime. But by the nineteenth century this hope was also being undermined as a result of critical self-examination. Reason was no longer the neutral and infallible arbiter of the social world. Reason was increasingly revealed to be the slave of hidden factors of which the reasoner was unaware, an immediate consequence being that it could no longer be taken at face value. Apparently rational argument need not be either defended or refuted; only the "unmasking" of its origins and hidden purposes counted. For Hegel Reason was the work of the World Spirit, not the individual, which caused the reasoner to fulfill purposes and goals that could contradict his conscious intentions. For Wilhelm Dilthey (1833–1911) and German historicism generally what passed for Reason was completely relative to the historical situation.[4] For Marx the political applications of Reason by the ruling class was mere ideology, reflecting (often unconscious) class interests and the constraints imposed on the social order by the means of production; it was usually a form of what Engels and later Marxists called "false consciousness" which systematically misled the proletariat as to its true position. For Vilfredo Pareto (1848–1923) the motivation behind human actions are deep nonrational perhaps unconscious drives of various sorts which he called "residues." Reason functions merely as a rationalization or "derivation" of these actions. Still later for Freud "Reason" was a rationalization of neurotic complexes. By the twentieth century, according to Max Horkheimer, it seemed that Reason was now limited to the technical and scientific spheres where it served as a tool or means to achieve various ends, but not as an authority to judge them. Social goals were now beyond Reason.[5]

The various deconstructions of Reason we have just described converged in the new twentieth century discipline of sociology, and particularly in its subdiscipline of the sociology of knowledge (or as we have abbreviated it "SK.") We believe that SK in turn is the most important intellectual precursor of the "third stage" of the eclipse of Reason (and one unforseen by Horkheimer), i.e., the assault on Reason in its remaining fortress of science through the rise of PIS. Especially in light of the sociological flavor of many of the propositions (i)–(xv), we feel that its contribution to the PIS bouillabaisse is more significant than either the philosophical antecedents examined in Chapter 3 or any unresolved technical issues in mid-twentieth cen-

---

[4] Berger and Luckman (1966), p. 6.
[5] Horkheimer (1947), Chapter I.

tury philosophy of science. Although many specific ideas characterizing SK can be found in the general atmosphere of nineteenth century thought or in early sociologists such as Emile Durkheim (1857–1917), Ludwig Gumplowicz (1838–1909), Max Weber (1864–1920), Max Scheler (1874–1928),[6] and Lucien Lévy-Bruhl (1857–1939), its modern form was primarily the invention of one man, Karl Mannheim (1893–1947). Ironically, Mannheim was a great admirer of science and basically accepted the (admittedly exaggerated) epistemological contrast sketched above between science and mathematics on the one hand and the humanities on the other,[7] but his thought contained the germs of SSK which others could exploit.

Mannnheim's life and career cannot be understood (as he would have been the first to admit) independently of his social and historical situation. He was a typical Central European intellectual trying to make sense of (and survive) the political and intellectual chaos following World War I. He was born in Budapest of a prosperous secular Jewish family in a period of extraordinary cultural brilliance, simultaneously menaced by violence and decay. An extraordinarily talented student, Mannheim studied philosophy and the social sciences at the Universities of Budapest and Berlin, receiving his doctorate in Philosophy and German literature *summa cum laude* in 1918 from Budapest. Fortunately his fragile health had spared him military service in the catastrophe of 1914. His teachers included Georg Simmel (with whom he was especially impressed), Ernst Cassirer, and Ernst Troelsch. For two months in the spring of 1914 he even visited Paris and attended the lectures of Henri Bergson. During the War he joined a discussion group in Budapest organized by his personal friend the Marxist philosopher Georg Lukáks. After the implosion of the Austro-Hungarian Empire in 1918, Lukáks joined the newly organized Hungarian Communist Party and in 1919 became Commissioner of Education in the violently radical and short lived Hungarian Soviet Republic established by Béla Kun. Although Mannheim was broadly leftist in politics and probably sympathized with many of the goals of the Kun government, he did not join the Party and was essentially a moderate who did not approve of the revolutionary excesses of the regime, especially the contamination of scholarship

---

[6]The term "sociology of knowledge" or *Wissenssociologie* in fact was the invention of Scheler, but Scheler's thought had much less relativism than later SK. "If one may describe Scheler's method graphically, it is to throw a sizable sop to the dragon of relativity, but only so as to enter the castle of ontological certitude better." (Berger and Luckman (1966), p. 7).

[7]In fact, our characterization of the humanities at the beginning of the previous chapter is not far from Mannheim's "total conception of ideology" discussed below.

by political ideology. Any doubts he felt, however, did not prevent him from accepting an appointment from Lukáks as a lecturer in philosophy at the University of Budapest. This official entanglement with the regime made it impossible for him to stay in Hungary after its collapse in August 1919.[8] Under the ensuing counter-revolutionary and anti-Semitic regime of Horthy, Mannheim became one of 100,000 exiles driven out of Hungary. 75,000 other sympathizers of Béla Kun were imprisoned and 5,000 killed.[9]

By 1921 Mannheim had settled in Heidelberg where he was appointed an extraordinary lecturer in sociology in 1926. It was in this period (1929) that he completed the work for which he is most widely remembered *Ideologie und Utopie*. In 1930 he was awarded a Professorship and became Director of a newly created College of Sociology at Frankfurt. This position was unfortunately brief. Being a Jew, Mannheim was dismissed by the "Law of Restoration of Civil Servants to their Offices" promulgated in April 1933 by the new Nazi government and once again had to go into exile. He spent the rest of his life in Great Britain, becoming first a Lecturer in the London School of Economics and in 1945 a Professor at the London School's Institute of Education. His life was prematurely ended by a heart attack in 1947.

Twice driven into exile, Mannheim's life had been tempestuous. However, he was also lucky. He had survived some of the most violent episodes of the twentieth century, and unlike many other refugees he was not reduced to poverty. He was able to obtain recognition and after some delay decent academic positions. With this background it is not surprising that he devoted his academic life to the goal of understanding the nature and function of political ideology, especially in its messianic Left and Right forms. Mannheim is more of a talented synthesizer who mixes existing intellectual ingredients in a new way than a truly original thinker who creates his own foundations *de novo*. He freely borrowed from the insights of such earlier thinkers as Hegel, Marx, Nietzsche, Lukács, Weber, Dilthey, and Simmel, some of whose contributions to an intellectual atmosphere in which SK could flourish we have already described. We have also pointed out (in Chapter 3) the widespread questioning by historians and other thinkers after World War I concerning the possibility of "objective" social and political knowledge. All these intellectual currents together with the ideological chaos of the postwar

---

[8] Hungary was invaded by Czech and Romanian armies as well as a French-backed force commanded by Admiral Horthy.

[9] See Woldring (1987), p. 17. I am indebted to Chapters I–III of this work for the biographical information in this section.

period helped to shape Mannheim's thought. He was, for instance, particularly fascinated by the historicism of his former teacher Ernst Troeltsch which was the subject of an early (1924) essay.[10]

In Mannheim's first book *Ideologie und Utopie* (1929) and in subsequent essays he began to develop the new discipline of SK which in his hands explained all ideological systems in terms of contextual factors such as class, interest, status, social roles, historical position, etc. But although this point of view owes much to Marx, Mannheim's approach was far more general than Marxism since the latter views ideology as purely a "superstructure" of economic class. He was also politically much more of a Social Democrat than a Marxist. This moderation together with his recognition of the relativism present in ideologies including Marxism caused him to oppose the pretensions of the radical Marxism of his colleagues at the Frankfurt *Institut für Sozizlforschung* directed by Max Horkheimer. As a consequence, there seems to have been a certain rivalry between Mannheim's College and the Institute with which Mannheim shared office space; personal contacts were distant but friendly.[11]

For Mannheim the purpose and function of SK was essentially *oppositional*; he believed that its origins and characteristic political use were first revealed by certain forms of humanist thought in the eighteenth century Enlightenment. It is worth quoting Mannheim at length on this subject since we will meet a similar attitude towards science among some of his modern sociological heirs:

> The systematic as well as sociological core of this oppositional science was its hostility toward theology and metaphysics—it saw its main task in the *disintegration* of the monarchy with its vestigially theocratic tradition, and of the clergy which was one of its supporters. In this struggle, we encounter for the first time a certain way of depreciating ideas ... We mean the phenomenon that one can call the 'unmasking turn of mind'. This is a turn of mind which does not seek to refute, negate, or call into doubt certain ideas, but rather to *disintegrate* them, and that in such a way that the whole world outlook of a social stratum becomes disintegrated at the same time. ... In denying the truth of an idea, I still presuppose it as a 'thesis' and thus put myself upon the same theoretical (and nothing but theoretical) basis as the one on which the

---

[10] *Ibid.*, pp. 103–107. Troeltsch had concluded "that historical study, rather than providing understanding, had fatally undermined 'all stable norms and ideals of human nature'." (Novick (1988), p. 157.)

[11] *Ibid.*, p. 32. The personnel of the *Institut für Sozizlforschung* included Theodor Adorno, Walter Benjamin, Erich Fromm, and Herbert Marcuse.

idea is constituted. In casting doubt upon the 'idea', I still think within the same categorical pattern as the one in which it has its being. But when I do not even raise the question ... whether what the idea asserts is true, but consider it merely in terms of the *extra-theoretical function* it serves, then, and only then, do I achieve an 'unmasking' which in fact represents no theoretical refutation but the destruction of the practical effectiveness of these ideas.[12] [emphasis in original]

In Mannheim's opinion the modern version of SK unmasks not the monarchy, the church, or—as in the case of Marxism—bourgeois society, but ideologies in general.

In unmasking ideologies, we seek to bring to light an unconscious process not in order to annihilate the moral existence of persons making certain statements,[13] but in order to destroy the social efficacy of certain ideas by unmasking the functions they serve.[14]

Marx had argued that certain ideologies served the interests of dominant classes and corresponded to the economic substructure of the society. The ideology of Feudalism or of Capitalism pretended to be "objective," but actually was a distortion of reality whose purpose was to enhance and defend the power of the dominant classes in these systems. Such concealed purposes, according to Marx, needed to be unmasked in order to make way for a true accounting of society by the proletariat whose ideology alone had the potential of objectivity. But Mannheim extends Marx's conception so that not even Marxism is spared unmasking. To this end, he begins by making a distinction between a "particular" and a "total" theory of ideology. Both (as in the case of Marxism) proceed by explaining an ideology in terms of social interests it satisfies:

The ideas expressed by the subject are thus regarded as functions of his existence. This means that opinions, statements, propositions, and systems of ideas are not taken at their face value but are interpreted in the light of the life-situation of the one who expresses them. It signifies further that the specific character and life-situation of the subject influence his opinions, perceptions, and interpretations.[15]

---

[12] Mannheim (1952), pp. 139–140.
[13] As in the case of unmasking "lies" which Mannheim has previously discussed.
[14] *Ibid.*, p. 141.
[15] Mannheim (1936), p. 50.

But the two differ in the sense that the first "designates only a part of the opponent's assertions as ideology—and this only with reference to their content"[16] while the second "calls into question the opponent's total *Weltanschauung* (including his conceptual apparatus)."[17] Here one is interested in analyzing the structure of the mind of one's opponent in its totality, and not merely in singling out a few isolated propositions.[18] To put it another way, the particular conception of ideology leaves open channels of meaningful communication between the parties. There is the the possibility of a common frame of reference by which the distortions and hidden functions of the ideology may be revealed even to the believer. On the other hand, in the total conception the entire thought systems of the parties are contained in different intellectual worlds, either due to the fact that they live in completely disjoint historical epochs or in different social strata. Because the divergence is so profound no resolution is possible. To use a later term, such thought systems are *incommensurable*. But both the particular and total conceptions unmask the belief systems they study by revealing the interests behind them. Mannheim soon notices, however, that there is no reason to limit the total conception of ideology to any particular ideology. That would be merely a special form of the theory. Haunted by the memory of violent ideologies of the both the Left and the Right—especially in Hungary, Mannheim earned the wrath of Marxists by being willing to apply his total conception to their ideology, just as he did to fascism or to bourgeois ideology.[19] Even more disturbingly, Mannheim recognized the possibility of applying the methods of SK to SK itself. As Mannheim was well aware we can ask that SK be "unmasked" and the political motives behind it exposed in just the same way as for all other ideologies. Mannheim struggled to come to terms with this self-destructive paradox throughout his career. Although the problem was logically inherent in SK, he came to believe that it was possible for thinkers to compensate for it. The group best suited for this task—of understanding how social factors affect its own thought—was what he called the "free floating intellectuals." Ordinary people locked into a class or social milieu cannot see how their situation determines their ideology. But the intelligentsia, while it may have arisen from various classes,

---

[16] *Ibid.*
[17] *Ibid.*
[18] *Ibid.*, p. 68.
[19] This aspect of Mannheim's was particularly criticized by Horkheimer and others of the *Institut für Sozizlforschung* since it seemed to subvert Marx's distinction between true and false consciousness. If Marxism was just another ideology, how could its claim to provide an objective analysis of society be justified? See Jay (1973), pp. 63-64.

has escaped class-bound intellectual limitations by virtue of its education. Hence it is in the best situation to understand all ideologies, including the constraints imposed by its own historical situation, and sympathetically "translate" a position arising from one ideology so that another can understand it. Few, however, have found this solution convincing and the possibility of "self-unmasking" or as it was later called "reflexivity" continued to haunt, as we shall see, Mannheim's intellectual descendants who think of themselves as sociologists of scientific knowledge.

The total conception of ideology also appears to lead to a radical relativism since the concept of an absolute truth in the cultural or political sphere which is independent of historical and social factors becomes meaningless. Mannheim accepts this consequence, but argues that it does not entail relativism, at least in its traditional form. According to him, ordinary relativism presupposes the *existence* of such truths, but believes that *all* observers are necessarily contaminated by ideology arising from their social or class position which distorts their perception. Instead, Mannheim introduces a position he calls "relationism." In matters relating to culture and politics there *is* no absolute truth independent of the subject's historical or social position.[20]

> Even a god could not formulate a proposition on historical subjects like $2 \times 2 = 4$, for what is intelligible in history can be formulated only with reference to problems and conceptual constructions which themselves arise in the flux of historical experience.[21]

But this implies that an individual's views cannot be distortions, for there is no "objective" asocial or nonhistorical point of view to compare them with. Thus while a person's social or historical situation may condition his judgements, they may be valid and count as real knowledge, if they accurately reflect that situation. Thus:

> Knowledge, as seen in the light of the total conception of ideology, is by no means an illusory experience, for ideology in its relational concept is not at all identical with illusion. Knowledge arising out of our experience in actual life situations, though not absolute, is knowledge none the less.[22]

---

[20] An unfriendly critic might claim that "relationism" is simply more radical form of relativism, although Mannheim refuses to admit this.
[21] *Ibid.*, p. 71.
[22] *Ibid.*, p. 76.

On the other hand, not all such views are equally valid. Some may be manifestations of "false consciousness," which usually arises when someone invokes the ideology appropriate to a past epoch in a present situation where it no longer applies.[23] An example would be a land owner running his estate on modern capitalistic lines but who is still imbued with a feudal, patriarchal ideology in relation to his tenants.[24]

There were two exceptions to Mannheim's total theory of ideology. As David Bloor[25], Michael Mulkay[26], Mary Hesse[27], and others have noted, SK accepted the thesis that science and mathematics transcended ideology in that they served the interest of no particular social group and either accurately mirrored reality or were formally true. Since they constituted "knowledge as such," Mannheim had no interest in "unmasking" them, but only in pointing out how these exceptional areas differ from other more subjective realms of human cognition.[28] But some qualification is needed here; Mannheim and other progressives (especially Marxists) of his generation would admit that social considerations might help to explain many things about science and mathematics such as their rate of progress, their cultivation by one culture or lack of by another, their contemporary influence in society, the structure of scientific institutions, and the nature of scientific or mathematical careers.[29] SK might even explain the persistence of false theories and practices like astrology, Ptolemaic astronomy, phrenology, or circle squaring, but it can not explain valid scientific or mathematical *content*. It seems obvious that the content of correct scientific or mathematical theories is epistemologically unlike that of either false versions or of political ideologies, religions, historical interpretations, and so forth. The truths of science and mathematics do not require a sociological explanation. They are believed because of their rationality and the evidence for them. Bluntly speaking, correct science consists of objective truths that mirror properties of a real physical world. Similarly, correct mathematics consists of true theorems. One can check scientific claims by experiment and the proofs of

---

[23] *Ibid.*, pp. 85–87.
[24] *Ibid.*, p. 86.
[25] Bloor (1976).
[26] Mulkay (1979a).
[27] Hesse (1980).
[28] Mannheim does not dwell on the contrast between the exact sciences and areas subject to the analysis of SK. He just accepts it. A few remarks on the contrast may be found in Mannheim (1936), pp. 39, 145–147, and 263.
[29] A somewhat crude Marxist analysis of such factors may be found in Hessen (1931) which makes Newton a mere epiphenomenon of the rise of the bourgeoisie and the economic transformations this wrought.

mathematical theorems by logic. *These* tasks have nothing to do with the cultural or political factors that may influence the development of science and mathematics. The content of these disciplines is independent of "existential" factors and cannot be analyzed in the same manner as a social ideology or religion.[30] For instance, we value Newton's derivation of the planetary orbits from the inverse square law of gravitation because of its ability to account for Kepler's Laws and the many predictive successes it made possible in astronomy. These properties also account for its rapid acceptance by seventeenth century astronomers and have nothing to do with religious, sociological, or political factors in Restoration or Hanoverian England. On the other hand, while science tells us about reality the hermeneutic investigations of scholars in humanities tell us only about the minds of these scholars and their cultural, political, and social milieu. Besides mere dates and elementary facts, they contribute no knowledge of permanent significance about the world. Even worse, as we have already noted, since assertions in the humanities are seldom verifiable, they are considered by the logical positivists to be *meaningless*.

These views were basically seconded by Robert K. Merton, a founder of the sociology of science and from the 1930s to the 1960s perhaps the most influential of American sociologists. He studied the role of characteristic Puritan/Protestant ideas in the stimulation of seventeenth century English science and in the foundation of the Royal Society and also their continuing influence on the values of modern science.[31] Concerning science, as David Bloor has pointed out:

> Merton largely took for granted that, in the proper functioning of the institution, the rational appraisal of evidence and the testing of theories were autonomous processes. The inner, rational core of scientific thinking was not itself social. Thus he would routinely contrast the rational and social properties of science.[32]

In a fundamental way the Mannheim-Merton version of SK was a culmination of the Enlightenment project to rationally organize society. Dominant but historically contingent and irrational political beliefs, religions, and social structure would be "unmasked" and "disintegrated" not by directly

---

[30]See Mannheim (1936), pp. 261–275. Mannheim especially views mathematics in this category and takes as a typical example the statement $2 \times 2 = 4$.
[31]See, Merton (1980).
[32]Bloor (2004), p. 920.

refuting the arguments supporting them, but by uncovering their origins and revealing their social purposes. The way therefore would be open to replace them by the products of scientific reason.

It is also clear that both Mannheim and Merton shared the general picture of science prevalent among the educated public of the mid-twentieth century which we have outlined in Chapter 1. They were firmly embedded within modernism, doubting neither the objectivity of science nor its progress. To them also, it was one of the most glorious achievements of our civilization and a vital tool for the betterment of society. The slow unraveling of this consensus and its replacement by the application of the "unmasking" idea characteristic of SK to the *content* of science has been the main trend in the history and philosophy of science in the last forty years. It is what we have called the "third stage" in Max Horkheimer's eclipse of Reason, the first stage of which originating in the skepticism and epistemological problems revealed by seventeenth century philosophy and the second perfected by SK itself. To get a crude but essentially accurate idea of the form the third stage will take, one need only perform the thought experiment of replacing "ideology" by "science" everywhere in *Ideology and Utopia*. In the hands of PIS science will become just another ideology. Scientific theories will not merely be distortions of truths about reality; no such truths are possible since that would require a nonexistent "God's eye," point of view. Instead, theories, like all other beliefs, are relational constructions ultimately depending on the total "life-situation" of the scientist. But scientists, like Mannheim's "ordinary people" cannot comprehend this. The role of Mannheim's "free floating intellectuals," who can, will be played by the theorists of PIS. They alone understand the true nature of science.

Chapter 6

Thought Styles and Thought Collectives

It did not take long for Mannheim's refusal to apply SK to the content of science to be challenged. Among the first to do so was Ludwig Fleck (1896–1961), a little known Polish physician and medical researcher who specialized in serology and bacteriology.[1] Fleck's most fundamental work *Entstehung und Entwicklung einer wissenschaftlichen Tatsache: Einführung in die Lehr vom Denkstil und Denkkollective*[2] was published in 1935, a year after Karl Popper's *Logik der Forschung*. The two books encountered very different fortunes. Popper's book was noticed and caused quite a stir. While many in the Vienna Circle as well as Werner Heisenberg disliked it, Einstein, Carnap, and other philosophers praised it lavishly. The work was reviewed by leading journals and via the potent idea of falsifiability began to make Popper's reputation as a serious philosopher of science. Fleck's book, on the other hand, was almost ignored. Except for a few reviews in German journals it dropped "dead-born from the press" as David Hume described the fate of his *Treatise on Human Nature* in 1739. While Popper's book was translated into English in 1958, Fleck's had to wait nearly another forty years before an English language version was available.

Ludwik Fleck's extremely original ideas on the nature and epistemology of science are presented in the context of a historical examination of the changing definition, treatment, and diagnosis of syphilis, culminating in the Wassermann test. But unlike the later social constructivist successors to Mannheim, Fleck's goal was not to "unmask" science (although this was an inevitable byproduct of his analysis), but to gain for himself a better

---

[1] Fleck was Jewish. He survived the Holocaust (although many of his relatives did not). At one point during the war he imprisoned in Auschwitz where he was compelled to produce typhus vaccine for the Wehrmacht.

[2] For a translated modern edition of Fleck's book with a foreword by Kuhn see Fleck (1979). A biography of Fleck can be found in the same work, pp. 149–153.

understanding of its historical development. But what is remarkable about Fleck's work is that it contains in embryonic form many of the main themes of PIS as listed in propositions (i)—(xv) of Chapter 1.

Fleck's two fundamental tools in his analysis of science are the ideas of "thought style" and "thought collective." These were not original with him; they are similar to ideas in Mannheim's concept of "total ideology." Others who anticipated him include Emile Durkheim, Wilhelm Jerusalem, Ludwig Gumlorwicz, and Lucien Lévy-Bruhl.[3] By the 1930s ideas similar to Fleck's were definitely part of the intellectual atmosphere. What is almost new is their use in the interpretation of science.[4] The concept of thought style is very general. Fleck defines it as the readiness:

> ... *for directed perception, with corresponding mental and objective assimilation of what has been so perceived.* It is characterized by common features in the problems of interest to a thought collective, by the judgement which the thought collective considers evident, and by the methods which it applies as a means of cognition. The thought style may also be accompanied by a technical and literary style characteristic of the given system of knowledge.[5] [emphasis in original]

Thus a thought style might be an ideology, some quasi-philosophical system of thought characteristic of a period such as Renaissance occultism, astrology, or cabalism, etc. Applied to science it can strongly resemble a "paradigm," at least in one of Kuhn's definitions of the term as the totality of "beliefs, values, techniques" held in common by members of a scientific community,[6] or it can designate the general attitude of scientific naturalism characteristic of Western thought since the Enlightenment. The "thought collective," on the other hand, is:

> ... *a community of persons mutually exchanging ideas or maintaining intellectual interaction... it also provides the special 'carrier' for the historical development of any field of thought, as well as for the given stock*

---

[3] See Fleck (1979), pp. 46–48 for a discussion of his precursors.

[4] Fleck had been anticipated by German theorists who had, as is well known, in the 1920s and 30s used the concept of "thought style" in connection with science. But they identified a thought style as a product of race or nationality. Thus there was a German physics or mathematics with different and "better" characteristics than "formalistic" Jewish physics or mathematics. One can see similar developments in Marxist treatments of science with class substituted for race. Fleck's ideas, however, are very different.

[5] *Ibid.*, p. 99.

[6] Kuhn (1970a), p. 75.

of knowledge and level of culture. This we have designated thought style.[7] [emphasis in original]

In the case of science the thought collective has as its most natural reference a body of scientists articulating (to use Kuhnian language) a paradigm in the form of normal science. But as in the case of "thought style," the concept is very general. According to Fleck a thought collective can be a political party, a collection of astrologers, a religious sect, and so forth. Also the number of individuals in a thought collective need not be large. Two people sharing a thought style can constitute a thought collective, or perhaps even one individual having a dialog with himself. Also a person may belong simultaneously to several thought collectives.[8] Fleck has a deeply sociological perspective on thought styles and thought collectives. The individual mind is more or less the slave of the thought style possessing it:

> A thought style functions by constraining, inhibiting, and determining the way of thinking. Under the influence of a thought style one cannot think in any other way. It also excludes alternative modes of perception.... A thought style functions at such a fundamental level that the individual seems generally unaware of it. It exerts a compulsive force on his thinking, so that he normally remains unconscious both of the thought style as such and of its constraining character.[9]

As this quotation hints, for Fleck observation is conditioned by the observer's thought style. Different thought styles give rise to different perceptions. Like Thomas Kuhn he compares these different perceptions to different *gestalten*. A seventeenth century anatomist like Caspar Bartholin (1673) literally sees and evaluates structures in the body differently than modern writers who exemplify a different thought style. Similarly, the physician Joseph Löw in a 1815 treatise on physiological and pathological aspects of urine *sees* a "lack of coction" in urine as a visible property. For us it is "theoretically constructed gestalt" which is no longer possible to see.[10]

---

[7] Fleck (1979), p. 39.
[8] *Ibid.*, p. 44.
[9] *Ibid.*, p. 159.
[10] *Ibid.*, p. 128 ff. His examples are clearer in the writer's opinion than those of either Hanson or Kuhn. For an extensive analysis of how the thought style of a period influenced clinical and anatomical perceptions see *Ibid.*, Chapter Four, Section Five, and especially p. 133 ff.

The essential characteristic of different thought styles is what was later called their "incommensurability." Fleck apparently does not use this Kuhnian term but his descriptions of the non-communication, mutual non-intelligibility and non-translatability of key concepts in different thought styles is very close to Kuhn. How for example can we make sense of "ichorous fluid with mummy-like corrugation" in Löw's book?[11] We no longer possess such concepts and are unable to translate the words describing them into terms appropriate to our own thought style. The further apart in time or space two thought styles are, the more difficult the communication becomes. A modern physicist may be able to communicate with a modern biologist (unless the latter is a vitalist) but not certainly with a cabalist or Chinese traditional physician. In such a case, to both parties:

> The principles of the alien collective are, if noticed at all, felt to be arbitrary and their possible legitimacy as begging the question. The alien way of thought seems like mysticism. The question it rejects will often be regarded as the most important ones, its explanations proving nothing or as missing the point, its problems as often unimportant and meaningless trivialities.[12]

Finally, what is "truth" and what are "facts"?[13] Fleck stresses the historical contingency and continuing development of both. "Truth" is not the result of a static correspondence of some assertion with the facts. Rather it is a solution to a given problem which in form and content is acceptable to a thought style: *"Such a stylized solution and there is always only one is called truth."*[14] But this does not imply that truth is relative or subjective. Within the same thought collective, the same assertion will be true or false for all. For different thought collectives, assertions or thoughts will be *different* and incomparable, so that a given assertion cannot be true for one collective and false for another. Thus

> Truth is not a convention, *but rather (1) [is] in historical perspective, an event in the history of thought, (2) in its contemporary context, [a] stylized thought constraint.*[15] [emphasis in original]

---

[11] *Ibid.*, p. 132.
[12] *Ibid.*, p. 109.
[13] Since Fleck begins his book with the question "what is a fact?", obtaining an answer to it must be important. However, Fleck's efforts are quite murky. We can only offer here an interpretation which we believe agrees with most of the passages where "facts" and "truth" are discussed.
[14] *Ibid.*, p. 100 [emphasis in original].
[15] *Ibid.*

To explain the nature of facts, Fleck makes a distinction between "active" and "passive" associations. The former are concepts that are open to choice and which can be explained by factors such as the prevailing thought style, individual or collective psychology, or the cultural-historical situation. The grouping of all venereal diseases together "under the generic concept of carnal scourge," which to our thought style erroneously conflates syphilis, gonorrhea, soft chancre, and other disagreeable conditions, as was done by Renaissance physicians, would be an example of an active association.[16] Passive associations, on the other hand, are elements of knowledge that do not seem to be explicable in terms of such factors and seem therefore "to be 'real,' 'objective' and 'true' relations."[17] In many cases they are derived from observations, e.g., "sometimes mercury does not cure the carnal scourge, but makes it even worse."[18] To employ later terminology, an active association is a *construction* by a particular thought collective, whereas a passive association is "discovered" although its significance and interpretation is also a function of the thought style. Fleck observes that passive and active associations are often mixed together or that one is necessary for the other in the same statement. For instance, the observation concerning mercury "could not even be formulated if it were not for the concept of carnal scourge and that, incidentally the very concept of carnal scourge contains active and passive elements."[19] With this distinction in hand Fleck defines a scientific fact as:

> ... *a thought-stylized conceptual relation which can be investigated from the point of view of history and from that of psychology, both individual and collective, but which cannot be substantively reconstructed in toto simply from these points of view.*[20] [emphasis in original]

Elsewhere he says that facts are a "signal of resistance" to the active associations of the collective[21] and that they are those concepts of the thought collective whose active element is minimized and whose passive element is maximized.[22] "Facts" are therefore in a certain sense relative to the collective and will change when the thought style of the collective changes

---

[16] *Ibid.*, p. 10.
[17] *Ibid.*
[18] *Ibid.*
[19] *Ibid.*
[20] *Ibid.*, p. 83.
[21] *Ibid.*, p. 101.
[22] *Ibid.*, pp. 95, 98.

because the active associations bound up with the facts will change. In this sense facts are collectively constructed; they are not "out there" waiting to be discovered and independent of the thought style; consequently, their development can be studied by historical methods. But they are not completely relative either; they have an objective aspect as well because of their passive ingredients. Even so, however, the interpretation of the passive ingredients and therefore also of the "facts" depends on the active associations bound up with them which can change. Clearly there is a subtle and not completely transparent mixture of relativism and objectivity in Fleck's conception of "facts" and "truth."

Chapter 7

The Reluctant Revolutionary

Ludwik Fleck was certainly a precursor of SSK and the postmodernist turn of mind generally, and would have probably agreed with many of the PIS-like theses (i)—(xv) we have listed. But he was nearly invisible, having as we have remarked no initial influence. An effective rebuttal to Mannheim and Merton's separation of the substance of science and mathematics from SK by their intellectual descendants had to await the 1970s when it was conjoined with a much broader attack on the entire Enlightenment project which privileged science as the only reliable source of knowledge. The process by which this happened and eventually triumphed in many humanistic disciplines is our central theme. It was certainly foreshadowed by Berger and Luckmann's influential 1966 book *The Social Construction of Reality* which implied that all knowledge including the most fundamental conceptions of reality was the product of human social interactions. Berger and Luckman, however, like other contemporary sociologists were following the tradition of Mannheim and Merton; unlike Ludwik Fleck they did not apply their ideas to explain the content of science. The primary (but quite unintentional) catalyst which changed the situation and opened the door to a *total* sociological interpretation of science was supplied not by a sociologist, but by a mathematical physicist turned historian of science. The physicist was Thomas Kuhn and the work was his famous *The Structure of Scientific Revolutions*. This book written in 1962 has almost certainly had more influence on the historiography and philosophy of science than any other individual work in the history of these disciplines; in fact, it appears to have been the single most cited book on *any* subject in the twentieth century.[1] By the time of Kuhn's death in 1996, sales had exceeded a million copies,[2]

---

[1] According to Keith Windshuttle who checked the *Arts and Humanities Citation Index*. (Stove (2001b), p. 2.)

[2] Heilbron (1996).

and the book had been translated into at least nineteen languages including Korean and Serbo-Croatian.[3] Unlike many historians and philosophers of science Thomas Kuhn (1922–1996) had a solid training in a scientific discipline. In 1949 he was awarded a Harvard Ph.D. in Physics under John H. van Vleck, who later (1977) shared a Nobel Prize. In 1947, while a graduate student, Kuhn was asked by the President of Harvard James B. Conant (1893–1978), who had been the director of the National Defense Research Council during the war and who was soon to become High Commissioner to occupied Germany, to assist in the teaching and organization of a science appreciation course for undergraduates in the humanities which Conant had wished to establish. The course was intended to illustrate the nature of science by analyzing historical case studies of its development, and it had a serious political purpose:

> As one of the masterminds behind the transition to Big Science, Conant was concerned that the public might have become suspicious of science if they understood its subsequent development exclusively through the horrific effects of the bomb.[4]

In particular, Conant wanted the humanist nonspecialist who might later rise to a position of power to have some insight into the scientific method, to be able "to abstract a distinctive scientific mind-set that has remained constant" despite vast changes in science during the last three hundred years, and to have the ability to distinguish good from bad science. He thought that the achievement of this goal was vital to the future of American democracy in the Cold War era.[5] To achieve this, Conant's method was to historically examine the rise and fall of old scientific ideas.[6]

In preparing one of these case studies on the history of mechanics from Aristotle to Galileo, Kuhn tells us that he was trying to understand Aristotle's views on physics. In reading Aristotle, he wondered *how* this great philosopher could be so wrong. He seemed to be a terrible physical scientist who knew no mechanics at all and his writings seemed "full of egregious errors, both of logic and observation."[7] But how was this possible? Aristotle was a profound logician, and skilled observer in biology. How could he have

---

[3] Hoyningen-Huene (1993), n. 2.
[4] Fuller (2000), p. 379.
[5] *Ibid.* See also pp. 9–10 and pp. 212–224 for a discussion of Conant's motives. Fuller also argues that another of Conant's goals was to divert attention from the actual connections of modern science to modern capitalism.
[6] Heilbron (1996), p. 507.
[7] Kuhn (2000b), p. 16.

been so incompetent in physics? One afternoon, in an sudden illuminating revelation, it dawned on Kuhn that Aristotle's work was not just incorrect Newton, but represented an entirely different world view which, taken on its own terms and involving key ideas in Greek philosophy, was just as rational as that of modern science even if it was incompatible with it.[8] This experience was the seed and inspiration of Kuhn's future work. Together with his exposure as a Junior Fellow to such writers as Alexander Koyré, Emile Meyerson,[9] Anneliese Maier, and Jean Piaget plus a developing personal dissatisfaction with physics, it caused him to change his research area to the history of science. Previous historians of science such as William Whewell or George Sarton had been content to write "Whig history," that is, to present the history of science as a chronicle of cumulative progress towards the truth as incarnated by contemporary science while simultaneously pointing out the errors of discarded theories. In contrast, throughout his career Kuhn tried to get inside the heads of past scientists in order to be able to think as they had thought without the distorting effects of modern developments. An early effort in this direction was his *The Copernican Revolution* (1957) with its sympathetic account of Ptolemaic astronomy and its gradual displacement by the heliocentric system.

In *Structure*, which synthesized views that had been maturing for at least a decade, Kuhn argued that a large part of the history of science consists of periods of varying duration he called "normal science." Here the conventional Whig account of science as cumulative progress is essentially correct. Scientists are guided by a "paradigm" which can be either an "exemplar," that is, a model—as in a text-book—of a correctly posed problem and its solution or "the entire constellation of beliefs, values, and techniques, and so on shared by a given community" of researchers.[10] In this environment they add to knowledge in a cumulative fashion by applying accepted theories, tools, and methods prescribed by the paradigm to outstanding problems. Their procedure resembles the solving of puzzles under

---

[8] See Kuhn's discussion of Aristotle's fundamental physical concepts in *Ibid.*, pp. 16–20.
[9] Especially Meyerson's *Identity and Reality* (1908).
[10] Kuhn (1970a), p. 175. Kuhn was notoriously vague in defining "paradigm." In *Structure* its primary meaning was that of exemplar. The second definition as a common body of technique and belief becomes more prominent in later work beginning with the "Postscript" to the second 1970 edition of *Structure*. The term is also sometimes replaced by "disciplinary matrix." See also Kuhn (1970b), Chapter 12 and Kuhn (1974). Nor are these the only meanings of "paradigm." Margaret Masterman managed to distinguish 21 *different* senses in which the word was used! (Masterman (1970), pp. 61–65). In all versions, however the paradigm defines what kind of questions are to be asked and the nature of the expected answers to them.

prescribed constraints and rules. To put it another way, normal science in essence is an attempt by specialists to "force" nature to conform to the theoretical commitments implied by the paradigm. What makes this activity possible is that a physical theory never quite "works" in all of its possible applications. There are always phenomena or "anomalies" which either are not explained by it or seem not to agree with its predictions. For instance, anomalies arising from the Newtonian paradigm as applied to astronomy would include irregularities in the orbits of Uranus and the Moon, and the precession of the perihelion of Mercury. Chemistry prior to Lavosier faced the problem why the calx of a metal or burnt sulphur *gained* weight when the process of calcination or combustion was supposed to *release* the element "phlogiston." Aristotelian physics had difficulty explaining the fact that an arrow continued to move in a semi-horizontal direction after losing contact with the bowstring instead of immediately seeking its "natural place" at the center of the earth. Similar contradictions between nature and theory can be found even in modern physics where they frequently arise out of routine research activity. Niels Bohr, for example, in the process of trying to improve results on another subject, realized that the classical Maxwell-Lorentz electromagnetic theory implied that the 1911 Rutherford model of the atom was unstable and should collapse. Again, the discovery of low level microwave radiation in the early 1960s seemed inconsistent with the then dominant steady state model of the universe and eventually reinforced the idea of the Big Bang. Fortunately, given sufficient ingenuity, most such anomalies can be incorporated into the accepted paradigm, and the effort to do so constitutes the career of most scientists in a period of normal science. Perhaps the anomaly is only a matter of a poorly designed experiment, inaccurate apparatus, or an inadequate mathematical model, and a more sophisticated procedure can clarify the situation. In the case of Uranus, the mathematical analysis of the perturbations of its orbit suggested a new planet and led in due course to the discovery of Neptune in the approximate location predicted. We could easily list many similar examples.

To Kuhn's many critics this account of the usual business of science was shocking. Strictly speaking, anomalies "falsify" the current theory which they contradict. According to Karl Popper and his school this means that the theory should be rejected. As we have seen, good science to Popper consists of daring conjectures, preferably overturning the very foundations of the subject, combined with rigorous efforts to falsify them by clever experiments. Further, since a theory makes infinitely many predictions we can never verify all of them. Hence we can never be confident that it is true; however, we *can* falsify it and find a better one. In this way scientists learn from their mistakes and scientific progress is guaranteed. This vision of science was probably dominant in the philosophy of science by the early 1960s and coincided with the popular image of scientists, exemplified by Einstein, as dynamic revolutionaries who completely redefine science.

In contrast, according to Kuhn, such a portrayal of science is most of the time utterly mistaken. Historically, scientists have simply not followed Popperian methodology. Educated and initiated into the reigning paradigm by an extremely narrow and restrictive system (comparable to training in orthodox theology), scientists have no interest in criticizing fundamentals; they simply want to make the paradigm work. Failure to do so in the case of a particular anomaly is usually more a criticism of the scientist involved than of the theory; the carpenter is blamed rather than his tools. In fact, the behavior that Popper celebrates is mostly found at a pre-scientific, immature stage of a subject. Then investigators form different "schools"; they quarrel over methodology, the nature of legitimate questions, acceptable answers, and the basic assumptions and definitions in the field. There are no common paradigms, no puzzles to solve, hence no normal science and "progress." All questions remain open and subject to endless dispute. All successful areas of science have managed to grow out of this stage. But some fields have not. One thinks of psychiatry with its various sects or sociology which often aspires to ape physics while remaining the arena of vicious ideological conflicts.

Revolutions, however, *do* happen in science. New paradigms replace older ones which vanish and almost become invisible. How is this possible given the conservatism of most practicing scientists? Kuhn's answer is that the very nature of normal science guarantees its demise. The successful incorporation of anomalies cannot go on forever. The "easy" cases are disposed of first and the anomalies become more challenging. Eventually one or more of them resists the best efforts of scientists. No accepted procedure or solution seems to work. After continued fruitless effort to explain the anomaly in terms of the old paradigm a "crisis" results. At this time the situation is ripe for an entirely new and revolutionary theoretical structure to be introduced to handle the problem. Sometimes the process occurs in steps and is drawn out. It took more than 150 years and the successive contributions of Copernicus, Kepler, Galileo, and Newton for a new paradigm to replace Ptolemaic astronomy conjoined with Aristotelian physics. On the other hand, the revolutionary transformation of classical physics by quantum mechanical ideas and relativity and a new conception of the atom by Bohr, Planck, Einstein, Rutherford, Heisenberg, Dirac, and others took less than three decades.[11]

Although the practice of normal science is necessary for the introduction of new paradigm, the occasion marks a profound break with the conventional science of the period—it is not merely a case of obtaining new explanations or discoveries by novel methods, but the whole foundation of previous scientific practice is torn up. Kuhn claims that the new paradigm is "incommensurable" with the old one. What exactly does this term mean? Kuhn chose it because of a mathematical analogy. As the ancient Pythagoreans are said to have shown, the hypotenuse H of a 45° right triangle with a unit side S is incommensurable with S. That is to say, there is no "common measure" or line segment of length l such that $H = ml$ and $S = nl$ where m and n are integers.[12] Similarly, to say that two paradigms or theories are "incommensurable" is to say that there is no common standard of "truth" or evidence by which they can be measured. Kuhn expresses this by saying that the proponents of each paradigm live in "different worlds". The new and old paradigms may differ not only in their fundamental assumptions and methods, but also in the *meaning* of these assumptions and methods as well as the nature of both "correct" questions a scientific theory is supposed to ask and what constitutes acceptable answers to them. In particular, the problems solved under the old paradigm may become meaningless and their "solutions" vacuous from the point of view of the new one.[13] Examples of such competing paradigms where these differences come into play include Ptolemaic versus Copernican astronomy, the plogiston theory versus the oxygen theory of combustion of Lavosier, the caloric fluid versus the dynamical theory of heat, and late nineteenth century classical physics versus relativity theory and quantum mechanics.

---

[11] How rapidly the transformation could occur is illustrated by the famous 1913 paper of Bohr who realized that an *ad hoc* adoption of revolutionary "quantum" ideas recently introduced by Planck would stabilize the previously unstable Rutherford atom and that these ideas in turn would explain Mendeleev's periodic table and with modifications also Balmer's formula predicting the wave lengths associated with the spectral lines of the hydrogen atom. See Heilbron and Kuhn (1969) for an interesting and thorough account of the origin and consequences of Bohr's theory.

[12] In modern mathematical language this is equivalent to saying that $\sqrt{2}$ is irrational, i.e., cannot be expressed as a fraction.

[13] For example, the successful elimination of equant points needed by Ptolemy to model planetary motion by the medieval Persian astronomer Naṣīr al-Dīn al-Ṭūsī (1201–1274) and his school via an ingenious compound of uniform circular motions was a brilliant solution to a long standing "problem" in this astronomical tradition, but it became totally irrelevant for post-Keplerian astronomy. For the details of Ṭūsī's clever geometric construction see Kennedy (1966), p. 377.

For Kuhn the tension between the old and new paradigms often resembles that between traditional and revolutionary political ideologies. In both cases:

> ... society is divided into competing camps and parties, one seeking to defend the old institutional constellation, the others seeking to institute some new one. And once that polarization has occurred, *political recourse fails* ... the parties to a revolutionary conflict must finally resort to the techniques of mass persuasion, often including force.[14] [emphasis in original]

As in the political case there is no way by rational argument in a revolutionary scientific situation to decide the issue—the contenders simply talk past each other. They share no common criterion or standard of decision; in particular, there can be no final appeal to observation and experiment since both are "theory-laden," the significance of and even *what* the observer sees being conditioned by the paradigm in which he is embedded.[15] There is no neutral observation language, acceptable to all sides, in which the results of observation can be expressed. In the words of Kuhn:

> Like the choice between competing political institutions, that between the competing paradigms proves to be a choice between incompatible modes of community life. ... the choice is not and cannot be determined merely by the evaluative procedures characteristic of normal science, for these depend in part upon a particular paradigm, and that paradigm is at issue. ... Each group uses its own paradigm to argue in that paradigm's defense.[16]

In his writings composed in 1969 and after Kuhn weakened his comparison of scientific and political incommensurability put forth in *Structure*. Now the incommensurability of two paradigms may no longer be total, but only "local"; only some of the elements of the old paradigm perish after a scientific revolution. Many problems, techniques, facts, etc., are left unscathed. But:

---

[14] Kuhn (1970a), p. 93.
[15] *Ibid.*, p. 109 and pp. 111 ff. Here Kuhn is influenced by Norwood Russell Hanson who was one of the earliest (1958) to argue that the visual experience of the observer is not merely an image on his retina mirrored in his brain, but in large part depends on his theoretical expectations. See Hanson (1958), Chapter I. Another influence may have been Ludwik Fleck who, as we have already observed, had a point of view very similar to Hanson's.
[16] Kuhn (1970a), p. 94.

> Though most of the same signs are used before and after a revolution—
> e.g., force, mass, element, compound, cell—the ways in which some of
> them attach to nature is somehow changed.[17]

That is, incommensurability is now almost totally a semantic rather than a political phenomenon. Influenced by Quine's *Word and Object* (1960),[18] by the Sapir-Whorf thesis that languages determine the metaphysics and world picture of the speaker, and also possibly (see below) by Ludwik Fleck, he argued that the incommensurability between two paradigms amounts to a change in language. Key words in the old paradigm may correspond to those in the new or may even be identical to them. But their meaning is different; they cut up nature conceptually in a different way, with the result that the language and core concepts of one paradigm may not be fully translatable into those of the other. The practical effect is that when arguments are given by supporters and opponents of the new paradigm, they miss the point. Their "conceptual lexicons" or scientific vocabularies are different. Neither side really understands the key concepts of its rival. In particular, those brought up in the new paradigm mistranslate and fail to comprehend the old. They find the previous paradigm not merely wrong but almost unintelligible. The same thing often happens when a modern scientist or historian tries to interpret an archaic scientific text. The following passage written in 1993 long after *Structure* is illuminating. Kuhn is considering the mutual incomprehension existing between certain Aristotelian physical propositions "regularly misconstrued as asserting the proportionality of force and motion or the impossibility of a void" and Newtonian mechanics:

> Using our conceptual lexicon, these Aristotelian propositions cannot be
> expressed—they are simply ineffable—and we are barred ... from access
> to the concepts required to express them. It follows that no shared
> metric is available to compare our assertions about force and motion
> with Aristotle's and thus to provide a basis for a claim that ours (or, for
> that matter his) are closer to the truth.[19]

---

[17] Kuhn (2000a), p. 267.
[18] Kuhn's view of Quine, however, can be critical. To Kuhn, Quine took for granted that anything expressible in one language is expressible in another. The problem in Kuhn's view is not indeterminacy of translation or inscrutability of reference but the fact that no translation at all may be available between radically different scientific lexicons. See e.g., Kuhn (2000d), pp. 61–62.
[19] Kuhn (1995), p. 330. See also Kuhn (2000a), pp. 266–278. and the essays "Commensurability, Comparability, Communicability" and "The Road Since Structure" in Kuhn (2000b), Chapters 2 and 4 for a discussion of the translatation issue.

How can the gap be bridged? Kuhn does not deny, the possibility of a historian (if not a scientist) reaching some understanding of an ancient scientific text which conforms to some non-modern paradigm. But this understanding cannot be reached by a process of simple one-to-one translation. The shifts in meaning are too profound. Instead, it requires hermeneutics, the art of interpretation. The historian must somehow learn the ancient science behind the text. Its concepts, not translatable into modern language, must be learnt not in isolation of each other but all together as part of a complicated but foreign cognitive network, much like the basic concepts of Newtonian mechanics such as force and mass. The problem of gaining access to old science is difficult and can never be solved exactly; it is rather like the difficulties found in translating poetry in a foreign language where some words have no exact equivalent in English. Choices must be made and different translators may make different decisions.

Because of its apparent implication (especially in *Structure*) that paradigm choice is in a fundamental sense irrational and that two paradigms cannot be compared, the incommensurability concept generated enormous controversy especially over the issue of whether or not "semantic" incommensurability really existed. At least one conference was devoted to just this question.[20]

If an individual abandons the old paradigm for a new one the change often is like conversion experience by which, say, a once ardent Catholic or Communist becomes a Calvinist or political conservative. Psychologically, it may be compared to a Gestalt switch; the world suddenly makes sense in a radically different way according to the new paradigm—in much the same fashion that the *same* picture may "flip" in a psychological Gestalt test, becoming to an observer an group of rabbits rather than a flock of ducks or vice versa. In the case of science the only difference in the situation is that the scientist cannot as easily flip back and forth between the two modes of perception. In the large, however, as we have seen, the victory of one paradigm over another to Kuhn resembles political change. If the adherents of the traditional paradigm are overcome, it is by a sociological transformation. The proponents of the new gain strength and numbers; the objections and arguments of the traditionalists are increasingly ignored, and eventually the opponents simply die out—fading into professional exile and oblivion just as the losers in a political revolution fall into political exile (or worse). The supplanting of a prior paradigm Kuhn calls a "paradigm

---

[20] See Devitt (2001) and the other papers in the proceedings volume of the conference.

shift," and if sufficiently important in the future development of science a "scientific revolution." The victors in such a paradigm shift will then rewrite history—usually in textbooks—to make their victory seem *necessary*; the entire scientific past is seen as inexorably leading up to it. Thus the writing of the history of a given field by scientists is invariably Whig history.

According to Kuhn the presence of a paradigm not only separates a mature science from its pre-scientific stage, but also marks the boundary between science and other pseudo-scientific belief systems. What, for example, is the essential difference between, say, astronomy and astrology? To Kuhn it does not result from the fact that astronomy and astrology are respectively "true" and "false" systems. The real difference is that the latter is a *craft* resembling "engineering, meteorology, and medicine as these fields were practiced until little more than a century ago" or, to take a modern example, present day psychoanalysis:

> In each of these fields shared theory was adequate only to establish the plausibility of the discipline and to provide a rationale for the various craft-rules which governs practice. these rules had proved their use in the past, but no practitioner supposed that they were sufficient to prevent recurrent failure.[21]

Practitioners certainly desired a stronger theory and more adequate rules, but since these were not available they had to make do with what they had. The outcome was that:

> ... neither the astrologer or the doctor could do research. Though they had rules to apply, they had no puzzles to solve and therefore no science to practice.[22]

Especially in his later writings, Kuhn agrees that there may be good and rational arguments for or against a disputed paradigm. Scientists will often appeal to common values such as consistency, simplicity, comprehensiveness, explanatory power, accuracy, fruitfulness, and so forth to justify

---

[21] Kuhn (1970c), p. 8.

[22] *Ibid.*, p. 9. Here Kuhn is attempting to refute Popper's assertion that astrology is a pseudo-science because the predictions of astrologers were so vague that they were unfalsifiable. Kuhn, however, points out that astrologers were well aware that many of their predictions failed but explained this much as contemporary meteorologists do by pointing out the extreme sensitivity or ill-conditioning of the system to small errors in the data necessary for the casting of horoscopes.

a paradigm.[23] But much also depends on rhetoric and persuasion together with straightforward argument. Unfortunately, the methods of mathematical proof are not applicable to paradigm choice. Premises and rules of inference may not be stipulated in advance and accepted by both parties; furthermore, detecting an error in an argument is not like detecting an error in a proof which will immediately compel an admission that a mistake has been made.[24] To use an historical analogy not found in his writings but one with which Kuhn would probably have agreed, there is no algorithm or "universal characteristic" for paradigm choice whereby both parties (as Leibniz hoped that Lutherans and Catholics would) can settle their differences by sitting down and saying "let us calculate." Very often the arguments on both sides are logically correct, but the problem lies in the nature of the premises which are assumed and in their application, the relative emphasis assigned to the above mentioned values, or even their proper definition.[25]

These views which Kuhn attempted to verify by case studies in the history of science render the relation of a scientific theory to the "reality" it represents ambiguous, and has profound consequences for the interpretation of science. If paradigm choice can result from an ill-defined mixture of political or religious conversion and "rationality," and if radical paradigm change can essentially negate the "progress" achieved in a previous period of normal science, then it seems that science fundamentally is not "cumulative," nor at times fully rational. All that can be said is that there exists an endless sequence of paradigms which may not converge to any final "truth." The way, therefore, seems open to relativism and a belief that science does not progress in the sense of "mirroring" reality in an ever more accurate way. Kuhn realized these implications and his view of them was complicated and not terribly clear. He seems constantly to oscillate between emphasizing rational and relativistic aspects of science, sometimes in the same paper. The conflicted, almost contradictory, nature of his thought is obvious from the passages we quote below. In the Rothschild Lecture "The Trouble with Historical Philosophy of Science,"[26] for instance, given

---

[23] See e.g., Kuhn (1970b), pp. 321–329.

[24] Kuhn (2000a), p. 261.

[25] See the 1969 postscript in Kuhn (1970a), pp. 199–200, for a discussion of these issues. Here Kuhn seems to be struggling to correct the impression that paradigm change is the product of irrational "mob psychology" that emerges in the original 1962 edition of his book. As Keith Parsons (Parsons (2005), Chapter 2) and others have noted, Kuhn in his later writings increasingly emphasizes the rational if nondeterministic aspects of theory choice and edges away from a comparison with political revolution which is so prominent in 1962.

[26] Reprinted in Kuhn (2000b), pp. 105–120.

in 1992 he denounced the extreme relativism of some of those who claimed inspiration from him for:

> ... claiming that power and interest are all there are. Nature itself, whatever that may be, has seemed to have no part in the development of beliefs about it. Talk of evidence, of the rationality of claims drawn from it, and of the truth and probability of those claims has been seen simply as the rhetoric behind which the victorious party cloaks its power. What passes for scientific knowledge becomes, then, simply the belief of the winners.[27]

Kuhn finds this vision of science "absurd: an example of deconstruction gone mad."[28] On the other hand, in the same lecture, he thinks that the relativists may feel:

> ... that the traditional philosophy of science was correct in its understanding of what *knowledge* must be. Facts must come first, and inescapable conclusions, at least about probabilities must be based on them. If science doesn't produce knowledge in that sense, they conclude, it cannot be producing knowledge at all.[29]

But perhaps the relativists are mistaken in their view of the nature of knowledge. It may simply be "the product of the very process these studies describe." Kuhn believes that scientific progress is real, but it does not consist of an ever more precise correspondence of theory with reality. This would require an "Archimedian platform" for the possibility of "neutral observation." But this would be "outside of history, outside of time and space" and is "gone beyond recall."[30] Only such a "fixed, rigid" platform:

> ... could supply a base from which to measure the distance between current belief and true belief. In the absence of that platform, it's hard to imagine what such a measurement would be, what the phrase 'closer to the truth' can mean.[31]

Kuhn admits that it is a common perception that there has been scientific progress in the sense of shaping an ever better fit between scientific theory and reality, and that this feeling is often reinforced by viewing an earlier paradigm as a a limiting or special case of a latter one. Newtonian mechan-

---
[27] *Ibid.*, p. 110.
[28] *Ibid.* Kuhn was referring to the Edinburgh Strong Program discussed below.
[29] *Ibid.*, p. 111.
[30] *Ibid.*, pp. 113–115.
[31] *Ibid.*, p. 115.

ics, for example, is a limiting case of Einstein's mechanics when v/c is near 0. But according to Kuhn this is an illusion. The mathematical derivation may be correct:

> But the argument has still not done what it purported to do. For in the passage to the limit it is not only the forms of the laws that have changed. Simultaneously we have had to alter the fundamental structural elements of which the universe to which they apply is composed.[32]

In other words, the entire *meaning* and world view engendered by the two theories differ. Einstein may improve on Newton in the ability to solve certain puzzles, but in fundamental respects he marks a return to conceptions of Aristotle so his replacement of Newtonian mechanics marks no "coherent direction of ontological development."[33]

What then is the nature of scientific progress? Kuhn believes that it consists only in the ability of scientists to solve ever more refined puzzles via increasingly sophisticated paradigms. Again, however, he emphasizes that this ability does not mean progress towards truth, for he maintains

> All past beliefs about nature have sooner or later turned out to be false. On the record, therefore, the probability that any currently proposed belief will fare better must be close to zero.[34]

Consequently, while Popper regarded a discarded, falsified theory as a "mistake" from which we can learn, Kuhn's anti-realism makes him reluctant to use such language to describe past theories: "what mistake was made, what rule broken, when and by whom in arriving at, say, the Ptolemaic system?"[35] Elsewhere, he says that in an intra-theoretic context one can talk about "truth" by applying the word to experimentally verified consequences of a shared theory. Similarly, the theory's application can point out "false" claims; but again he refuses to maintain that one theoretical structure or paradigm is better than another in the sense of being a better representation of "what is out there."[36] As we have observed, later in his

---

[32] Kuhn (1970a), p. 101.
[33] *Ibid.*, p. 206.
[34] Kuhn (2000b), p. 115.
[35] Kuhn (1970c), p. 10.
[36] Kuhn (2000a), p. 265. I have found no evidence that Kuhn read Mannheim. Yet, his position resembles Mannheim's "relationism." For Kuhn there seems to be no paradigm or body of scientific truth that mirrors the world accurately, just as for Mannheim there is no universal non-historical social truth.

career Kuhn came to emphasize the role of rational argument more than in *Structure* in the decision between competing paradigms, but this has little to do with deciding whether or not the claims of either paradigm are true. It is possible to compare their value without talking about "truth." One paradigm might be better than the other in the sense of being:

> ... *more* accurate, *more* consistent, broad*er* in its range of applicability, and also simpl*er* without for those reasons being any tru*er*. Indeed, even the term 'truer' has a vaguely ungrammatical ring: it is hard to know quite what those who use it have in mind.[37]

Or again if we consider a paradigm as a way of describing the world:

> Some ways are better suited to some purposes, some to others. But none is to be accepted as true or rejected as false; none gives privileged access to a real, as against an invented world.[38]

Also, Kuhn is careful to point out that there are losses as well as gains in the "progress" resulting from paradigm changes. Ignorance increases along with knowledge. Old questions and old solutions furnished by the previous paradigm are forgotten or dismissed. Newtonians accepted gravity as an unexplained feature of nature whereas both the Aristotelians and Cartesians had explanations for it. To take a second example, phlogiston, a key explanatory concept of the qualitative oriented chemistry of the eighteenth-century, was supposed to be present in all metals and explained their shared qualities. No such explanation was possible in post-Dalton chemistry which emphasized combining weights and proportions.[39] In summary, science does not compare theory with "reality", but only one paradigm with another. Paradigms become more complex and are able resolve previously insoluble puzzles, but they do not evolve in the direction of a final objective truth.

In considering the evolution of science Kuhn finds a fascinating parallelism to Darwinian biological evolution. Both processes are directionless and do not progress towards any predetermined goal. Anomalies and environmental challenges play similar roles in stimulating paradigm and species change. As in biological evolution the process of scientific change is contingent. There is no necessity to it. Both our modern paradigms

---
[37] Kuhn (2000b), p. 115 [emphasis in original].
[38] *Ibid.*, p. 104.
[39] *Ibid.*, pp. 190–91.

and *homo sapiens* are presently dominant. But both the history of science and evolution might have had a different outcome. The world might now be populated by intelligent dinosaurs with mammals either nonexistent or playing a subservient role. Entirely different paradigms, unknown to us, might have arisen which solve their creators' puzzles, but not ours. The analogy between the two kinds of evolution goes even deeper, however. In the case of science, "normal science" resembles periods of evolutionary stasis in which species change slightly if at all. Scientific revolutions, on the other hand, "are much like episodes of speciation in biological evolution."[40] The precise evolutionary model, in fact, that Kuhn's theory seems to resemble most is the theory of punctuated equilibria due to Stephen J. Gould and Niles Eldredge which was published in 1972.[41] Attempting to evade the relativism that this evolutionary model, his skepticism concerning the "truth" of successive theories, and his view of paradigm change seem to imply, Kuhn argues that later theories differ from earlier ones in that they are more highly evolved. They are more complicated and specialized than their predecessors. Just as in the case of evolving species, it is easy to give criteria that allow us to separate theories later in the sequence from the earlier. "In this sense scientific development is a unidirectional and irreversible process and that is not a relativistic view."[42]

Underlying Kuhn's interpretation of the history of science is a certain neo-Kantian ontological duality which we now outline and which is found in some form in many of his successors.[43] Kuhn divides "reality" into two parts. There is first of all the phenomenal world, the set of all possible sense data. In *Structure* Kuhn maintains that those of concern to the scientist are primarily visual since the scientist "can have no recourse above

---

[40] *Ibid.* p. 98. Ideas comparing biological to scientific evolution are developed in Kuhn (1970a), Chapter 13 and in Kuhn (2000b), pp. 96–99.
[41] Eldredge and Gould (1972). Gould admits that *Structure* was:

> ... if I was to cite any one factor as probably the most important among the numerous influences that predisposed my own mind in joining Niles Eldredge in the formation of punctuated equilibrium.

(Gould (2007), p. 283.) The resemblance was also noted by Bird in Bird (2000), p. 212. The structural similarities and differences between these two theories would be interesting to explore further.
[42] Kuhn (1974), p. 508.
[43] The following discussion owes much to Hoyningen-Huene (1993), especially Chapters 2, 3, and the Epilogue. This book is almost certainly the most elaborately detailed account of Kuhn's work yet written, and was "warmly recommended" by Kuhn himself in the preface.

or beyond what he sees with his eyes and instruments."[44] This world is essentially constituted by the scientist's paradigm. It is a truism of course that the meaning, interpretation, or significance of an element in the phenomenal world depends on the preconceptions or mental categories of the observer. The interpretation of a falling stone would be quite different for an Aristotelian than for a follower of Newton. But Kuhn (following Hanson and Fleck) has a stronger thesis in mind. What the scientist actually *perceives*—in other words his visual sensations—may depend on all sorts of factors: his education, the group he belongs to, his prior expectations, and the paradigm which guides his research. To use an example of Hanson,[45] Tycho de Brahe and Simplicius gazing at the sun at dawn actually *see* a rising sun and a stable earth while Kepler and Galileo looking at the same dawn sun *see* a fixed local star and a rotating earth.[46] Given this state of affairs, the function of scientific training is to initiate the student into the paradigm and thus enable him to "become an inhabitant of the scientist's world, seeing what the scientist sees and responding as the scientist does."[47] Thus:

> Looking at a contour map, the student sees lines on paper. Looking at a bubble-chamber photograph, the student sees confused and broken lines, the physicist a record of familiar subnuclear events.[48]

When the paradigm changes after a scientific revolution what is perceived *changes*. Scientists now see the world differently, just as Kepler sees differently from Brahe. Consequently:

> In so far as their only recourse to that world is through what they see and do, we may want to say that after a revolution scientists are responding to a different world.[49]

---

[44] Kuhn (1970a), p. 114.
[45] Hanson (1958), Chapter 1.
[46] Why do both Hanson and Kuhn reject the idea that two observers may see the same thing and merely *interpret* it differently? It is tempting to indulge in an SK explanation here. If perception involves interpretation of an underlying reality, one's interpretation may be incorrect. But if perception is a Gestalt-like act, then no perception can be called incorrect and there is no need of truth language in science. You just see an ugly crone; I see an attractive young woman (á la Toulouse-Lautrec (cf. *Ibid.*, p. 11). Again, as we have noted above, this position is evidently similar to Mannheim's "relationism" in relation to ideological issues.
[47] Kuhn (1970a), p. 111.
[48] *Ibid.*
[49] *Ibid.*

To clarify these changes in perception Kuhn uses, as we have seen, the language of Gestalt psychology. The changes amount to a Gestalt shift:

> What were ducks in the scientists world before the revolution are rabbits afterwards. The man who first sees the exterior of the box from above later sees its interior from below.[50]

The second element in Kuhn's division of reality is the "actual" world that scientists are attempting to understand and describe. Kuhn feels that underlying the conceptual changes and differentiations brought on by paradigm shifts "... there must, of course, be something permanent, fixed, and stable. Its existence is a necessary postulate made to avoid "both individual and social solipsism."[51] This world is independent of the observer and his paradigm; even though the scientist enters a "different world" after a paradigm change, it does not change. The actual world, moreover, is responsible for the "stimuli" that give rise to the visual sensations of two observers (like Brahe and Kepler) who are standing in the same place and looking in the same direction. But we do not have access to stimuli: "people do not see stimuli; our knowledge of them is highly theoretical and abstract."[52] Moreover, the same stimulus may produce different perceptions or representations in different observers and conversely different stimuli may produce the same perception. In a late essay[53] Kuhn calls his concept of the actual world "a sort of post-Darwinian Kantianism." But like Kant's *Ding an sich*, the actual world is "ineffable, undescribable, and undiscussible."[54]

One consequence of Kuhn's two world theory is nominalism. Since the criteria we use to classify groups of related things in the world of phenomena come from our paradigm and must be taught to us, it seems that they are our constructions and do not reflect the inherent structure of the world. A second consequence is non-uniqueness of the phenomenal world, since it it is given form, structure, and meaning by the varying paradigms scientists have invented to interpret it. And since there is a potentially endless sequence of paradigms, there is also a potentially endless sequence of corresponding phenomenal worlds, each as legitimate as any other. This view leads Kuhn to oppose a thesis concerning the differences between the

---

[50] *Ibid.* The example is taken from Hanson (1958), p. 9.
[51] Kuhn (1970a), p. 193.
[52] *Ibid.*, p. 192.
[53] Kuhn (2000b), p. 104
[54] *Ibid.*

human and physical sciences put forth by Charles Taylor. To Taylor the understanding of human actions requires "hermeneutic interpretation" according to criteria which will differ from culture to culture or even from person to person. The night sky with its stars and planets—or more generally the physical world, on the other hand, is the "same for all cultures, say, for the Japanese and for us."[55] Its meaning, if any, is the same for all. Hermeneutics, therefore, is not required to understand it. Kuhn disagrees. Whatever may be the case with the Japanese and us, the heavens for the Greeks were "irreducibly different from ours."[56] We have seen repeatedly that in making statements like this, Kuhn does not mean merely that the Greeks had different theories or interpretations of the same night sky. They were really living in a *different* world from us:

> The nature of the difference is the same as Taylor so brilliantly describes between the social practices of different cultures. In both cases the difference is rooted in conceptual vocabulary. In neither can it be bridged by a description in a brute data, behavioral vocabulary. And in the absence of a brute data vocabulary, any attempt to describe one set of practices in the conceptual vocabulary, the meaning system, used to express the other, can only do violence.[57]

Although he denied the charge of relativism since, for example, just as for evolving species, latter paradigms are more complex and sophisticated than earlier ones, Kuhn is certainly an epistemological relativist since no paradigm is "truer" than others. At this point he differs from Kant. Both agree that the phenomenal world is in part caused by the imposition of *a priori* concepts on the noumena. But for Kant these concepts are common to the human race, so that there is only *one* phenomenal world for all times, places, and cultures. Epistemological relativism, therefore, is not a consequence of Kantian metaphysics although it is for Kuhn and his successors because the paradigms, perspectives, etc., are historically situated and change from group to group.

Kuhn wrestled with the metaphysical problems inherent in his neo-Kantianism throughout his career without giving particularly coherent answers to them. Presumably the actual world exists. But we can say nothing about it. Aside from being an abstract cause what is its relation to the phenomenal world? To have any function in some fashion it *together* with the

---
[55] Kuhn (2000c), p. 218.
[56] *Ibid.*, p. 220.
[57] *Ibid.*

paradigm must determine our representations. But how does it do this? Replying to a question by Dudley Shapere as to whether our classifications are discovered or constructed, Kuhn dances around the issue:

> ... learning a similarity relationship is learning something about nature that there is to be found. It informs you of a way to group stimuli in natural families and only some groupings will do.
>
> But in another sense the group does put them there *or finds them already there*, and what I want most to resist about the question is the implication that it must have a yes or no answer. We have no recourse to stimuli as given, but are always—by the time we can see or talk or do science—already initiated to a data world that the community has divided in a certain way. ... the group might have done the division job in other ways ... But it has done so in one particular way and initiates are stuck with it. It is a good way, or the group would not have survived, but there is no way to get outside it—back to stimuli and to specify the respects in which it is good ...[58]

Kuhn should not be blamed for this incoherence; it is a flaw in all "two world" doctrines going back to Kant and it is also occurs, as we shall see, in the SSK followers of Kuhn who attempt to extend the Mannheim SK program to science.

There is, finally, a sociological turn to Kuhn's thought which is implicit in *Structure of Scientific Revolutions* and is further emphasized in later writings.[59] It is not that Kuhn feels that individual scientists are insignificant. But the "analytic unit" for historians of science should be:

> ... the practitioners of a given speciality, men bound together by common elements in their education and apprenticeship, aware of each other's work and characterized by the relative fullness of their professional communication and the relative unanimity of their professional judgement.[60]

It is this community that socializes junior scientists into the paradigm and is the final arbiter of the practice of normal science and theory-choice. Writing in 1970, Kuhn is even clearer on the sociological factors in science:

---

[58] Kuhn (1974), p. 509 (Discussion).
[59] See e.g. the "Postscript" to the 1970 edition, pp. 176–90 or pp. 147–50.
[60] Kuhn (2000a), p. 147.

> Some of the principles deployed in my explanation of science are irreducibly sociological, at least at this time. In particular, confronted with the problem of theory-choice, the structure of my response runs roughly as follows: take a group of the ablest available people with the most appropriate motivation; train them in some science and in the specialities relevant to the choice at hand; imbue them with the value system, the ideology, current in the discipline ... and, finally, let them make the choice. If that technique does not account for scientific development as we know it, then no other will.[61]

A still more explicit passage stressing sociological factors is the following. Kuhn claims that the explanation of the development of science should:

> ... in the final analysis, be psychological or sociological. It must, that is, be a description of a value system, an ideology, together with an analysis of the institutions through which that system is transmitted and enforced. Knowing what scientists value, we may hope to understand what problems they will undertake and what choices they will make in particular circumstances of conflict. I doubt that there is another sort of answer to be found.[62]

It is just these sociological aspects of Kuhn's thinking that leads Barry Barnes to view him as a precursor of the SK approach to science:

> By introducing a social dimension and relating the status of scientific knowledge to the contingent judgements of specific communities of people, Kuhn undermined a whole range of philosophic arguments designed to secure a privileged epistemological or ontological status for science.[63]

Be this as it may, it is certainly believable that by (i) refusing to regard scientific theories as either "true" or progressing to "truth" in the sense of mirroring nature, (ii) arguing that paradigm choice was a not entirely objective or rational sociological process, (iii) asserting that we can know only our representations which are conditioned by our paradigm rather the mysterious "stimuli" which cause them, Kuhn supplied at least three of the ingredients required for an incorporation of science into SK. Obviously if his ideas are correct, science can no longer be viewed as "knowledge as

---
[61] *Ibid.*, p. 131.
[62] Kuhn (1970b), p. 290.
[63] Barnes (1982), p. 12. Barnes was the co-creator of the SSK generalization of the Mannheim program to science.

such" and, instead, would only be a contingent, historically situated way of looking at the world similar to other belief systems generated by a culture, and not epistemologically superior to them. To use a term of which Kuhn probably would not have approved, but which has become a cliché among his successors, scientists, by imposing their paradigm on the physical world, "construct" scientific reality; they do not "discover" it. Hence science is effectively detached from any kind of mind-independent reality. So-called "scientific progress" becomes merely a byproduct of the activity of normal science; since the business of normal science is to solve puzzles, some will indeed be solved—which (by definition) is "progress." Paradigm change, in particular, does not represent progress except in the writings of Whig historians; instead, as critics such as Lakatos maintained it was a form of "mob rule." Such conceptions of science, however disdained by conventional philosophers of science, were delightful to humanists; it freed them from the hitherto prevailing positivist views that "privileged the hard sciences at the expense of other departments of the university.... [Now] they too were respectable knowledge producers laboring under paradigms."[64]

Kuhn's theory also helped in another way to topple science from its conceptual throne and put it on the same level as other forms of cultural activity. He says concerning its reception in the postscript to the second (1969) edition of *Structure* that:

> ... A number of those who have taken pleasure from it have done so less because it illuminates science than because they read its main theses as applicable to many other fields as well. ... But they should be, for they are borrowed from other fields. Historians of literature, of music, of the arts, of political development, and of many other human activities have long described their subjects in the same way. Periodization in terms of revolutionary breaks in style, taste, and institutional structure have been among their standard tools. If I have been original with concepts like these, it has mainly been by supplying them to the sciences, fields which had been widely thought to have developed in a different way.[65]

This point of view, of course, was like manna from heaven to humanists who had long suffered from an inferiority complex vis-à-vis the hard sciences. And if it was objected that the punctuated equilibria in the humanities do not represent cumulative progress, they were now licensed to give the *tu quoque* reply that according to Kuhn the same was true of science.

---

[64] Fuller (2000), pp. 2–3.
[65] Kuhn (1970a), p. 207. Also see Gould (2007), pp. 285 ff. for the quotation and an interesting discussion of Kuhn's impact on other disciplines.

As we have seen, Kuhn himself rejected many of the later extrapolations of his ideas as far too extreme. He suffered the fate of many who begin revolutions by living long enough to be appalled by the excesses committed in his name.[66] One feels that Kuhn neither intended or foresaw the revolution that he unleased.[67] He was merely trying to point out the weakness both in the traditional positivist account of science and in its revisions due to Karl Popper. These views of science were largely the work of philosophers not of working scientists which Kuhn had once been. To Kuhn the ideas in *Structure* reflected the way science had historically actually *worked*. His goal was just to correct the record. He was a reluctant and initially probably an unconscious revolutionary. But unfortunately, the ambiguities in his work, the excellence of his writing, and the general Zeitgeist made revolution inevitable. There were also personal factors which had a "multiplier effect" on the impact of his theories. Kuhn was *not* primarily an academic philosopher, historian, or some other type of humanist. Instead he was a first rate mathematical physicist of excellent pedigree who criticized traditional realist conceptions of science and said things that sociologists found agreeable.[68] Had Kuhn been a humanist with little scientific training in some second-tier university, he probably would not suffered the fate of Fleck. But his book might not have enjoyed the enormous success it did.

While Kuhn's thought was highly original it did not spring up *ex nihilo*. Many of the intellectual strands in *Structure* had been anticipated in some sense by others and by 1962 were "in the air." His achievement was one of synthesis. He could blend earlier ingredients into a new intellectual whole greater than the sum of its parts. As we have previous noted and as Kuhn states in the Preface to *Structure of Scientific Revolutions*,[69] he encountered many of the writers who were to influence him as a Junior Fellow at Harvard (1949–51). Jean Piaget's portrayal of the world view of a child and its changes as the child matures enabled him to understand the naturalness of Aristotle's cosmology and physics. His debt to Gestalt psychology and to the Sapir-Whorf thesis concerning the effect of the structure of a lan-

---

[66] Somewhat hurt, sociologists complained that he had rejected his children.

[67] In this respect Kuhn's historical position resembles that of Luther or of the Abbé Sieyès whose explosive pamphlet *"Qu'est-ce le tiers état?"* was published in 1789 at the convocation of the Estates General. Luther did not anticipate the religious extremism unleased by his Reformation, nor did Sieyès foresee the Terror.

[68] Had Kuhn not become a historian of science, he would have almost certainly been successful as a physicist. If the reader disagrees, he invited to plumb Kuhn's deeply mathematical treatment in his book *Black-Body Theory and the Quantum Discontinuity, 1894–1912* (1978).

[69] Kuhn (1970a), p. vi.

guage on a culture's world view is clear. In the history of science there was Anneliese Maier's sensitive treatment of Galileo's medieval precursors *Die Vorläufer Galileis im 14 Jarhundert* (1949) and Alexandre Koyré whose *Études Galiléennes* (1939) and *From the Closed World to the Infinite Universe* (1958) strongly contrasted the Aristotelian and Galilean world views and argued that observational data especially in mechanics was conditioned by the abstract mathematical Platonism of Galileo. Other writers he encountered included Emile Meyerson, Hélène Metzger, and A. O. Lovejoy. All of the above he says were "second only to primary source material in shaping my conception of what the history of scientific ideas can be."[70]

But his most important predecessor was Ludwik Fleck. The relation between Kuhn and Fleck is an ambiguous one. Fleck was one of the writers whom Kuhn encountered at Harvard and he admits that Fleck "anticipated many of my own ideas."[71] Later, however, in a foreword to an English translation of Fleck's book which he commissioned, Kuhn states that several of Fleck's notions had already been "on my mind" when he discovered them.[72]. In fact, Kuhn's frequent praise and mention of Fleck is probably the main reason the latter was rescued from oblivion.[73] But this may not be evidence of some kind of atonement to Fleck. Since Fleck wrote in German and in a style which Kuhn (whose German at this time was weak) found difficult, it is not clear what if anything Kuhn directly borrowed from him. Nevertheless there are some startling parallels in fundamental ideas between the two. Especially noteworthy is the similarity between Fleck's "thought style" and Kuhn's "paradigm." The former of course is a much more general notion. It could amount to nothing more than the Zeitgeist of a historical period; but suitably narrowed it could be a paradigm. In fact, the various thought styles Fleck describes in his history of syphilis could well have been included as paradigm examples in *Structure* or *Essential Tension*. It is also the case that Kuhn's concept of the semantic incommensurability of paradigms is close to Fleck's analysis of the conceptual differences between two "thought styles,"[74] especially that (i) the same word may have different meanings as it passes from one thought style to another, and (ii) key terms

---

[70] *Ibid.*
[71] *Ibid.*, p. vii.
[72] See Fleck (1979), p. ix.
[73] Having discovered that in twenty-six years only two people Edward Shils ("who had apparently read everything") and Mark Kac (who had known him) had read Fleck "independently of my intervention," it was Kuhn who arranged in 1976 that Fleck be translated into English in order to provide him "with an audience at all," (*Ibid.*, p. vii).
[74] See *Ibid.*, p. 109.

of past thought styles are unintelligible from the point of view of later thought styles.[75]

We will not speculate on the origins of the above similarities between Fleck and Kuhn; they may very well have been serendipitous coincidences. In any event, there are elements in Kuhn which Fleck does not anticipate. Fleck has no well-developed notion of paradigm shifts, nor of scientific revolutions. His thought styles develop and change with time; there can be radical changes in thought styles in a short period, but these tend to be correlated with periods of "general social confusion" such as the early Renaissance or just after World War I.[76] There is, however, a vague (and very brief) anticipation of Kuhn's distinction between normal science and a scientific crisis due to the emergence of anomalies. Fleck observes that all comprehensive theories pass

> ... first through a classical stage, when only those facts are recognized which conform to it exactly, and then through a stage with complications, when the exceptions begin to come forward.[77]

In the classical stage anomalies or contradictions are either unthinkable, unseen, or if seen, kept secret or ignored, or "laborious efforts are made to explain an exception in terms that do not contradict the system."[78] Fleck calls these alternatives "preserving the harmony of illusions." The last activity seems to correspond to Kuhnian puzzle solving within normal science, but the idea is not very well developed. Conversely, there are aspects to Fleck which Kuhn rejects. The notion of thought collective Kuhn believes is too close to an individual mind writ large and is a "source of recurrent tensions" in Fleck's work.[79] Also Kuhn finds the idea of passive and active aspects of knowledge unenlightening.[80]

---

[75] Ibid., pp. 125–145.
[76] Ibid., pp. 177-178, n.4.
[77] Ibid., pp. 28–29.
[78] Ibid., p. 27.
[79] Ibid., p. x.
[80] Ibid., p. xi.

Chapter 8

Anything Goes

Kuhn still, however, had a high regard for science. He admired the power of increasingly sophisticated paradigms to solve puzzles, and I have found no evidence that he questioned the prominence and prestige of science in modern western societies. Paradigm choice may not be a *completely* logically determined process like proof in mathematics; but Kuhn, as we have seen, came to believe in the virtue of rational argument and stressed the importance of "internal" factors in scientific change. In none of his published writings does he express the view that a legitimate scientific theory mirrors in any simple fashion a political ideology or originates in the class interest of a certain group, although he probably would have admitted that nonscientific aspects especially connected with philosophy may be present.[1] From his student days at Harvard until his death, he was also a man deeply committed to finding what he conceived to be the truth, and not merely to winning arguments.[2] These barriers would have to be demolished before the sociological deconstruction of science could proceed further.

The most significant subsequent agent in this work of demolition was the Berkeley philosopher Paul Feyerabend (1924–1994). Unlike Kuhn, Fey-

---

[1] Kuhn did point out the importance of external philosophical currents in the development of paradigms such as neo-Platonism in the case of Copernicanism. He was also extremely impressed by the Ph.D. thesis "The Environment and Practice of Atomic Physics: A Study in the History of Science" by his and Hunter Dupree's student Paul Forman who argued that the noncausal viewpoint of quantum mechanics was in deliberate conformity to the political climate of the early Weimar republic. After Forman had completed the thesis under Dupree because Kuhn had moved to Princeton and sent it to him, Kuhn claimed that it was the most exciting paper he had read since discovering Alexandre Koyré. (Kuhn (2000b), p. 304). Forman's approach, however, does not seem to have actually influenced his interpretation of science.

[2] When he was a graduate student at Harvard he was asked at a cocktail party by a young woman what he did. "As he answered the party chanced to fall silent. Everyone heard the answer 'I just want to know what truth is'." (Heilbron (1996), p. 506).

erabend was not chronologically a precursor of SSK. His most influential works were written well after the movement was established. The founders of SSK were certainly inspired by Kuhn, but they—as we shall see—were perfectly good at demolition on their own. Rather, it is the case that Feyerabend provided a volatile magazine of powerful explosives for the mature development of SSK and PIS generally after about 1975. If we could ignore chronology and attempt a "rational reconstruction" of the history of the process in the spirit of Lakatos, Feyerabend would emerge as its most important role model other than Kuhn.

At the same time, it is difficult to describe Feyerabend as clearly and accurately as we can Kuhn since he was a much more complicated and variegated thinker. There is a Jekyll-Hyde aspect to his writings which tax our faculties of both charity and appreciation. On the one hand, he won a reputation well before Kuhn for brilliant and original technical articles in the philosophy of science. Many, especially those published prior to the late 1960s, are reasonably mainstream. Although provocative, they fall within the accepted boundaries of the discipline. On the other hand, books such as *Against Method* (1975), *Science in a Free Society* (1978), *Farewell to Reason* (1987), *Three Dialogues on Knowledge* (1991), as well as the unfinished, posthumously published *Conquest of Abundance* (1995) won him a reputation as the "worst enemy of science," and excited blind rage and astonishment among critics.[3]

Throughout his career Feyerabend combined fairly sane analysis with rhetorical exaggeration and violent personal abuse of opponents—who were sometimes simultaneously his friends. A common verdict was that he combined brilliance with irresponsibility. In the judgement of John Watkins, an off-and-on friend:

> Paul Feyerabend was one of the most gifted, colorful, original, and eccentric figures in postwar academic philosophy—irreverent, brilliant, outrageous, life-enhancing, unreliable and, for most who knew him, a lovable individual.[4]

Feyerabend loved to astonish and offend more conventional colleagues and administrators, but was genuinely hurt if they retaliated, complaining that

---

[3] For example, consider the following reaction of David Stove to *Against Method*: "Of all the productions of the human mind, this must rank as one of the most curious. It is impossible to convey briefly the unique absurdity of the book." (Stove (2001a) p. 14.)

[4] Watkins (1999), p. 56.

people took his writings and actions too seriously and could not recognize a joke. This complaint has some justification, but it is tied to the fact that Feyerabend's writing is often unclear. One has the impression of an out of focus brilliance and it is sometimes hard to tell just *what* is being asserted or—whatever our interpretation of it is—whether or not he really means it. However, because of his enormous influence, we have no choice except to take his work more or less literally; after all, even a Berkeley philosopher should be held responsible for what he writes.

At first (c. 1947) after returning from service in the Wehrmacht with a spinal wound that made him semi-paraplegic and impotent, Feyerabend was a Vienna Circle logical positivist. Then in the early 1950s he became the student and follower of Karl Popper. By the end of the decade he had achieved a considerable recognition in the philosophy of physics. His ideas were sometimes controversial, but were clearly expressed and backed by ingenious argument. Like Popper he now rejected characteristic theses of the Vienna Circle such as the verifiability criterion of meaning and the abolition of metaphysics represented by philosophers like Rudolf Carnap or A. J. Ayer. He opposed positivism and instrumentalism. He was a scientific realist. The goal of scientific theories according to Feyerabend is to explain the way the world *is*, not just to make predictions and "save the appearances." He gives no ontological arguments for realism, but considers (like Popper) that it is the most productive approach since it especially encourages the development of many contending theories which he considers a positive feature of science. Straight instrumentalism is barren. Had Galileo and others followed Osiander's advice in the Preface to *De Revolutionibus*,[5] the new physics and optics needed to refute Aristotelian objections to Copernicanism would never have been developed. But also in papers anticipating his future radicalism he goes beyond the thesis, found in the work of Fleck, Hanson, and Kuhn, that observation statements are theory-laden[6] and argued that they are "fully theoretical," i.e., "observation statements have no 'observational core'."[7] And as early as 1962 we can find an early version of a completely constructivist position, probably obtained by extrapolating Kuhnian ideas as expressed in *The Structure of Scientific Revolutions* to

---

[5] Osiander, a Lutheran theologian, had argued that the goal of astronomy was merely to devise convenient models that allowed planetary motions to be calculated easily and accurately. Since the true causes of astronomical phenomena cannot be known, the astronomer must not regard his models in any sense as real. Osiander recommended Copernicus' book on this basis.

[6] See note 15, Chapter 7.

[7] Feyerabend (1981), Vol. 2, viii.

their logical limit. He writes that:

> ... scientific theories are ways of looking at the world; and their adoption affects our general beliefs and expectations, and thereby also our experiences and our conception of reality. We may even say that what is regarded as 'nature' at a particular time is our own product in the sense that all the features ascribed to it have first been invented by us and then used for bringing order into our surroundings.[8]

In the same paper he arrives at a notion of incommensurability, but one which was less general than Kuhn's initial ideas. In *Structure*, as we have seen, Kuhn had spoken of incommensurability between two paradigms as incorporating global changes "in the standards governing permissible problems, concepts, and explanations."[9] Moreover, since this is the case scientists on each side simply talk past each other and the triumph of one paradigm over another had to be due to essentially political rather than strictly rational factors. In contrast, Feyerabend limits himself to a semantic incommensurability between rival theories. He argues that when a theory T_1 is replaced by a theory T_2 both the meaning of critical terms of T_1 and the ontology T_1 postulates changes. This in turn implies that it is impossible to establish logical relations between T_1 and T_2, and specifically to define terms of T_1 using T_2.[10] We have observed that Kuhn in his later writings began to weaken his early quasi-political concept of incommensurability, and focused on the difficulty of translating (or even making sense of) the terms of an older paradigm using the concepts and vocabulary of a later one. Thus Kuhn arrived at a position not too dissimilar to Feyerabend's early concept of incommensurability. Feyerabend, however, went in the opposite direction; his vision of incommensurability began to become ever more expansive. By 1965 he goes beyond semantic questions of translation and argues that different theories are incommensurable in the sense that since no observation, permeated as it is by the theory in question, can decide between them:

---

[8]Feyerabend (1969), p. 45.
[9]Kuhn (1970a), p. 105.
[10]Feyerabend (1969), especially pp. 44–45, p. 59, and pp. 67–94. Feyerabend gives a number of examples purporting to show that even when it is commonly thought that $T_1 \subset T_2$, for example, Galilean physics in relation to Newtonian dynamics, it is impossible to deduce Galileo's results from Newtonian theory. See pp. 59–61. For additional remarks concerning incommensurability according to Kuhn and Feyerabend see Kuhn (2000b), n. 2, pp. 33–34 and Feyerabend (1978), pp. 65–70.

To express it more radically, each theory will possess its own experience and there will be no overlap between these experiences. Clearly a crucial experiment is now impossible ... because there is no universally accepted *statement* incapable of expressing whatever emerges from observation.[11] [emphasis in original]

This claim, of course, is similar to some of more radical parts of *Structure* discussed in the previous chapter.[12] But how then does one decide between two such theories if crucial experiments or rational argument will not work? For Feyerabend the only option is a pragmatic approach. One theory can be superior to the other in the fruitfulness and accuracy of its predictions. Note that here Feyerabend is less revolutionary than Kuhn since he offers a rational method to choose the better of two incommensurable theories. For Feyerabend the victory of one over the other is not a matter of "mob rule," as Kuhn's analysis of paradigm change in *Structure* seemed to imply to some critics. Some of Feyerabend's early ideas concerning observation and incommensurability may not have been exactly mainstream, but if we consider the work of Hanson and Kuhn, they were certainly "in the air" at the time. Moreover, even if Feyerabend's papers have radical implications, they were still within the accepted boundaries of contemporary philosophy of science and exhibit immense technical sophistication (far more in my opinion than most of Kuhn's work). This relatively "conventional" phase in Feyerabend's career lasted until the late 1960s.

In 1958 Feyerabend accepted a visiting position in the Philosophy Department at the Berkeley campus of the University of California which was made permanent the next year. He officially remained there in spite of frequent leaves of absence, conflicts with colleagues and administrators, and a temporary resignation for the next thirty years.[13] What was the environment like in Berkeley at this time? In 1958 the campus was fairly

---

[11] Feyerabend (1965), p. 214.
[12] Later, as we shall see, Feyerabend will extend this notion of incommensurability much further, arriving at a position similar to Ludwik Fleck's overarching notion of "thought style."
[13] I frequently saw Feyerabend hobbling around campus on his crutches. But I did not know him, and since he seemed vaguely terrifying I dared not take any of his courses. I thought that they would probably be too difficult. This was a mistake. Feyerabend became famous among students (and infamous to colleagues and administrators) for the laxness of his teaching—he required essentially nothing of his students. The perpetual frown I detected, which scared me off, was probably caused by the pain arising from his war injury.

conservative. The faculty was liberal, even leftist.[14] Academic standards were high especially in the liberal arts, but fraternities and football were also important. The majority of students as elsewhere in the Eisenhower era were middle class and career oriented, majoring in subjects like business administration and engineering. What there was of student unrest took the form of "panty raids" on sororities, an especially violent episode occurring in 1955. The administration was politically conservative, and had a certain indifference to subsequent notions of academic freedom. For instance, that dangerous radical Adlai Stevenson was barred from campus by President Robert Gordon Sproul during the election campaign of 1956. Also, John Searle, then an assistant professor of philosophy, was forbidden by the vice-chancellor to lecture to a law school club concerning the House Unamerican Activities Committee (HUAC) film "Operation Abolition."[15] The atmosphere began to change with the anti-HUAC demonstrations in San Francisco in the spring of 1960 which resulted in several student arrests and with a steady infiltration of beatnik ideology from the San Francisco North Beach district. The fact that many students were sympathetic to or participated in the Mississippi Freedom rides in the summer of 1961 probably also contributed to a growing radicalism.[16] The situation culminated in the Free Speech student uprising led by Mario Savio and Jerry Rubin in the fall of 1964. This was the first major episode in a decade of student unrest which soon spread from Berkeley to the rest of the nation and which by 1966 had become fueled by an increasingly divisive Vietnam war.

Soon after arriving in Berkeley Feyerabend became friendly with and, as some of the passages quoted above show, strongly influenced by Thomas Kuhn. But his own thought evolved, paralleling the students, in a much more radical direction than Kuhn's.[17] He began first to rebel against his

---

[14] Several had been members of the Communist party or "fellow travelers" in the 1930s and 40s. The McCarthyite loyalty oath controversy of 1951 had cost several faculty their positions and had damaged the morale of many more.

[15] Personal recollection and Searle (1972), p. 175. This writer also recalls that around 1958 a rumor developed that the Chemistry Department made a job offer to an academically outstanding candidate. They had, however, somehow overlooked his photograph in his file. When it was noticed that the candidate was Afro-American, the offer was withdrawn. Whether or not the rumor was true is uncertain, but it illustrates the temper of the times.

[16] At first innocently signaled by an increasing density of guitars among the student population combined with a fondness for folk singers such as Pete Seeger and Woodie Guthrie.

[17] There is such a perfect correspondence between Feyerabend and the heady Berkeley scene of the Mario Savio era that had he not existed, he could have very well been a fictional character in a David Lodge novel. On occasion, Feyerabend's mode of living

former mentor Karl Popper. There are increasing criticisms of Popper starting in early 1960s. By the 1970s these were often personal and insulting, and not, as in the case of Kuhn, confined to a criticism of Popper's doctrines. From a psychoanalytic point of view they amounted to acts of intellectual parricide against the person who in the early years had been his advisor, mentor, and who had arranged his first academic appointment at the University of Bristol.[18] In the mid-1970s after the publication of his first book *Against Method* he became the *enfant terrible* of the philosophy of science. In this book, whose themes he continued to elaborate for the remainder of his career, Feyerabend attacked the idea, then accepted by most philosophers of science, that there exist methodological or structural characteristics of scientific practice which define science and separate it from non-science. Any attempt, Feyerabend believed, to codify a "scientific method" would in fact damage science and inhibit its development. In particular, he was critical of Popper's view that science in contrast to non-science consisted of falsifiable statements and that good practice in science consisted of efforts to falsify theories. Like Kuhn, Feyerabend argues that no interesting theory is consistent with all the facts. If a theory is sufficiently attractive, counterexamples to it are often put on hold for later treatment or even ignored; for example, in certain areas of quantum mechanics such as renormalization calculations are just tossed out if they disagree with the facts.[19]

Feyerabend supported his claims by a careful study of the actual historical development of science, especially Galileo's defense of Copernicanism. In defiance of the directives of Karl Popper, Galileo did not *allow* the fact that the heliocentric hypothesis violated common sense, established philosophic principles, well-known contemporary scientific "facts," or the evidence of the senses to falsify it.[20] Instead, Feyerabend argued that Galileo proceeded "counterinductively." He began by accepting a theory that by

---

was also appropriate to the times. In 1969/70, for instance, he shared his quarters with a rat who lived under his bed and was given the name Kautsky after the revisionist Austrian Marxist theoretician. (Feyerabend (1995b), p. 113.)

[18] Feyerabend even made negative remarks about the ugliness of Popper's wife. But Popper gave as good as he got. In a brutal putdown he wrote: "As far as my former student Feyerabend is concerned, I cannot recall any writing of mine in which I took notice of any writing of his." (quoted in Krige (1980), p. 107.)

[19] Feyerabend (1975), p. 61.

[20] Such as the absence of stellar parallax, the "fact" that a heavy object dropped from a tower would fall behind it, that there would be a 1,000 mile per hour East to West wind, the apparent lack of change in the sizes of Venus and Mars viewed from the Earth as they orbit the Sun, etc.

contemporary standards was clearly "falsified" and then proceeded to get around the objections by a rich stew of good and bad arguments, seasoned by intellectual dishonesty, rhetorical trickery, and bombast. According to our contemporary canons of scientific method, Cardinal Bellarmine and the Inquisition were right. At best the Copernican model was a convenient computational tool for astronomers. There was *no* evidence that it was true; all received facts and physical theory appeared to contradict it. An uncritical promulgation of this theory might, moreover, have serious social and political consequences for the Church at a a dangerous period in her history, given the spread of heresy in Europe.[21] However, if in the future compelling evidence were found for its truth, the Church was prepared to change its interpretation of biblical texts which implied a stationary earth, just as she had earlier done with passages that supported the flatness of the earth. In the meantime, it was reckless and unscientific for Galileo to argue otherwise. Yet, ultimately Galileo's way of looking at the world prevailed, and this was because he was not limited by a sterile "scientific methodology," but helped to *force* a change in all sorts of issues that created the intellectual space for Copernicanism to flourish.

Some philosophers of science like Herbert Feigl were prepared to agree that the actual historical development of science is a messy affair where scientists do not use any recognizable "method," but they made a distinction between the "context of discovery" and the "context of justification." The latter should clean up science and justify its discoveries by the rules (promulgated, of course, by philosophers) of good science. Feyerabend asserted that this possibility is an illusion. No such rules ever existed, have been followed, or can be given; any attempt to legislate or enforce them would ruin science. If, however, the need for such rules are still insisted upon, the only one possible is Cole Porter's "anything goes."

Feyerabend coupled his denial of the existence of a scientific method with an attack on traditional scientific realism and objectivity, together with the affirmation of a thorough relativism. Scientists may believe that atoms, say, were *discovered* and have an independent existence and that:

> ... there were atoms long before the scintillation screen and mass spectroscopy; they obeyed the laws of quantum theory long before those laws

---

[21] The report of the Inquisition on Galileo's opinions reports his praise of William Gilbert, "a perverse heretic and a quibbling and quarrelsome defender" of Copernicanism. (Santillana (1958), p. 247.) Another suspicious aspect to Galileo in the eyes of the Church may have been his correspondence with the Lutheran heretic Kepler.

were written down and they will continue to do so when the last human being has disappeared from the earth ...[22]

But this is an illusion. Feyerabend observes that the Olympian gods according to the ancient Greeks had the same kind of independent existence as atoms. Do atoms exist together with the gods? Why should the gods be treated differently from atoms? If it is argued that the the existence of the gods is not a "reasonable" belief or is incompatible with the "scientific world view," Feyerabend is happy to agree. But this just means that such belief is not compatible with the present historically conditioned Zeitgeist. Essentially, the dismissal of the gods is a circular argument. We simply appeal to the scientific world view which has historically replaced belief in the gods to justify disbelief in the gods; no *independent* justification exists. Atoms were substituted for the gods not by "reason" or good arguments, but rather as a result of profound historical and social changes: " ... leading to a new attitude, new standards, and new ways of looking at the world: history, not argument, undermined the gods."[23] Besides, present physics will probably be as "unreasonable" from future perspectives as the gods are now from our perspective. Perhaps then, we should say that neither the gods or atoms have an existence independent of human beings. The truth of the matter is that:

> *Scientific entities* (and for that matter all entities) *are projections and thus tied to the theory, the ideology, the culture that postulates and projects them.* The assertion that some things are independent of research, or history, belongs to special projecting mechanisms that 'objectivize' their ontology; it is not valid outside the historical stage that contains such mechanisms.[24] ... We now say that scientists, being embedded in constantly changing social surroundings, used ideas and and physical equipment to *manufacture* first metaphysical atoms then crude physical atoms, then complex systems of elementary particles out of material lacking all these features. Scientists are sculptors of reality ...[25] [emphasis in original]

It follows that the Olympian gods and the subatomic particles of modern physics are on the same epistemological level; the same is true Feyerabend

---

[22] Feyerabend (1975), p. 262.
[23] *Ibid.*, p. 263.
[24] *Ibid.*, p. 265.
[25] *Ibid.*, pp. 269–270.

will argue of Western science in general and other belief systems that develop out of other traditions and cultures. The appeal to "objectivity," "reason," or "truth" in defense of science is really quite parochial and subjective; it is part of and depends upon the scientific tradition it is trying to support. Politically it amounts to nothing but sophisticated propaganda designed to suppress rival traditions.

In opposition to nineteenth century figures such as Whewell or modern logical positivists who believed that science had a secure epistemological foundation Feyerabend views its ideology simply as a dogma, no better or worse than any other and not so different from that of the Roman Catholic Church. Both are authoritarian and wish to suppress deviant "projections" arising from competing traditions.[26] As an example he considers the extreme hostility of astronomy to astrology, manifested by a 1975 denunciation of astrology signed by 186 leading scientists including 18 Nobel Prize Winners. No real arguments are given refuting astrology; rather, the article is an anathema against a scientific heresy similar to a bull against witchcraft by Pope Innocent VIII in 1484 which was a preface to *Malleus Maleficarum*, an infamous manual to aid in the discovery and punishment of witches, issued by the Inquisition in the same year. However, this manifesto against astrology was less literate and intelligent than either Innocent's bull or the book, since it exhibited extreme ignorance concerning astrology and ignored the possibility of real physical influences by the planets. At least the Pope and the Inquisition knew what they were talking about concerning witchcraft.[27]

During the Enlightenment science may have been a liberating force against clerical and Old Regime obscurantism. Now it is the dominant ideology in the modern world. It is power hungry and ruthlessly suppresses all competitors. But the victory of science is not the result of its excellence or:

> ... *comparative merits but because the show has been rigged in its favour.*
> ... *the apostles of science were the more determined conquerors*, because they *materially suppressed* the bearers of alternative cultures. There was no research. There was no 'objective' comparison of methods and achievements. There was colonization and suppression of the views of the tribes and nations colonized.[28] [emphasis in original]

---

[26] Feyerabend once taught a history of Catholic dogma, making explicit its parallels with science.
[27] See "The Strange Case of Astrology" in Feyerabend (1978), pp. 91–98. Feyerabend gives a list of some real physical and chemical phenomena which he claims may indirectly depend on the position of the planets.
[28] *Ibid.*, p. 102.

The ruthless snuffing out of the traditional knowledge of other cultures such as shamanism, tribal medicine, or yin/yang duality, acupuncture, and the theory of the chi—has meant the loss of much of potential value.[29] The sterile and abstract scientific rationalism characteristic of our own culture has blinded us to certain aspects of nature to which the pre-rationalist past and non-western cultures were more open. For instance:

> We have learned that there are phenomena such as telepathy and telekinesis which are obliterated by a scientific approach and which could be used to do research in an entirely novel way (earlier thinkers such as Agrippa of Nettesheim, John Dee, and even Bacon were aware of these phenomena).[30]

These facts together with the danger in the modern West of a tyranny of scientific "experts" means that science should be separated from the State just as religion is. Just as the State is not allowed to establish a religion or a political doctrine, why should it be permitted to "establish" science? Why should students, for example, be forced (as they are now) to study astronomy rather than astrology, science rather than magic, history rather than the study of legends, or evolution rather than creationism?[31] This situation is unjust and anti-democratic! To correct it, Feyerabend proposes that all intellectual traditions be considered equal and have an equal right to be taught in schools. In no case should science be given an educational monopoly. Scientists, moreover, should be subject to the close supervision of committees of lay citizens and their claims and proposals decided upon by democratic vote. We may on occasion consult them and sometimes the technology resulting from their work is useful. But we must recognize that scientists usually are limited, uncreative types, "devoid of

---

[29] Feyerabend felt that traditional Chinese medicine "has methods of diagnosis and therapy that are superior to those of Western scientific medicine" and celebrated the decision of the Chinese communist regime to reinstate this medicine into its hospitals. (*Ibid.*, p. 103.) When asked by a critic if having a child diagnosed with leukemia, he would consult Sloan Kettering or "his witch-doctor friends," Feyerabend answered that he would definitely prefer a witch-doctor whose medicine he considers superior to that of soulless western rationalism. There is a strong component in Feyerabend of Rousseau's "noble savage" ideology. Like both eighteenth century French aristocrats and some contemporary radical-chic academics, Feyerabend usually judged the customs and beliefs of traditional third world societies to be superior to those of the West.

[30] Feyerabend (1999), p. 186.

[31] "Three cheers to the fundamentalists of California who succeeded in having a dogmatic formulation of the theory of evolution removed from the text books and an account of Genesis included." (*Ibid.*, p. 187.)

ideas" and "slaves of institutions and slaves of 'reason'," and we will ensure that they "will not play any predominant role in the society I envisage. They will be more than balanced by magicians, or priests, or astrologers."[32] Feyerabend calls this political program "Democratic Relativism," and in a passage revealing his low opinion of the average scientist Feverabend argues that such a separation between science and the state will not have any adverse consequences:

> We need not fear that [it] will lead to a breakdown of technology. There will always be people who prefer being scientists to being masters of their fate and who gladly submit to the meanest kind of (intellectual and institutional) slavery provided they are paid well and provided also there are some people around who examine their work and sing their praise. Greece developed and progressed because it could rely on the services of unwilling slaves. We shall develop and progress with the help of numerous willing slaves in universities and laboratories who provide us with pills, gas, electricity, atom bombs, frozen dinners and occasionally, with a few interesting fairy tales. We shall treat these slaves well, we shall even listen to them, for they have occasionally some interesting stories to tell ... [33]

Feyerabend, especially in this later years, proudly identified with relativism. Two of his cultural heros in this respect were Protagoras and John Stuart Mill. Protagorean relativism was primarily *cultural*. Therefore, it is:

> ... *reasonable* because it pays attention to the pluralism of traditions and values. And it is *civilized* for it does not assume that one's own village and the strange customs it contains are the navel of the world.[34] [emphasis in original]

Beyond this, Feyerabend is not prepared to interpret Protagoras' famous dictum that "Man is the measure of all things" as implying a simple epistemic relativism, that is:

> ... all traditions, theories, ideas, are equally true or equally false or, in an even more radical formulation, that any distribution of truth values over traditions is acceptable.[35]

---

[32] Ibid., pp. 189–190.
[33] Feyerabend (1975), Concluding Chapter of the first edition.
[34] Ibid., p. 244.
[35] Feyerabend (1978), pp. 82–83.

His views were more complicated. Feyerabend appears to admit, say, that "physics may be objective" *within* the tradition in which it is embedded, but the tradition itself is not objective. There is no such thing as a tradition-independent belief system:

> ... opinions not tied to traditions are outside human existence, they are not even opinions while their content depends on or is 'relative to', the constituting principles of the traditions to which they belong. Opinions may be 'objective' in the sense that they do not contain any references to these principles. They then *sound* as if they had arisen from the very essence of the world while merely reflecting the peculiarities of a particular approach: the values of a tradition recommending absolute values may be absolute, but the tradition itself is not ... [36]

Thus judgements of belief systems, theories, etc., are relative to "traditions" which simply "are" and are neither "good or bad."[37] These traditions, moreover, cannot be rationally evaluated in a tradition independent manner since "rationality is not an arbiter of traditions, it itself is a tradition or an aspect of a tradition."[38] This relativism also holds for moral aspects of a tradition: " 'Objectively speaking', i.e. independently of participation in a tradition there is not much to choose between humanitarianism and anti-Semitism."[39] Of course, we are free to make value judgements, but "value judgements are not 'objective' and cannot be used to push aside the 'subjective' opinions that emerge from different traditions."[40] These considerations reinforce Feyerabend's opinion that all traditions whether cultural, political, religious, or intellectual are in a fundamental sense equal and should be tolerated and treated equally by the state. Astrology and Astronomy, Alchemy and Chemistry simply represent different traditions and should be free to develop and compete. This "pluralism of ideas and forms of life" would enrich society and even benefit science since it would actually have to compete with the other traditions and could not rely on authoritarian dogma enforced by the State. Although he claimed not to be a supporter of political anarchy,[41] he certainly believed in fostering intellec-

---

[36] Feyerabend (1987), p. 73.
[37] Feyerabend (1975), p. 243.
[38] *Ibid.*, p. 243.
[39] *Ibid.*
[40] Feyerabend (1978), p. 83.
[41] As usual there is ambiguity here. John Watkins (Watkins (1999), pp. 54–55) claims that Feyerabend *insinuated* that violence and anarchy were "beneficial for the individual" while not actually claiming this.

tual and cultural anarchy and thought that the result would be a realization of the ideals of Mill's *On Liberty* which encouraged many experiments in living in a society.

These views pursued over a twenty year period clearly contain the main ideas of PIS and helped pave the way for and reinforce the application of SK to science, which is our major theme. He was both a precursor and "fellow traveler" of what was to follow. Historically he stands in much the same relation to Kuhn as wild Anabaptist thinkers such as John of Leyden or Thomas Münzer did to Luther. Just as religious radicals reacted to Luther by pushing some of his central theses to extremes while rejecting others, Feyerabend supported but exaggerated the anti-methodological and relativistic implications of Kuhn's thought. We have already seen, for instance, that both Kuhn and Feyerabend had discovered incommensurability as early as 1962, with Feyerabend's version being the more technical and less general. By the early 1970s, however, he had generalized the idea so that it went beyond anything envisaged by Kuhn. For Kuhn Copernican and Ptolemaic astronomy or the chemistry of phlogiston and Lavoisier may represent incommensurable paradigms, but they are within recognizably the *same* science committed to the *same* values of rational argument and empirical research. On the other hand, for Feyerabend incommensurability is found at the level of, say, Western scientific rationalism versus Greek mythology or third world shamanism. These global cognitive structures, which represent entirely different ways of encountering the world, Feyerabend calls "traditions." To Feyerabend traditions have their own epistemological standards and ontologies which are opaque to each other; consequently, (like Kuhn in relation to paradigms) he argues that no "objective" standard or "common measure" exists to judge the superiority of one tradition over another.[42] We are all locked into whatever tradition supports our cognitive system. We may use reason and logic to judge claims within our tradition, but since the particular form reason and logic take is relative to the tradition; neither applies (except as propaganda weapons) to judgements in other traditions. It immediately follows—since one cannot show "objectively" otherwise—that traditions such as science and witchcraft or astronomy and astrology are merely different, but equally valid "ways of knowing." They are comparable in society to different religions or political ideologies. Thus while Kuhn believed there may exist no *completely* ra-

---

[42] In fact "tradition" for Feyerabend functions in much the same way as "thought style" for Ludwik Fleck; but this may be a serendipitous coincidence rather than evidence of direct influence.

tional method to decide between incommensurable paradigms, Feyerabend abandoned any global appeal to rationality altogether. Not for nothing did he sympathize with intellectual anarchism and dadaism.[43]

However, Feyerabend throughout his career remained enough of a Popperian to strongly dislike the Kuhnian notion of "normal science." To Feyerabend normal science with its concentration on "tiny puzzles" was a form of pedantic "professional stupidity," and he thought it dogmatic and authoritarian.[44] Like Popper, he believed that science should constantly challenge its theories and did not regard the uncritical acceptance of received paradigms characteristic of normal science either as a feature distinguishing science from non-science or as a guarantee of scientific progress. This attitude is evident as early as 1967 in a paper sharply attacking Kuhn.[45] He argues that the identification of science with puzzle solving does not really distinguish it as an activity. Many activities other than science involve puzzles. Organized crime, for instance, as practiced by professional safe crackers or by John Dillinger (who built full sized models of banks he intended to rob on his farm and conducted dress rehearsals) also perfects itself by solving puzzles. By defining normal science as he does, Kuhn misses the aim of science. The aim of safe cracking or bank robbery is money. What is the aim of science? This question is missing from Kuhn's analysis.[46] Also, by stressing the narrow, dogmatic, and authoritarian characteristics of normal scientists Kuhn makes it difficult to understand how competing theories can ever arise.[47] But what if the claimed existence of normal science is a myth? Feyerabend tries to show historically that "normal periods if they ever existed, cannot have lasted very long, and cannot have extended over large fields either."[48] In the last third of the nineteenth century, for ex-

---

[43] *Ibid.*, pp. 23, 286–7.
[44] Feyerabend, although he acknowledges a debt to Kuhn, can be rather insulting:

> Kuhn's ideas are interesting but, alas, they are much too vague to give rise to anything but lots of hot air. If you don't believe me, look at the literature. Never before has the literature on the philosophy of science been invaded by so many creeps and incompetents. Kuhn encourages people who have no idea why a stone falls to the ground talk with assurance about scientific method. Now I have no objection to incompetence but I do object when incompetence is accompanied by boredom and self-righteousness. And this is exactly what happens We do not get interesting false ideas, we get boring ideas or words connected with no ideas at all. (Feyerabend (1999), p. 186.)

[45] Feyerabend (1967).
[46] *Ibid.*, pp. 133-4.
[47] *Ibid.*, p. 139.
[48] *Ibid.*, p. 141.

ample, astronomy, thermodynamics, and electrodynamics as developed by Faraday and Maxwell were all guided by mutually incompatible paradigms, much as Kuhn argues happens in pre-science. The presence, conflict, and interaction of "different, incompatible, and occasionally even incommensurable disciplines" is what generates real progress such as the invention of quantum theory rather than the practice of so-called "normal science."[49] All this is not too different from Popper. But whereas Popper believed in the virtues of competition of recognizably *scientific* theories, Feyerabend, as we have seen, came to believe in the virtues of complete theoretical anarchy. In Feyerabend's view, the more competing theories—preferably drawn from radically different traditions such as high-energy physics, New Age crystal worship, or witchcraft—the better!

Despite his extremism Feyerabend was extremely successful professionally. He received offers from Atlanta, New Zealand, Sussex, Oxford, Freiberg, Berlin, Kassel, and London. At one time he held positions at Berkeley, the University of London, the Free University in Berlin, and Yale simultaneously. In Berlin he had fourteen assistants, most of a revolutionary persuasion and two wanted by the police.[50] He obviously suited the Zeitgeist of the 1960s and 1970s. A Mannheimian analysis of Feyerabend's success is tempting. Clearly he would have been impossible before the 1960s. Can his thought be explained by "existential factors" such as the moral and intellectual anarchy flourishing on the Berkeley campus in the age of Mario Savio and Jerry Rubin, or the post-Vietnam disillusion with science on the part of the New Left?[51] It is also tempting to think that another explanatory factor may be connected to the vast expansion of higher education in the 1960s together with the inclusion of minority groups for whom the University had never before been a realistic possibility. Feyerabend relates that:

> In the years 1964 ff. Mexicans, Blacks, Indians entered the university as a result of new educational policies. There they sat, partly curious, partly disdainful, partly simply confused, hoping to get an "education". What an opportunity for a supposed prophet in search of a following! What an opportunity, my rationalists friends told me, to contribute to the spreading of reason and the improvement of mankind! I felt very differently. For it dawned on me that the intricate arguments and the

---
[49] *Ibid.*
[50] See Feyerabend (1995b), Chaper 11.
[51] As contrasted with the Marxist-oriented Old Left of the 1930s who viewed themselves as the political heirs of the *scientific* Enlightenment.

wonderful stories I had so far told my more or less sophisticated audience
might just be dreams, reflections of the conceit of a small group who had
succeeded in enslaving everyone else with their ideas.[52]

For Feyerabend personally this experience had the same effect as the exposure to the strange customs of Egypt and Persia had on Heroditus—it pushed him in the direction of relativism. Certainly, the view that he might only be teaching what "white intellectuals had decided what was knowledge" must have helped form his attitude that the belief systems of other groups, cultures, and traditions were of equal value to western science, thus making him a pioneer of "multiculturalism" as an ideology. More cynically, one might speculate that the huge clientele now present on campus was in need of an intellectual solvent to dissolve the difficult hard sciences many of whom resented and lacked the educational background to understand.

We shall not pursue the above analysis further. Many sociological factors obviously combined to make Feyerabend's philosophy attractive to such disparate groups as perpetual graduate students in philosophy of science, the lumpenproletariat intellectuals who infest the squalid rooming houses that surround the modern mega-university, faculty in English, sociology, and cultural studies, or perhaps even the drug saturated panhandlers who live in the rough in alleys off Telegraph avenue. All of the above and others in the counterculture opposing the hegemony of science, technology, and rationalism would appreciate being informed that there is little to choose between science, voodoo, and astrology and that:

> Science is not a closed book that is understood only after years of training. It is an intellectual discipline that can be examined and criticized by anyone who is interested and that looks difficult and profound only because of a systematic campaign of obfuscation carried out by many scientists ...[53]

Towards the end of his life, however, there were signs that Feyerabend's ideas were changing. In the Third Dialogue (1989) of the late *Three Dialogues on Knowledge* he seems to disown his previous fascination with astrology, implicitly comparing his present feelings with the slow estrangement from an old friend that can arise from boredom.[54] The multiculturalism

---

[52] Feyerabend (1978), p. 118.
[53] Feyerabend (1999), p. 187.
[54] Feyerabend (1991), pp. 127–129.

that in the 1970s he defended was by the 1990s an official rigidly enforced ideology on US campuses, and the tyrannical rationalism and scientism he had denounced was replaced by an equally barren "political correctness." Ever the gadfly, Feyerabend tended to oppose all reigning orthodoxies. In the past he notes that pluralism was:

> ... called irrational and expelled from decent society. In the meantime it has become the fashion. This vogue did not make pluralism better or more humane; it made it trivial and, in the hands of its more learned defenders scholastic...[55]

In the same passage he accuses these defenders of the new orthodoxy of preparing "a New Age of ignorance, darkness, and slavery." Likewise, Feyerabend began to have mixed feelings about the relativism he had helped make fashionable: "If on almost every university toilet door there are relativistic theses, then it is time to distance oneself from relativism."[56] In 1990 he admits to once having been a relativist "at least in one of the many meanings of this term" but now only regards it as a "useful and, above all, humane approximation to a better view" which he has not yet found.[57] At the same time, much of the unfinished *Conquest of Abundance* continues to pursue the same themes as *Against Method*. His conclusions are similar; but now "Traditions" are called "World Views." And in the First Dialogue (1990) of *Three Dialogues on Knowledge* he proposes the esoteric doctrine that knowledge is the product of interactions between an incomprehensible, almost person-like, indeterminate Being and humanity. When approached spiritually Being responds by revealing the gods to mankind, "not just ideas of them but real visible gods whose actions can be followed in detail."[58] When approached scientifically:

> Being gives us, one after the other, a closed world, an eternal and infinite universe, a big bang, a great wall of galaxies, in the small an unchanging Parmenidean block, Democritean atoms and so on until we have quarks, etc.[59]

---

[55] Feyerabend (1995b), p. 164.
[56] Quoted in Hoyningen-Huene (1999), p. 14.
[57] Feyerabend (1991), pp. 156–57.
[58] *Ibid.*, pp. 43–44.
[59] *Ibid.*, p. 43.

Any of these responses of Being are as "objective" as any other. How seriously Feyerabend took these ideas is unclear, but they do entail a relativism that "leaves more leeway than assumed by realism." He also softened his attacks on science and was willing to admit that it had many positive features. In a late article (1992) he writes concerning science:

> How can an enterprise depend on culture in so many ways, and yet produce such solid results? Most answers to this question are either incomplete or incoherent. Movements that view quantum mechanics as a turning point in thought—and that include fly-by-night mystics, prophets of a New Age, and relativists of all sorts—get aroused by the cultural component and forget predictions and technology.[60]

How all this would have played out had he lived another decade or two is a fascinating question. Perhaps, given the contemporary dominance of political correctness, he would have become a neoconservative and a sympathizer with David Horowitz's campaign for a compulsory conservative presence in academia.[61] Perhaps he would have even detected intelligence in the mind of George W. Bush. One can only speculate, but with Feyerabend one has the feeling that "anything goes."

---

[60] Feyerabend (1992), p. 28.
[61] In the early 1960s Horowitz was an enthusiastic socialist activist at Berkeley. Now an intellectual hero to conservatives, he is the leader of a national movement to ensure proper balance between conservatives and liberals in academia. Since Horowitz thinks that conservatives are seriously under-represented in universities, he feels that the ideology of candidates should be a factor in hiring, promotion, and tenure.

Chapter 9

The Sociological Attack

The final stage in the process of scientific deconstruction inadvertently begun by by Thomas Kuhn and continued by Paul Feyerabend was the achievement of Barry Barnes, David Bloor, Steve Shapin,[1] and John Henry[2] at the University of Edinburgh's Science Studies Unit[3] who, beginning in the early 1970s, created the so-called "Edinburgh Strong Program." Over the next twenty years closely related ideas were advanced by Harry Collins and Trevor Pinch[4] in Bath ("The Bath School"), Michael Mulkay[5] at York, Simon Schaffer,[6] Karin Knorr-Cetina[7] at the University of Pennsylvania, Steve Woolgar[8] at Brunel, Michel Callon and Bruno Latour[9] in Paris, and several others. Their collective approach became a new and fashionable academic discipline, usually called "sociology of scientific knowledge" (SSK), some form of which is now extremely influential in the historiography of science and in STS generally. SSK is not monolithic; as in any academic field there are discordant points of view and disputes, sometimes vicious, among its practitioners. We will briefly discuss some of these and attempt to point out some of the differences among the major figures in the movement. But

---

[1] Now at Princeton.
[2] See e.g. Barnes and Bloor (1982), Barnes, Bloor, and Henry (1996), Bloor (1976), and Bloor (1982a).
[3] This organization was founded under the Labor government of Harold Wilson. Wilson's motives were similar to Conant's. He wanted a better mutual understanding between science and society. One of the central books studied in the first few years of the Unit was Kuhn (1970a). See Fuller (2000), p. 318.
[4] Collins and Pinch (1988).
[5] Mulkay (1979a).
[6] See in particular Shapin and Schaffer (1985).
[7] Knorr-Cetina (1981) and Knorr-Cetina (1999).
[8] Woolgar (1988a).
[9] Latour (1987).

an exhaustive discussion of SSK and its various strands is beyond our scope, requiring at the very least a much larger book than the present one.

Delicious morsels of SSK's characteristically jaundiced view of science have been consumed and reprocessed by a variety of groups such as feminists, Afrocentrists, the New Left, anti-Darwinian creationists, and Christian Evangelicals. With the exception of the last two, all have been on the left—sometimes extreme left of the political spectrum. This aspect of SSK is its most important feature. Had it remained an esoteric academic cult of little social or political importance—like, say, K-theory in mathematics—this book (and many others) would not have been written. The political excesses of SSK have been thoroughly analyzed and debunked by authors such as Paul Gross and Norman Levitt, Alan Sokal and Jean Bricmont, and Noretta Koertge whose books we have already mentioned, as well as by the Conference Proceedings *The Flight from Science and Reason* (1996) edited by Gross, Levitt, and M. L. Lewis. These books comprise the heavy artillery of one side of the Science Wars. But our interest here is in the intellectual structure and origin of SSK together with the philosophic doctrines it espouses rather than polemic. Only at the end of the chapter will we briefly consider the political "applications" of SSK to feminism and Afrocentrism.[10] Criticism will be presented later.[11]

A minimal common core of belief characterizing SSK has been well expressed by Michael Mulkay:

> ... the factual content of a science should not be treated as a culturally unmediated reflection of a stable external world. Fact and theory, observation and presupposition, are inter-related in a complex manner; and the empirical conclusions of science must be seen as interpretive constructions, dependent for their meaning upon and limited by the cultural resources available to a particular social group at a particular point in time. Similarly, general criteria for assessing scientific knowledge-claims cannot be applied universally, independently of social context, as most sociologists have previously assumed.[12]

As stated these ideas seem fairly harmless and have a kernel of truth. Who, for example, would not agree with the judgement that "fact and theory, observation and presupposition are inter-related in a complex manner"?

---

[10] The political consequences of SSK and PIS will be discussed more thoroughly in Chapters 13 and 14.
[11] Chapters 11, 12, and 14.
[12] Mulkay (1979a), p. 119.

Trouble, however, arises from the multiple possibilities of their interpretation and application. In their mildest versions they are not too different from Kuhn's views. But practitioners of SSK, as we shall see, often push one or all of them to extremes. Especially when they are applied to case studies of scientific development, a fair judgement is that SSK has united the anti-realist epistemology common to both Kuhn and Feyerabend to the "unmasking" turn of mind of Mannheim's SK. These two perspectives form the two components of a powerful weapon with which SSK tries to undermine the claims and cultural prestige of science.

Anti-realism and some ontological, and methodological issues

Anti-realism is hinted at in the first two sentences of Mulkay's quotation. The weakest interpretation (especially of the beginning clause) of the second sentence is that SSK endorses Hanson's thesis that observation is "theory-laden," and more generally Kuhn's view that the relationship of a scientific theory or "fact" to the *actual* world is problematic, and not usefully characterized by words such as "truth." Here SSK generally pushes Kuhn's skepticism to its limits. Not only is there a denial of any kind of correspondence theory of truth, but words such as "knowledge" and "truth" cannot refer to the way the world really *is* and only designate strongly held beliefs or positions of a "local culture" whether that of scientists, witch doctors, organic chemists, the Roman Inquisition, etc.; what is advocated is a purely consensus theory of truth. Science may in some sense "work" for its believers (just as witchcraft does) and yields control over nature, but it does not offer "objective" knowledge about the world. Its claims to knowledge, moreover, may not even be influenced by the world. Harry Collins, for instance, speaks approvingly of a school which "embraces an explicit relativism in which the natural world has a small or non-existent role in the construction of scientific knowledge."[13] But the strongest example of anti-realism is provided by Steve Woolgar, most notably in his book *Science the Very Idea* (1988). Using arguments that go back to Berkeley and Kant Woolgar claims that there is no mind-independent reality which is the basis of scientific knowledge. Objects in the world do not determine their representations; on the contrary, our representations determine the object. In other words, the direction of the causal arrow which is ordinarily considered to point from the natural world to our scientific representations

---

[13]Collins (1981a), p. 3.

should be reversed. Further, even in the unlikely case that objects exist independently of our representations of them, we have no contact with them. We are locked in a world of representations which we call "reality," and all we can do is to compare one representation with another, not a representation with an object as it is "in itself." Applied to science, this means that our "representational practices" and our "scientific knowledge" respectively *construct* the natural world, not the other way around. An important consequence of this view is that so-called "discovered" objects do not exist prior to their "discovery" by scientists. What we call "scientific discovery" is merely a network of representations that is the end result of a sociological process of "fabrication," leading to a consensus by scientists that the discovered object indeed exists and has the features that scientists assign to it. One of Woolgar's examples given to support this claim is the "discovery" of pulsars in 1968. To Woolgar, a pulsar is no more than an elaborate construction devised by scientists over a time span of several years to explain "an anomalous trace on routine chart recordings" of radio signals from space.[14] Woolgar's attitude concerning the external world, then, amounts to pure idealism; his "representations" are simply Berkeley's "Ideas." But in fairness, we should note that many SSK theorists are unwilling to go *quite* this far. They admit that the external world exists independently of us and in some sense constrains scientific conclusions. To Barry Barnes, for example:

> Occasionally, existing work leaves the feeling that reality has nothing to do with what is socially constructed or negotiated to count as natural knowledge, but we may safely assume that this impression is an accidental by-product of over-enthusiastic sociological analysis, and that sociologists as a whole would acknowledge that the world in some ways constrains what is believed to be.[15]

And for Michael Mulkay although nature may not "uniquely determine the conclusions" of scientists, it is "self-evident" that it "exerts constraints" on them.[16] However, Karin Knorr-Cetina, apparently agreeing with Woolgar, believes:

---

[14] Woolgar (1988a), pp. 31–33 and Chapter 4, especially pp. 54 ff. For the analysis of pulsar discovery see pp. 61–65.
[15] Barnes (1974), p. 7.
[16] Mulkay (1979a), p. 61.

> ... reality not to be given but constructed. There are ... no initial, undissimulatable 'facts': neither the domination of workers by capitalists, nor scientific objectivity, nor reality itself.[17]

This becomes obvious she thinks from a close deconstructive analysis of knowledge production. The "solid entity" of scientific fact is the product of intricate labor applied to "countless non-solid ingredients from which it derives [and] confusion and negotiation that often lies at its origin ..."[18] But, on the other hand, although the "real-world entities represented by scientific descriptions" are constructed, there is a "material reality" which interacts with us; facts are not made "by pronouncing them to be facts but by being intricately constructed against the resistances of the natural (and social!) order."[19] The position of Barnes, Knorr-Cetina, and Mulkay evidently is a vague and unstable compromise between the full-blown idealism of Woolgar, Collins, and (possibly) Bruno Latour (discussed below) and ordinary scientific realism. But purists in the SSK movement will have none of it! Woolgar, for instance, who likes his 100 proof Berkeley taken neat, thinks it a cowardly compromise which is "epistemologically relativist and ontologically realist."

Whatever their ontological commitments, however, all parties in the SSK camp de-empathize the role of observation and experiment in determining scientific theory. Like Kuhn or Hanson they argue that observation is "theory-laden." Scientists see more or less what their theory dictates that they should see. This position in turn leads them to deny the possibility of falsifying a theory by experiment which according to Karl Popper demarcates science from non-science. In general, partisans of SSK reject falsifiability, in part because the contamination of observation by theory undermines the possibility of a neutral observation language which seems necessary to evaluate experiments. They also appeal to a thesis originally due to Pierre Duhem and later elaborated by Quine. The Duhem-Quine thesis asserts that the negative result of any experimental test of a theory can be always explained away, since one can argue that there is experimental error (a common phenomenon in the complex technically delicate experiments of modern science), some failure in an auxiliary theory, or that there exists some unknown perturbing influence. Much of the SSK literature attempts to support such claims not just by abstract argument, but by historical ex-

---
[17] Knorr-Cetina (1981), p. 148.
[18] *Ibid.*
[19] *Ibid.*

ample. Barnes, Bloor, and Henry, for instance, argue that Robert Millikan used the first strategy to explain away observations of charges which were *fractions* of e in his oil drop experiments.[20] Another example according to Farley and Geison would be Pasteur's rhetorical denunciations of Pouchet's experiments which apparently falsified his theories on the impossibility of spontaneous generation.[21] An explanation of the "perturbing influence" type would be the positing of a hypothetical planet distorting the orbit of Uranus in order to explain the failure of the inverse square law to accurately predict the motion of Uranus; this was the preferred alternative to simply "adjusting" the inverse square law.[22]

If falsification of a theory is impossible, so also is its verification. Following Quine, SSK often argues that no finite set of data can uniquely determine a theory. We can always construct a new theory, logically distinct from the old one which explains the same set of data. In fact the same is true regarding all *possible* observations:

> Theories can still vary though all possible observations be fixed. Physical theories can be at odds with each other and yet compatible with all possible data even in the broadest sense. In a word they can be logically incompatible and empirically equivalent. This is a point on which I expect wide agreement.[23]

Quine, however, does not explain clearly just why he expects "wide agreement." But he does argue that this position implies the indeterminacy of translation. Consider the problem of translating "a radically foreign physicist's theory":

> If our physical theory can vary though all possible observations be fixed, then our translation of his physical theory can vary though our translations of all possible observation reports be fixed.[24]

---

[20] Millikan was trying to refute Ehrenhaft's thesis that such fractional charges were possible. Barnes, Bloor, and Henry (1996), pp. 18–30.
[21] Farley and Geison (1982).
[22] A similar but less successful strategy of supposing the existence of rings of unobserved matter between the orbits of Mercury and Venus to account for the precession of the perihelion of Mercury was proposed by Newcomb and Seeliger in 1895–96 (Newcomb (1911)).
[23] Quine (1970), p. 179.
[24] *Ibid.*, pp. 179–80.

The Strong Program

Having shown that observation then plays little role in the "interpretive constructions" of science, we can ask what does? The answer SSK offers is hinted by the final part (after the word "dependent") of the second clause of the second sentence and the third sentence of Mulkay's quotation, and points to the second ingredient in the SSK synthesis: it is the sociological context that determines a scientists conclusions. It is at this point that the corrosive acids of Mannheim's SK makes their appearance. In general scientific theories and observations are said by SSK to be "socially constructed" (a term which as we shall shortly see has several imprecise meanings). Whatever their connection to reality, scientific "facts" are a form of socially agreed upon "convention." They come into existence by being "constructed", "socially negotiated" or even "invented" rather than "discovered." Thus, according to a notorious judgement of Shapin and Schaffer:

> As we come to recognize the conventional and artifactual status of our forms of knowing we put ourselves in a position to realize that it is ourselves and not reality that is responsible for what we know. Knowledge, as much as the state, is the product of human actions.[25]

The most systematic, ambitious, and influential interpretation of science from a sociological point of view is the so-called "Strong Program," (henceforth, SP) mentioned at the beginning of this chapter which was developed by Barry Barnes, David Bloor, and a number of colleagues at the University of Edinburgh in the early 1970s. SP is both a body of doctrine and a research program intended to apply the methods of SK in a unified way not only to science, but to any other "way of knowing" as well. It is characterized by certain goals and a methodology. The task of a sociological investigator according to Bloor[26] is to explain the "credibility" of scientific belief according to the following four directives:

(i) The explanation should be causal, "concerned with the conditions which bring about belief or states of knowledge."

(ii) "It would be impartial with respect to truth and falsity, rationality or irrationality, success or failure." Both "require explanation."

(iii) "The same types of cause would explain, say, true and false beliefs."

---
[25] Shapin and Schaffer (1985), p. 344.
[26] Bloor (1976), p. 7.

(iv) "In principle its patterns of explanation would have to be applicable to sociology itself."

Bloor calls (i)—(iv) respectively the requirements of "causality", "impartiality", "symmetry", and "reflexivity." The symmetry postulate (iii), in particular, is a radical departure from traditionally accepted "non-symmetric" ideas concerning the causes of true and false belief. For both epistemologists and sociologists in the tradition of Mannheim and Merton, rationally supported true belief—or "knowledge as such"—in science or mathematics needs *no* explanation. One checks that the arguments supporting the scientific claim are sound or that the proof of the theorem is correct, and that is the end of the matter. No other causal explanation is required. False, irrational, or ideological belief, on the other hand, such as the witchcraft mania in the sixteenth century, human sacrifice in a primitive tribe, or twentieth century utopian politics requires a *different* kind of analysis; in this case we can and should draw on sociology, anthropology, or even perhaps abnormal psychology. Bloor will have none of this asymmetry. *All* beliefs including those relating to the content of science, whether true or false, rational or irrational, he thinks, should have the *same* kind of explanation. One kind, Bloor admits, may be biological deriving from our evolutionary mental hard wiring. In the future it may be possible to show that a person's belief is a consequence of the biochemical state of his brain which neurologists can unravel. At present, however, for Bloor the most important type of cause is *social*. Every belief reflects a person's social position, interests, or other social forces acting on him. The arguments and "rational" explanations he gives are secondary; they are functions of far more fundamental causal factors. By uncovering them one can "understand" the belief by—in true Mannheim fashion—unmasking the extra-theoretical functions it serves. While Mannheim fruitfully employed this technique to the content of political and religious ideologies, he declined to use the same approach concerning science. Bloor intends to correct this omission. Both witchcraft and high-energy physics will be found to have the same general cause of belief, and knowledge of the cause will reveal the social function served by these beliefs. But, as Bloor repeatedly stresses, the cause *cannot* include the belief's purported "truth" or "rationality." This would not only would contradict the symmetry requirement, but it would introduce a non-naturalistic teleological aspect into the explanation of human belief. We should have to postulate the existence of a "teleological" property of the human mind, a mysterious attraction to "Truth" and "Rationality"—both of which amount to a Platonic mysticism. Real science allows only material causes, and thus any appeal to transcendental mental properties would be unscientific. Unlike some other practitioners of SSK, Bloor claims *not* to be hostile to science. Although he puts science on the same epistemological plane as any other belief such as Azande magic (see below), somewhat paradoxically he considers SP the very epitome of a scientific approach designed to allow science to know itself:

> Throughout the argument I have taken for granted and endorsed ... the standpoint of most contemporary science. In the main science is causal, theoretical, value-neutral, often reductionist, to an extent empiricist, and ultimately materialistic like common sense. This means that it is opposed to teleology, anthropomorphism and what is transcendant. The overall strategy has been to link the social sciences as closely as possible with the methods of other empirical sciences. In a very orthodox way I have said: only proceed as the other sciences proceed and all will be well.[27]

To use an analogy (which Bloor does not give but which seems to follow from his position) explanations of a teleological or transcendent kind would be similar to vitalism in biology.[28] In the specific case of science Bloor's approach implies that in explaining the acceptance of some past theory, now deemed correct, we must not invoke the "good evidence" or "properties of nature" which its discoverers may have found—or even the arguments they give. This, as remarked above, would be an appeal to teleology; furthermore, what we regard as "good evidence" or "properties of nature" is sociologically dependent. What counts as "evidence" or "properties of nature" differs from believer to believer and culture to culture and in all cases is a result of social processes.[29] One should not therefore merely point to good evidence; a sociological analysis is required to discover why it was (or is) regarded as "good."

Bloor also offers an analysis of what he means by "knowledge" and "truth." Rather than defining knowledge as "justified true belief" in the manner of philosophers, he maintains that:

> ... it is whatever people take to be knowledge. It consists of those beliefs that which people confidently hold to and live by ... which are taken

---
[27] *Ibid.*, p. 157.
[28] *Ibid.*, p. 11; cf. Bloor (1982a), p. 44.
[29] See Barnes and Bloor (1982), pp. 28–29. Barnes and Bloor also want to rule out appeals to "evidence" or "rationality" to judge the *validity* of a belief. There seems to be little difference for them between *credibility* and *validity*.

for granted or institutionalized, or invested with authority by groups of people.[30]

"True beliefs" furthermore are defined as either (1) those that work for the believers or (2) those affirming "how the world stands" according to the "ultimate schema with which we think,"[31] or (3) what we "affirm and count as true whatever is the outcome" of our investigative and socially explained processes.[32] It follows that the "truth" of the obsession with witchcraft and poison oracles on the part of the Azande[33] is not the "truth" of high-energy physics; all that can be said is that each is equally credible according to the standards of the culture in which each is cultivated. But, having given these definitions, Bloor, echoing Kuhn, argues that the notion of "truth" is really superfluous:

> ... why not abandon it altogether? It should be possible to see theories entirely as conventional instruments for coping with and adapting to our environment.... What function does truth or talk of truth play in all this? It is difficult to see that much would be lost by its absence.[34]

"Truth" for Bloor is thus a humanly constructed attribute that denotes either those of our theories or beliefs that "work" or serves a rhetorical function in argument; in particular, invoking "truth" is a claim of authority and transcendence especially when one wants to contradict received opinion. What "truth" does *not* do is denote a correspondence between a belief and "reality" in any absolute sense transcending local cultural norms.

As we have seen, for Bloor and SP the causation of scientific belief has nothing to do with the "truth" of the belief or the arguments scientists give for it. These are mere propaganda and self-persuasion which mask

---

[30] Bloor (1976), p. 4.
[31] *Ibid.*, pp. 41–44. "Of course this schema is filled out in many different ways. The world may be peopled with invisible spirits in one culture and hard, indivisible (but equally invisible) atomic particles in another." (*Ibid.*, p. 42.)
[32] *Ibid.*
[33] SSK is fond of exhibiting this collection of Sudanese tribes living in Central Africa, investigated in the 1930s by the anthropologist Evans-Prichard, as a counterpoint to Western scientific rationalism. The Azande answer questions about the future by giving a certain poison to a chicken and observing whether or not the chicken lives or dies. They believe that bad events are caused by witchcraft which can be diagnosed *postmortem* by a substance in the belly which is inherited by the same sex descendants of the male or female witch making them witches also. See the discussion in Barnes (1974), pp. 27–29, or in Bloor (1976), pp. 138–46.
[34] Bloor (1976), p. 40.

the real sociological causes of the belief. Indeed, although Bloor does not explicitly claim this, they are almost a manifestation of false consciousness in the same sense as the abstract and rational arguments given, say, by defenders of bourgeois economics or feudal privilege. But what are the sociological causes? At this point there is a fundamental split in the SSK community. In numerous historical case studies, some of which we shall analyze at length below, the adherents of—or those influenced by—SP tend to stress the external, often political or ideological nature of these causes. Thus, for example, Newton or Boyle's theories concerning matter are related to Restoration political concerns, Pasteur's theories on spontaneous generation to the reactionary politics of the Church and the Second Empire, quantum mechanics to the decline in cultural status felt by German scientists after World War I, etc., etc. Another possibility, however, is to analyze the sociology of belief in terms of the mutual interactions of scientists themselves in the laboratory. In this interpretation scientific argument again is not to be taken seriously as an attempt to discover the facts; rather it is a form of propaganda or rhetoric in the service of laboratory politics and struggle among scientists.

The French approach

The most uncompromising vision of science as the fruit of a Hobbesian *bellum omnium contra omnes* in the laboratory is to be found in the work of Bruno Latour, a French anthropologist. In his famous book *Laboratory Life* published in 1979 and coauthored with Steve Woolgar, Latour described the "culture" of the laboratory of Dr. Roger Guillemin at the Salk Institute much as in his previous anthropological career he had described the culture of African tribes in the Ivory Coast. The conclusions of this book were later elaborated in *Science in Action* (1987) and also in *The Pasteurization of France* (1988).

The task of Guillemin's laboratory—for which in 1977 he shared a Nobel Prize with two other scientists—was to isolate and determine the chemical structure of a substance known as Thyrotropin Releasing Factor (Hormone) or TRF. This hormone when released by the hypothalamus regulates the release of a thyroid-stimulating hormone from the pituitary. The procedure Guillemin's laboratory had to follow was extremely delicate. There were many false starts and technical difficulties to be overcome. It finally took 300,000 sheep hypothalami extracted from 500 tons of sheep brains to obtain 1 mg of the pure substance in 1968. The following year the chemical formula was determined by mass spectrometry. As a result a synthetic version of TRF could be produced in unlimited quantities. It has been shown to be active in every vertebrate species and especially in man.[35]

All of the scientific justifications and explanations by laboratory workers of their goals and procedures Latour and Woolgar treat as so many rituals and mythologies—in principle no different from the rituals and mythologies of some tribal shaman; they are to be reported but not believed by the anthropologist. The scientists may naively think that TRF is an actually existing chemical compound which they have discovered by sophisticated bioassays and whose structure and properties they have determined. The results of their activities are thus new "facts," which are independent of the scientists who discovered them. The anthropologists of science Latour and Woolgar, however, know better. In a manner reminiscent of Feyerabend's views on the existence of atoms, they maintain that Guillemin's laboratory has *constructed* TRF, and in general that scientists socially construct their so-called "facts." In the case of TRF they argue that this substance *entirely depends* on the apparatus and various "inscription devices"[36] used to find it. It has no existence outside of these instruments:

> The bioassay is not merely a means of obtaining some independently given entity. The bioassay constitutes the construction of the substance. Similarly a substance could not be said to exist without fractionating columns... It is not simply that phenomena *depend on* certain material instrumentation; rather the phenomena are *thoroughly constituted by* the material setting of the laboratory, The artificial reality, which participants describe in terms of an objective entity, has in fact been constructed by the use of inscription devices.[37] [emphasis in original]

Not only however are such facts "created" by the equipment, but they are the product of a thoroughly *political* process. Latour and Woolgar observe that the validation of scientific theories or facts are almost always preceded by controversies. The data is inconclusive and the apparatus and experiments yield contradictory and obscure results. Interpretations and decisions are necessary to decide what to keep and what to throw

---

[35] Guillemin (1977), p. 366.
[36] "I will call an instrument (or *inscription device*) any set-up, no matter what its size nature and cost, that provides a visual display of any sort in a scientific text." (Latour (1987), p. 68.)
[37] *Ibid.*, p. 64.

out. Different factions in the laboratory quarrel with each other over these issues. But to proceed and establish "facts" these disagreements must be resolved. How does this happen? Latour and Woolgar argue that it is not a matter of disinterested observation, experiment, or rational argument. Rather, it has everything to do with the "agonistic field," that is, the small group politics, power networks, shifting alliances, and rhetoric found in the the laboratory. The settlement of scientific facts is thus a form of warfare. Describing the methods of this warfare, Latour writes:

> The rules are simple enough: weaken your enemies, paralyse those you cannot weaken... help your allies if they are attacked, ensure safe communications with those who supply you with indisputable instruments... oblige your enemies to fight one another... if you are not sure of winning be humble and understated. These are simple rules indeed: the rules of the oldest politics.[38]

The weapons in the controversy are "texts," that is, article literature. As a controversy grows more intense both sides assemble greater and greater quantities of increasingly technical papers and reports. A dissenter is thus faced with the task of wading through and discounting more and more paper, often created by prestigious authors and institutions. Those who become the losers in the dispute are those who collapse first from exhaustion. As one side prevails its statements undergo "splitting" and "inversion":

> *The statement becomes a split entity.* On the one hand, it is a set of words which represents a statement about an object. On the other hand, it corresponds to an object which itself takes on a life of its own. It is as if the original statement had projected a virtual image of itself which exists outside the statement.... Before long, more and more reality is attributed to the object. Consequently, an inversion takes place: the object becomes the reason why the statement was formulated in the first place.[39] [emphasis in original]

By this process a claimed result is stabilized so that it becomes a "fact" or (to borrow a cybernetic term) a "black box"; that is, it is used and accepted without inquiry into its origins like a complicated but useful "inscription device" or piece of electronic apparatus whose operation and function are taken for granted, even though the device may be very complex and at

---

[38] Latour (1988a), pp. 37–38.
[39] Latour and Woolgar (1979), pp. 176–178.

one time controversial. In this setting the "fact" looks like an objective property of nature and neither it or the apparatus are "opened" and the circumstances of their origins critically analyzed. If the dissenters do not capitulate and accept it, they are driven into oblivion and their careers destroyed. Future researchers will use the new fact uncritically and the messiness and political factors in its construction will have been forgotten.[40] "Nature," Latour and Wolgar emphasize again and again plays no role in the process; rather it is the "outcome" of the process as expressed in the their well-known quotation given in proposition (v) in Chapter 1. When the "facts" are determined and embedded in the scientific canon, all these political factors are suppressed and forgotten so that the "fact" looks like it has a real existence independently of the scientists who have constructed it.

A critic of this nihilistic thesis might argue that one needs to consider what stands behind the "texts," rhetorical tricks, and laboratory politics. Are there not experimental results supporting the claims of scientists that exist independently of such factors? In reply, Latour supposes that a dissenter goes to the laboratory to see for himself. The defender of the claim, called in Chapter 2 of *Science in Action* "the Professor," shows the dissenter the equipment, explains the experiment, and demonstrates the results. This is done in terms of a hierarchical sequence of "black boxes." At each point, the critic may be tempted to open the black box and criticize it; if so, more black boxes are brought into the fray, and so on *ad infinitum*. Rather as in the case of texts he is overpowered in the end by the number of black boxes arrayed against him and the enormous labor needed to open them all up. His only hope is to obtain his own "counter laboratory" and introduce his own black boxes, texts, allies, etc.

At no point in their account of scientific controversy, do Latour or Latour and Woolgar speak of "truth" to describe the claims of the victorious side. For them there are *only* winners and losers and the main weapon in scientific warfare is Rhetoric, the ancient and despised art of persuasion devised by the sophists. In science as in politics the worse can be made to seem the better cause. The resolution of scientific quarrels is arbitrary and subjective. That this is not obvious is a consequence of the fact that victory gives the dominant party the right to suppress, ignore, or explain away incompatible data, i.e., to write Whig history. As a corollary, it fol-

---

[40]To give a specific example, the double helix structure of DNA is such a black box for molecular biology. Like a computer or some other complicated instrument it is routinely used, but not "opened."

lows that the outcome of scientific research is contingent. At any point in time, beginning with such fundamental discoveries as the double helix different facts *could* have arisen and science might have taken a different path.

This interpretation of the genesis of scientific facts or discoveries has (at least to the unsophisticated) some odd consequences. For instance, according to Latour, to say, as some French egyptologists did, examining the mummy of Rameses II in 1976, that he died of tuberculosis is as senseless as saying he died of machine gun fire since the machine gun did not exist in c.1213 B.C.E., and neither did the TB bacillus because Robert Koch "constructed" it only in 1882; prior to Koch "the bacillus had no real existence."[41] In a similar vein Latour gives elaborate arguments which we will sketch below that Pasteur's microorganisms did not exist before Pasteur "discovered" them. To take a second example, in 1903 a French physicist working in Nancy René-Prosper Blondlot claimed to have discovered a new type of ray, distinct from X-rays. He named them "N-rays" after his home town. He published several papers in reputable journals about them, and they seemed on their way to being accepted as a new fact of nature. An American physicist Robert Wood, however, doubted Blondlot's claims and visited his laboratory. During a demonstration of N-rays by Blondlot, Wood secretly detached the aluminum prism which was supposed to generate the N-rays from Blondlot's apparatus. Nevertheless, Blondlot continued to report N-rays projecting themselves on a screen. This, of course, completely discredited N-rays! No one believed Blondlot any longer and the explanation for his "results" became psychological self-deception. The consensus of science became and remained that Blondlot was discredited because he was *wrong*; there are no N-rays in nature.[42] Even Latour does not tell us by what rhetorical device or political strategy Blondlot could have recovered from this "trial of strength" and prevailed against Wood. But we have the impression that he feels that it would have been possible. Modern physics was in part the *result* of the settlement of this controversy. For contemporary historians of science to use it to claim that Blondlot was wrong is to write:

---

[41] Sokal and Bricmont (1998), pp. 88–89, n. 123.
[42] The best account of the N-ray episode may be found in Nye (1980). Blondlot detected N-rays by seeing a slight increase in intensity on a faintly illuminated screen. He was operating at the limits of perception and explained the failure of critics to detect the rays as a result of the insensitivity of their vision.

> ... Whig history, that is, a history that crowns the winners, calling them the best and the brightest and which says that losers like Blondlot lost simply *because* they were wrong.[43] [emphasis in original]

Apparently had things turned out differently, N-rays would have had exactly the same status today as X-rays; they would have been a fact. Who knows what valuable applications (N-ray scanners?) to medicine and other fields they might have had!

From the above account and (especially!) his Third Rule of Method (see Chapter 1, proposition (v)) the reader might be excused for concluding that Latour is a full blown and rather extreme social constructivist who believes that the content of science is entirely a human fabrication. However, in a later work *Pandora's Hope: Essays on the Reality of Science Studies* (1999) he denies that he has *ever* been guilty of such opinions and affirms in reply to a psychologist's doubting question that he has always believed in "reality." It is unfortunate, he complains that:

> ... science warriors too often waste their time attacking someone who has the *same name* as mine, who is said to defend all the absurdities I have disputed for twenty-five years: that science is socially constructed; that all is discourse; that there is no reality out there; that everything goes; that science has no conceptual content ... that the mightiest, manliest, and hairiest scientist always wins provided he has enough allies in high places; and such nonsense.[44]

Both scientific realism and the opposite pole of constructivism are according to Latour bad consequences of the classical subject-object dichotomy due to Descartes and still maintained by the "modernist settlement." This sets up an impenetrable gap between the observer and the observed. Realists who accept this settlement propose that Reason bridges the gap and can produce enduring knowledge of reality. Constructivists, on the other hand, argue that the mind can know nothing but its own "representations." Therefore naive scientific realism is impossible and the world of science is our own fabrication. Since this fabrication is not anchored in an external world, it can and will change as our social and historical situation changes; such a position obviously implies relativism. The solution to this conundrum, proposed by Kant, of the mind's *a priori* categories supplying

---

[43] Latour (1987), p. 100.
[44] Latour (1999), pp. 299–300.

structure to a noumenal world which exists independently of ourselves but is unknowable in itself is also unsatisfactory:

> Kant invented this science fiction nightmare: the outside world now turns around the mind-in-a-vat, which dictates most of the world's laws, laws it has extracted from itself without help from anyone else. A crippled despot now ruled the world of reality.[45]

Latour proposes that we abandon the subject-object distinction. He does not give technical arguments against it, but considers it a useless diversion and attempts to show that it actually plays no role in the creation of science. The resulting synthesis will occupy a middle ground between constructivism and realism. How it does is shown in a number of case studies. In 1991 Latour joined an expedition to Brazil to determine the behavior of the boundary between the rain forest of Boa Vista and the surrounding savanna. The question under investigation was "is the savanna advancing into the forest or is it retreating before an advancing forest"? The expedition begins with a vast, hot, confusing, nearly impenetrable forest; it ends with a thin neat scientific report filed in Paris. There is a tremendous gap between the forest and the report. How is it bridged? The answer is by constructing a long chain of "transformations," each across a tiny gap. In the end the forest and the soil beneath it is "transformed" into the report which replaces it and allows us to hold Boa Vista in our hands. At each stage link n is transformed into link $n+1$; link $n+1$ imposes "form" on link n which is its "matter." Each step, moreover, is a hybrid involving both objects and human constructions in the form of equipment and concepts to bridge the subject-object gap. Thus there are soil samples and vegetation to be preserved which are objects and graph paper, wooden boxes and instruments like the pedocomparator[46] used to store and organize the soil samples which are constructions. As the chain lengthens "locality, particularity, materiality, multiplicity, and continuity" present in the chain is lost while "compatibility, standardization, text, calculation, circulation, and relative universality" is gained. Key aspects are extracted and amplified while the sheer quantity of information is reduced.[47] The chain is reversible; one can go in either direction. Each link is valid so the chain is unbroken and truth can circulate in it as electricity does in a circuit.

---

[45] *Ibid.*, p. 6.
[46] A flat square divided into a matrix of small cubical compartments.
[47] *Ibid.*, pp. 70–71.

Later in *Pandora's Hope* Latour considers the question: "Did Pasteur's microbes exist before Pasteur?" His answer is "No!"—which certainly *seems* like a constructivist position. Latour, however, denies this and justifies the claim of previous nonexistence by proposing a new ontology. Everyone admits that human culture changes in time; the sequence of these changes we record as history. What is present now may not have existed previously and may no longer exist in the future. To Latour the same is true of nonhumans; things have historicity and change from one thing to another in time. Consider, for example, the fermentation of lactic acid. The ferment of Pasteur is a yeast. Prior to him in the laboratory of Justus Liebig it had been a chemical phenomenon allied to decomposition. Had the yeast then always been a hidden and persecuted Cinderella, finally unveiled by the Prince Charming—Pasteur? No, but by a chain of small transformation as in the case of the Boa Vista expedition Pasteur brought the yeast into being; it had not existed before. Its fabrication was the joint effort of the nonhuman ferment and Pasteur; both were changed as a result—Pasteur wins medals and the ferment is granted life, becoming a yeast:

> ... we should be able to say that not only the microbes-for-us-humans changed in the 1850s, but also the microbes-for-themselves. Their encounter with Pasteur changed them as well. Pasteur, so to speak happened to them.[48]

It is a product of the modernist subject-object settlement to think of the yeast and other microbes of Pasteur as being permanently "out there" before he discovers them. But what was "out there" had different attributes, and relationships to the rest of the universe and hence was a different object for Pasteur and Liebig:

> Pasteur is not Liebig. Lille is not Munich. The year 1852 [the year of Liebig's investigations] is not the year 1858 [the year of Pasteur's experiments]. Being sown in a culture medium is not the same as being the residue of a chemical process, and so on.[49]

We think differently because the dichotomy between the human subject and the nonhuman object makes us believe that only subjects change and participate in history. Objects are only acted upon. Without human intervention they are timeless and enduring. For this reason we want to think

---
[48] *Ibid.*, p. 146.
[49] *Ibid.*, p. 150.

of unchanging substances as supporting or standing under the attributes of an thing. But a substance in this sense does not really exist; a thing is defined only by its collection of attributes and relations with other things. "A substance is more like the thread that holds the pearls of a necklace together than the rock that remains the same no matter what is built on it."[50] Because of the "historicity of things" these attributes and relations change in time. Thus what Pasteur viewed as microorganisms in his dispute with Pouchet over spontaneous generation had different relations to the Academy of Sciences, Napoleon III, Republican or conservative Catholic ideology, etc., before and after Pasteur. Just as in the case of ferments in the eyes of Pasteur and Liebig, these microorganisms had different attributes for Pouchet and Pasteur. It is Pasteur himself who makes the microbes into enduring "substances" and transforms Pouchet's spontaneous generation or Liebig's purely chemical ferments into phenomena which had *never* actually existed. He "retrofits" the past with the microbes, so it is psychologically impossible not to believe in their permanent existence. He "tries to convince the Academicians that his story is not just a story, but that it has occurred *independently* of his wishes."[51] In this way the microbes end up by *becoming* substances in the old fashioned sense; the process is similar to rewriting German history after 1945, so that it is seen as a preparation over the centuries for the rise of Hitler, or theological interpretations which view certain events in the Old Testament as harbingers of Christ's birth.[52]

Pasteur did his job so thoroughly that we live in *his* world, not Liebig's. But the job of maintaining this world is never complete. To ensure their continued existence the microbes require the permanent hard work of *institutions*, many of which were created by Pasteur, e.g., laboratories, methods of biochemistry, universities, etc. Should these disappear, so would the microbes. This triumph of Pasteur parallels triumphs in other areas of human activity. For instance, there is a:

> ... *homology* of the narrative of the spread of microbiological skills with one that would have described, say, the rise of the Radical party from obscurity under Napoleon III to prominence in the Third Republic, or the expansion of diesel engines into submarines.[53]

---

[50] *Ibid.*, p. 151.
[51] *Ibid.*, p. 125.
[52] *Ibid.*, p. 170.
[53] *Ibid.*, p. 155.

In the final chapters of *Pandora's Hope* Latour embarks on a political mission. He identifies the "Science Warriors"—i.e., those who identify science with Reason and believe that it discovers laws of nature that are "inhuman" in the sense that they have always been "out there" independently of human beings—with Socrates in the dialogue *Gorgias*. Both Socrates and his adversary the sophist Callicles agree in their contempt for Athenian democracy. Their only disagreement is on how best to control it. Callicles believes in the Nietzschean solution of aristocratic rule over the "human debris and assorted slaves" which constitutes the bulk of the *Demos*. Socrates, although he has equal disdain for the *Demos*, believes in the power of rational argument to control them. No demagogue or decree of the Assembly can refute either a geometrical proof or the philosopher's demonstration of what should be done to secure the Good. One man can defy the Many and be right. Stephen Weinberg, a Nobel prize winning Harvard physicist, who is a vehement scientific realist is, Latour thinks, an intellectual descendant of Socrates. Weinberg's belief that the discovery of impersonal natural laws strengthens "the vision of a rationally understandable world" which is needed "if we are to protect ourselves from the irrational tendencies that still beset humanity."[54] To Latour Weinberg's inhuman Science plays the same role as Socrates' Reason. Both are ramparts—"our great wall of China, our Maginot line against the dangerous unruly mob."[55] They are tools "to eliminate even more thoroughly the damnable tendency of the mob to discuss and disobey."[56]

Science as rhetoric

Another writer evidently influenced by Latour, but in some ways even more radical, is Alan Gross, a Professor of Rhetoric in a Department of the same name at the University of Minnesota. For Latour, as we have seen, the sophistic art of Rhetoric, which in the *Gorgias* Plato argued stood in the same relation to justice as cookery does to medicine, together with "the oldest politics of all" are the engines bringing about scientific change in the laboratory. The objects of scientific study may be artifacts of "inscription devices" or "fabricated" by scientists, but nevertheless they have some sort of concrete existence. Gross, on the other hand, in his book *The Rhetoric*

---

[54]quoted in *Ibid.*, p. 216.
[55]*Ibid.*, p. 217.
[56]*Ibid.*, p. 265.

of Science goes beyond Latour and suggests a novel (to say the least!) ontology. He argues that science just *is* the tissue of rhetoric scientists employ to persuade others of their theories: "the claims of science are solely the products of persuasion."[57] There is *no* difference between posited scientific objects and the rhetoric used to "describe" them; there is no "hard scientific core" after rhetorical analysis."[58] The concept, therefore, of "scientific discovery" is an oxymoron. Scientists do not discover "what is out there"; rather, their theories are purely rhetorical fabrications, a realization which, as Gross happily notes, challenges "the intellectual privilege and authority of science."[59] Consider, for example, Watson and Crick's "discovery" of the double helix structure of the DNA molecule. Latour would have probably argued that DNA did not exist apart from the complicated equipment used to study it, but Gross wants to:

> ... support a more radical claim: that the sense that a molecule of this structure exists at all, the sense of its reality, is an effect only of words and numbers, and pictures used judiciously with persuasive intent.[60]

Such a view Gross argues further implies "a sophisticated relativism in which truth depends not on a conformity to a substratum of reality, but on agreement among significant persons"[61] and states that bringing about such a consensus is a "natural effect of the persuasive process."[62] Apparently then, Watson and Crick persuaded their scientific peers of the reality of the double helical structure of DNA by techniques not dissimilar to the techniques by which Cleon persuaded the Athenian Assembly to order the massacre of the inhabitants of Mytilene or Cicero the Roman Senate to proscribe Catiline. Gross links his rhetorical thesis to a variety of the antirealism which we have already encountered in many strands of SSK/PIS. Scientific terms and theories refer, but not to a mind-independent reality; they refer only to objects which do not exist independently of our conceptual schemes and whose validity is testified to by a consensus of scientists produced by rhetoric. Realists may object that science, after all, "works"; it makes correct predictions time after time, which would be a miracle unless

---

[57] Gross (1990), p. 3.
[58] *Ibid.*, p. 33.
[59] *Ibid.*, p. 7.
[60] *Ibid.*, p. 54.
[61] *Ibid.*, p. 21.
[62] *Ibid.*, p. 204

"scientific theories really referred to the causal structure of the world."[63] But to Gross this inductive argument is mistaken:

> It assumes that science is a generally successful enterprise, a methodological stance that virtually guarantees the production of theories that are largely true. But the history of science is not like that: "overwhelmingly, the results of the conscious pursuit of scientific inquiry are failures: failed theories, failed hypotheses, failed conjectures, inaccurate measurements, incorrect estimations of parameters, fallacious causal inferences, and so forth. ... the problem of the realist is how to explain the *occasional success* of a strategy that *usually fails*. [emphasis in original][64]

In view of these crushing arguments against realism, why do scientists believe that their concepts refer to real objects, existing independently of rhetoric? Gross believes that scientists have to think this way to motivate their work. For these benighted souls realism is just a psychological crutch. Would they work so hard just to produce *words* (which in fact is what they mainly produce)?

Science as a golem

A similar but milder point of view than that found in Latour's earlier books *Laboratory Life* and *Science in Action* has also been espoused by Harry Collins, formerly at Bath and more recently a Professor in the Department of Sociology at Cardiff. The neo-Kantian or even idealist metaphysics present in Kuhn, Feyerabend, Woolgar, and other advocates of SSK is not much emphasized by Collins. Despite his radical early statement quoted above, he now allows that in some sense nature exists, has something to do with science, and that scientific facts are more than "texts."[65] But Collins is also a cultural relativist reminiscent of Feyerabend. He believes that just as scientists operating in different paradigms do not "see" the same things, the same is true of cultures having radically different world views; hence "their perceptions and usages cannot be fully explained by what the world is really like."[66] Collins' favorite metaphor to describe

---

[63] *Ibid.*, p. 195.
[64] *Ibid.*, p. 195. The quotation is from Fine (1984), pp. 89, 104.
[65] For instance, "In so far there are facts of science, the relationships between matter and energy put forth by Einstein are facts." (Collins and Pinch (1988), p. 28.)
[66] Collins (1981a), p. 16.

science is "a golem." In Eastern European Jewish tradition, the golem was a powerful essentially stupid creature fabricated of clay which could without evil intention cause great damage if it escaped the control of the Rabbi who created it. Consequently, the word came to describe "any lumbering fool who knows neither his own strength nor the extent of his clumsiness and ignorance."[67] Thus in his 1988 book *The Golem* written with Trevor Pinch, Collins declares that his aim is:

> ... to show that [Science] is not an evil creature but is a little daft. Golem science is not to be blamed for its mistakes; they are our mistakes. A golem cannot be blamed if it is doing its best. But we must not expect too much. A golem powerful as it is, is the creature of our art and our craft.[68]

Like the golem who was animated by having the Hebrew word for truth inscribed on its forehead, science may be motivated by a search for truth. "But this does not mean that it understands the truth—far from it."[69] To demonstrate his thesis that science resembles the the golem, Collins proposes to show the essential indeterminacy and ambiguity of scientific experiment. He argues that the actual behavior of scientists puts into doubt the possibility advocated by traditional notions of scientific method (or even by otherwise revolutionary philosophers of science such as Karl Popper) of cleanly falsifying or supporting a theory or even replicating experimental results. Agreeing with Latour and SSK in general, Collins believes that decisions by scientists are fundamentally social processes, reached by negotiation (though not necessarily as for Latour by "warfare"), but are not "compelled" by the properties of nature.

In his books *Changing Order* (1985), the previously mentioned *The Golem*, and *Frames of Meaning* (1982) (also coauthored with T. Pinch) Collins offers several case studies in support of his views. Take, for example, the idea that experiments should be repeatable, i.e., that others should be able to replicate an experiment and verify its claims, which is basic to standard discussions of scientific method. This sounds like a straightforward procedure which is necessary for a claimed result to be added to the scientific canon. Collins first makes the obvious point that even in "simple" cases replication of experiments can be very difficult to carry out—even in

---
[67] Collins and Pinch (1988), p. 28.
[68] *Ibid.*
[69] *Ibid.*

situations where the problems are technological rather than scientific. He notes that in the early 1970s British scientists in several laboratories tried to copy and improve a TEA Laser which had first been successfully built at a Canadian defense laboratory in 1968. The theory was well understood and the British were eventually successful. But the construction of a TEA laser was surprisingly tedious and problematic, taking many months longer than expected. Everything (and more) that could go wrong did go wrong. Polarities were reversed; parts were defective; short circuits occurred; capacitors failed, and so forth. Diagnosis of what exactly was wrong was much of the time almost impossible. But in this case, criteria of successful replication or failure were obvious and accepted by all. If the device, say, melted concrete, then the experiment was successful. If it exploded, then it had failed.[70] Next, Collins examines the task of gravitational wave detection. This form of radiation is predicted by General Relativity and should be detectible on Earth by tiny variations in the value of g. Such detection was reported by Professor Joseph Weber at the University of Maryland in 1969. This claim in the early 1970s generated considerable interest. Several other laboratories, some using slightly different methods, attempted to replicate Weber's results. They all failed and by 1975 the nearly universal conclusion was that Weber's gravitational fluxes were the result of experimental error. As stated, this seems a textbook case of Popperian falsification of a theory. Collins, however, attempts to demonstrate that matters were not nearly so simple. As for the TEA laser, many things could and did malfunction in the complicated apparatus that Weber and his competitors built. But more significantly the experiment was exceedingly delicate. Gravitational waves would be very weak and difficult to detect. All sorts of "noise" could contaminate the results and elaborate computer algorithms were required to extract the signal from the noise. This situation led, Collins claims, to a situation he calls "the experimenter's regress." Because of the extreme complexity of the experiment and weakness of the signal, Weber could and did claim that scientists who obtained negative results did so only because something was wrong with their apparatus or technique. Conversely, since the radiation "discovered" by Weber was much stronger than the theory predicted, skeptics could and did claim that the same was true of *him*. No matter what outcome was obtained the experiment was sufficiently "fragile" that it could be technically criticized at many points. How to resolve this impasse? New experiments and procedures would be necessary to test the

---

[70]Collins (1985), Chapter 3.

accuracy of the device. But these in turn were just delicate and problematic as the original experiment. Cogent criticisms could be made of them also, and so on, *ad infinitum*. In the end all that can be said was that the experiment was sound if it validated the "proper" result which was in the eye of the beholder. To Weber and his supporters a "good" experiment was one that found gravitational waves and "failure" could be convincingly explained away; to his skeptics this judgement was reversed. The settlement of the controversy, therefore, was not a simple function of the "observed" results and depended on "sociological" circumstances which Collins examines in great detail.[71] Collins' analysis is clearly a radical "sociological" version of the Duhem-Quine thesis. Although he is less cynical than Latour, he like Latour feels that the settlement of this and other controversies is never a simple function of observation or experimental results, but depends on the sociological circumstances and interests involved. This implies that the settlement might have been different had these factors been different. In Weber's situation a different consensus might have been reached on the outcome of his experiments if he had been luckier in his opponents (the opposition of one charismatic individual in particular was very damaging to Weber). In such a case it is possible that Weber might today be celebrated as the discoverer of gravitational waves.

Similar phenomena are present, according to Collins, when a theory is "verified" and its claims accepted. He offers as an example Sir Arthur Eddington's verification in 1919 of a prediction of Einstein that the gravitational field of the Sun should bend light rays from stars close to the Sun's disc by about 1.7 seconds of arc versus about .8 seconds as predicted by Newtonian theory. This difference at best would be very difficult to measure. In the first place, it had to be done during a solar eclipse. Otherwise the brightness of the Sun would totally obscure nearby stars. Secondly, the predicted difference is very small, a second being 1/3600 of a degree. This, in particular, would mean that changes in temperature, vibrations, and a myriad of other factors (including cloudy weather) could wreck the accuracy of the observation. Additionally, the solar eclipse of 1919 could only be seen from remote and rather primitive tropical areas. Finally, no one was sure that Einstein's prediction was a correct derivation from his theory. Nevertheless, the experiment, organized by Eddington was carried out. Two parties were sent, one to Brazil, the other to an island off the West African coast. In Brazil eight "good" photographic plates gave a mean of 1.98",

---

[71] *Ibid.*, Chapter 4; Collins and Pinch (1988), Chapter 5.

while 18 "bad" plates using a different telescope gave a mean of .86." The African expedition obtained a mean of 1.62" from two "poor" plates. To an unbiased observer this data seems inconclusive. But by rather complicated "massaging" of the data, Eddington and the Astronomer Royal threw out the 18 Brazilian plates while keeping the two "poor" African plates. When this was done, a result more or less agreeing with Einstein was inevitable. Collins does not explain just *why* Eddington and the Asronomer Royal did this. They indeed gave complicated technical reasons for their action, but a possible inference is that they had a prior commitment to the "truth" of Relativity. The consequence of their report, essentially due to their prestige and authority, gave enough credibility to Einstein to "license" other observers to look for additional confirmation of the theory. For instance, it was soon noticed that if one had the ingenuity to ignore (a lot of) contrary evidence, there was an observable red shift in light from the Sun as also predicted by Einstein's theory. Looking backward, it was also clear that the Michelson-Morley experiment which failed to detect an "aether wind" was consistent with Einstein since light as seen from the Earth should have the same velocity in any direction relative to the Earth's orbit. The scientific climate became so receptive to Einstein that almost no experiment could shake confidence in his theory. Such efforts as a very carefully designed experiment by Miller in 1933 which *did* detect an aether wind (hence inconsistent with Einstein) were ignored or explained away as was also the fate of a whole series of measurements taken before 1952 which were similar to those done in 1919 and which gave results disagreeing with *both* Newton and Einstein.[72] Collins concludes his discussion of Eddington's experiment with the observation:

> What we have seen are the theoretical and experimental contributions to a cultural change, a change which was just as much as a licence for observing the world in a certain way as a consequence of those observations."[73]

It follows that Relativity "was a truth brought on by agreement to agree about new things. It was not a truth forced on us by the inexorable logic of a set of crucial experiments."[74]

---

[72] They were far too high, in one case 2.24 seconds of arc.
[73] *Ibid.*, p. 52. The account of the Eddington experiment is given in Chapter 2.
[74] *Ibid.*, p. 54.

Science as a social construction

Whether or not proponents of SSK believe that scientific theories derive from external political causes or from micro-sociological processes among scientists, almost all assert that "scientific theories, methods, and acceptable results are "social constructions."[75] Let us revisit this slippery and elastic term. In a trivial sense the practice of a great deal of science—especially the "big science" of high-energy physics and the like—consists of interactions of groups of scientists in laboratories and universities trying to make sense of experimental phenomena. The results of such interactions are interpretations and theories "constructed" by the groups and published in the form of scientific articles (often having dozens of authors) in journals such as *Nature*. All this is certainly a social process. In this pedestrian manner "social construction" is practically universal in science. But a much more radical thesis is intended by SSK. To be "socially constructed" means that the "science" is partially or entirely determined by contextual forces having little or nothing to do with nature. These forces, such as "laboratory politics" or the prevailing paradigm, may be internal to the discipline and determine which problems are to be studied and shape how phenomena are to be interpreted. Others may be "external" and derive from political factors often involving the scientist's class, ideology, gender, the requirements of the military industrial complex, and so forth. There are also overtones of unreality/falsehood/artificiality/contingency involved in the term. That is to say: (i) A social construction may be a concept mirroring what its creators feel is a real object, but in fact is either a distortion or does not correspond to anything *real* outside of itself at all—being entirely "fabricated" by the social group. As examples one thinks of the concepts of witchcraft or hysteria "constructed" by misogynistic sixteenth century inquisitors or nineteenth century physicians. That all science is socially constructed in this way is widely held by SSK. Such a view certainly follows from "idealist" variants of SSK such as Woolgar's and seems a central thesis of Latour (at least in some of his writings).[76] (ii) Since a particular social construction depends on historical circumstances, it is contingent; it need never have existed or might have been completely different. This attitude is strongly implied as we have noted by both Latour

---

[75] *Ibid.*, p. 43.
[76] Given his "two world" ontology and claim that science can only deal with the world of phenomena and not with the "actual" world it is even implicit in Kuhn's analysis.

and Collins. One of the most detailed arguments for such contingency has been given by Andrew Pickering in his book *Constructing Quarks: A Sociological History of Particle Physics* (1992) which presents a substantial and detailed history of the revolutionary transition from the old (pre 1974) hadron-centered physics to a "new world" focusing on quarks and leptons as the basic constituents of particles. Pickering claims that the change was not inevitable. The old physics might have continued to flourish or something completely different might have emerged. The transition according to Pickering was not compelled by the experimental evidence; rather, the significance, meaning, and even existence of such evidence was conditioned by the changing theoretical overview which depended on a series of free choices and judgements by the investigators. In Pickering's words:

> Historically, particle physicists never seem to have been *obliged* to make the decisions they did; philosophically, it seems unlikely that literal obligation could ever arise. This is an important point because the choices that were made *produced the world of the new physics*, its phenomena and its theoretical entities. ... the existence or nonexistence of pertinent natural phenomena was the product of irreducible scientific judgements.[77] [emphasis in original]

In the particular case of gauge theory, which was a necessary ingredient in the new synthesis:

> ... the adoption of the theoretical apparatus of gauge theory entailed the view that misfits between predictions and the data (and deductive lacunae) indicated grounds for further elaboration of the theory rather than for its rejection: again, an unforced and irreducible choice.[78]

Such contingent decisions were the results of interpretations deriving from complex sociological factors, especially the institutional interests of the various theoretical and experimental groups of physicists, rather than a straightforward imprinting from experiments. (iii) Lastly, there is often a pejorative connotation in the notion of social construction, especially found in "political" versions of SSK: not only might the construction have been different, but it *should* have been. Society would have been be better off without it, since it served some bad "extra-theoretical" (i.e. political) function. In this sense calling an idea a "social construction" is to "unmask"

---
[77] Pickering (1984), p. 404.
[78] *Ibid.*

it in the Mannheim sense. What is intended is "disintegration" of the idea rather than its refutation. Used uncritically (and too frequently) the word becomes a vague insult, rather like the epithet "fascist" applied to a political opponent in the 1930s.[79]

Some type of relativism is also necessarily a consequence of SSK. At its weakest this relativism is "methodological." Faced with conflicting scientific beliefs, the sociologist tries to understand each sympathetically "on its own terms," explaining each in the same general way without allowing subsequent judgements of "truth" or "falsehood" to interfere with the task. Most recent historians of science proceed in this way. To allow present standards of truth to influence our judgement of past science would be, they believe, to write Whig history, according to them the worst sin of historiography. A good example of this kind of historical impartiality would be Thomas Kuhn's path-breaking book *The Copernican Revolution* (1957) which treats Ptolemaic astronomy with the same sensitivity as the Copernican system. Until at least the mid-seventeenth century both were "rational" alternatives. As we have pointed out in our discussion of Feyerabend, a case can even be made that the Ptolemaic system was more "rational" than the Copernican according to the intellectual standards of the sixteenth century because of the physical and cosmological arguments in its favor. However, the account of the notions of "truth" and "knowledge" favored by SP seem on occasion to go much further and transforms methodological relativism into a full-blown *epistemological* relativism. "Knowledge" and "truth" are simply relative to the local culture that has produced them. We may prefer one set of "truths" to another, for example that produced by our own "scientific" culture; but we cannot say that our preferences are "more rational" or "truer" than others. Consider again the poison oracle of a primitive tribe such as the Azande. This ritual is considered efficacious by the Azande. It "works" for them and follows logically from their metaphysics. A modern scientist would say, however, that the oracle did not *really* work and therefore in the words of Barry Barnes:

> Must it not be irrationally held, because of its 'real' inefficacy? This question could only be answered affirmatively if we could demonstrate inefficacy by the power of reason alone. In practice we see their oracle as inefficacious because of our *theory* of their oracle. To relate rationality to efficacy as we define it is an undercover way of giving special status to our own theories. When the trick is exposed we are left without arguments

---

[79] These aspects of "social construction" language have been nicely developed by Ian Hacking to whose work the writer is indebted. See Hacking (1998), Chapter 1.

for we cannot justify the special status of our theory by an argument which assumes it.[80] [emphasis in original]

Relativism, in turn, tends to veer towards idealism. We have already shown that thinkers such as Woolgar and (possibly) Latour apparently hold that "science" determines or "constructs" "reality" (which has no independent existence) rather than the reverse. In fact, something like this point of view seems necessary for the SSK enterprise:

> If scientific representations were simply determined by the nature of reality, then no sociological accounts of the production and evaluation of scientific knowledge could be offered. Perhaps one might attempt to understand why certain features of reality were selectively attended to at different periods and different social settings, but of the resulting knowledge nothing of sociological interest could be said.[81]

We noted earlier, however, that most of the SSK community will not go as far as pure idealism, They believe that a mind-independent world exists and plays some role in determining scientific belief. Even if we can say nothing about it as Kuhn believes, it does offer "resistance" in the form of anomalies to the constructions of the paradigm. Stephen Cole who calls himself a "realist-constructivist" is willing to go further. He apparently believes that we have some access to the actual world and that both it and social processes have varying degrees of influence on science, the relative importance of each needing to be assessed in particular cases.[82] Bloor in a recent article[83] in part agrees, but criticizes Cole by claiming that the mixture of nature and social factors on science is not like a mixture of salt and water whose relative proportions can be calculated:

> But what if it is not like this at all? What if both ingredients are equally necessary, and both factors are fully engaged and taxed to their limit in all cases? ... One can ask *what* each contributes, but not *how much*.[84]

---

[80]Barnes (1974), p. 29.
[81]Shapin (1982), p. 159.
[82]Cole (1991), pp. 24 ff.
[83]Bloor (2004), p. 940.
[84]Bloor is concerned to reject the accusation that the Edinburgh Program is only concerned with social causation of belief which he views as a unfair misinterpretation. If it is a misinterpretation it is an easy one to make given the four directives of the programme discussed above; in fact, it is difficult to see what *else* besides "social factors" broadly construed could cause belief. The reference to "nature" as a co-determiner of belief in the present quotation is of little weight since throughout SSK (see e.g. the next quotation) *every* statement about nature is conditioned by socially given interpretations.

These arguments concerning the relative importance of social versus non-social causes of belief remind one of the ancient subject-object distinction which in one form or another has continuously perplexed western philosophy since Descartes. Does the mind directly know the external world (realism), or is the world an idea in our minds (idealism)? A third possibility (Kantianism) is that phenomenal world is *constructed* and given its form out of the raw material of sensation via the categories of the intellect. Both factors are necessary, but we cannot know the *res in se*, i.e., the thing as it really *is*. This type of analysis, of course, was anticipated by Kuhn, but for him the agent of construction is the paradigm as interpreted by the community of scientists. But in SSK the mental categories of Kant or Kuhn's paradigm are replaced by explicitly social factors and the Kantian compromise becomes:

> When people confront the experience of their senses, they do so within an already existing structure of knowledge given them by their community and within a structure of purpose sustained by that community.[85]

While SSK is willing to admit that some of of the social influences on the construction of a scientific theory have to do with internal factors such as scientific traditions, the status and interests of scientific groups involved, interpretations generated by a shared paradigm, or "negotiations" among networks of scientists, great emphasis is placed especially by SP and its sympathizers on external, often political causation of scientific belief. Uncovering these offers the clearest application of the "unmasking" technique applied to the pretensions of science. In proposition (ix) of Chapter 1 we have given a famous quotation from Schaffer and Shapin stating that solutions to epistemological problems "are solutions to the problem of social order." At the minimum, these "solutions" may provide the metaphors that scientists can exploit in their theories. Darwin, for example, was stimulated by the survival of the fittest competitive ideology of early capitalism, as well as Malthus's concept of the natural geometric growth of a population versus the arithmetic growth of its food supply.

To give a deeper example of social causation due to Paul Forman,[86] physicists in post World War I Germany faced a hostile intellectual climate consisting of a heady mixture of *Lebensphilosophie* and Spenglerism which rejected positivism, causality, materialism, and reductionist rationality. It

---

[85]Shapin (1995), p. 303.
[86]Forman (1971).

was felt that these qualities, enemies of life and feeling, were characteristic of the exact sciences and were the basic spiritual cause of Germany's defeat. To counter this and recover their lost social prestige, Forman believes that German scientists began to agree with their critics and disparage their own disciplines. In an opportunistic—even cowardly manner—they capitulated to the Zeitgeist. As a particular consequence, they abandoned determinism and causality which happened *before* the creation of the new quantum mechanics and was a critical factor in its invention. Forman concludes that the accommodating reaction of German physicists to the political pressures of their environment points "inescapably to the conclusion that substantive problems in atomic physics played only a secondary role in the genesis of this acausal persuasion."[87] If Forman is correct the new quantum mechanics, as useful as it has become, is really just a species of ideology in the Mannheim sense rationalizing and justifying the social interests of its creators. But frequently, the influence of overt political factors is even more direct according to SSK.

To Farley and Geison,[88] Pasteur in his arguments with Pouchet over spontaneous generation was biased against this concept because of his support (which Farley and Geison find unattractive) for Catholicism and the tyranny of Napoleon III. Both considered spontaneous generation dangerously materialistic and potentially disruptive of the social order because it was connected to the morally subversive Darwinian theory of evolution. During his career Pasteur had waffled on the issue. In his later work he had shown that fermentation was not a chemical process as Liebig and others had argued, but was caused by organisms introduced into the medium. This was consistent with opposition to spontaneous generation. However, earlier research on crystallography "led him not only to believe in the possibility of abiogenesis [spontaneous generation] but actually to attempt such a feat experimentally."[89] Therefore, Farley and Geison hint that Pasteur was a political opportunist: "Pasteur's public posture on the issue seems to reveal a quite high degree of sensitivity to reigning socio-political orthodoxies."[90] He suppressed his prior belief to please the establishment. Pouchet, on the other hand, consistently thought that, whatever the political consequences, he was doing good science. But whatever the political origins of Pasteur's position, is it not true that his clever experiments show

---

[87] *Ibid.*, p. 110.
[88] Farley and Geison (1982) and Geison (1995).
[89] Farley and Geison (1982), p. 79.
[90] *Ibid.*, p. 197.

that Pouchet was wrong? Not even this is correct according to Farley and Geison. The experiments were indecisive. The growth in Pouchet's apparently sealed flasks could be seen with equal cogency either as evidence of spontaneous generation or of Pouchet's incompetence. The eventual decision in Pasteur's favor was made by a government committee which shared Pasteur's political values and was biased against Pouchet. This rather than the superiority of his science was the real reason for Pasteur's triumph over Pouchet.

Shapin and Schaffer have argued in *Leviathan and the Air Pump* (1985) that Robert Boyle, son of the Earl of Cork (one of the richest men in England) and his supporters in the Royal Society of which he was one of the principal founders, fought with Thomas Hobbes over properties of the newly invented air pump chiefly to keep scientific authority in the hands of an elite of gentlemen like themselves with the proper credentials. Politically they were supporters of the Restoration settlement which preserved power in the hands of the propertied upper middle classes and the aristocracy. They feared the potential for anarchy and "leveling" in the atheistic materialism and libertinism they perceived in Hobbes. Thus Boyle viewed Hobbes' doctrine of the plenum "as morally pernicious and subversive of true Christian religion."[91] Hobbes, on the other hand, believed that science should be grounded on rational demonstration by philosophers not on unreliable so-called "experiments." A vacuum or partial vacuum that Boyle thought he had produced using his leaking (!) contraption of an air pump was logically impossible. Empty space was a self-contradiction. Besides, putting philosophical authority in the hands of an experimental rabble who were independent of the state would certainly undermine social peace in the same way putting religion in the hands of priests did. Only the Sovereign could lay down the truths of philosophy by demonstration in a manner similar to geometric demonstration. For Shapin and Schaffer the arguments of Boyle and Hobbes were equally cogent (with Hobbes even perhaps having the better of the dispute). They conclude that the victory of Boyle over Hobbes was contingent, having little to do with the facts of nature. Boyle won because he won the support of the political elites of the period who backed the Restoration settlement. Boyle was more in tune with the political realities of his time. Had the political history of England turned out differently, we might be living today in a totally different scientific environment descended from Hobbes rather than Boyle.

---

[91] Shapin and Schaffer (1985), p. 181.

To David Bloor, even Boyle's corpuscular philosophy which interpreted nature as a collection of inert particles subject to mechanical laws can only be understood in terms of the politics of his time. In the period between the the Civil War and the Restoration England had suffered a near breakdown in social order and "had also witnessed the proliferation of radical groups and sects, such as the Diggers or True Levellers, the Ranters, Seekers and Familiasts."[92] Now for the earlier vitalistic Renaissance philosophy, nature was alive and self-organizing, furnished with a soul or *anima mundi*. Subversive religious groups agreed with this vision; to them nature was divine and God was the soul of the world. But:

> To say that matter could organize itself carried the message that men could organize themselves. By contrast, to say that matter was inert and depended on non-material active principles was to make nature carry the opposite message.[93]

Bloor goes on to draw similar conclusions concerning the political purposes of Newtonian mechanics.[94] He does not claim that Boyle and Newton proceeded deliberately from a political stance and simply made up scientific propaganda to support it. Instead, their arguments were examples of self-deception. Their politics (perhaps unconsciously) caused these scientists to believe in the conclusions they drew from the "evidence." But their opponents who had a different political outlook could and did draw opposite conclusions from the *same* evidence.[95] The cause of controversies derived from differences in political values. Bloor concludes:

> Of course, neither Boyle or Newton, nor their free-thinking opponents, will be found saying that they believe what they do just because of its political implications, though they were deeply concerned with these. Both sides will believe what they do because experience, or reason, or the Bible makes it plain to them. Nevertheless, we know enough of the divergent interests of both sides to explain why all these sources of rational evidence lead to such opposing conclusions.[96]

---

[92] Bloor (1982b), pp. 284–288.
[93] *Ibid.*
[94] *Ibid.*, pp. 288–91.
[95] We have seen, for instance, that to Pouchet the presence of life in an apparently sterile flask was a proof of spontaneous generation, while to Pasteur it was proof of Pouchet's inability to properly conduct an experiment.
[96] *Ibid.*, p. 290.

Not even mathematical issues are free of politics. Still another of Bloor's many theses[97] is that William Rowan Hamilton's (1805–1865) quarrel with Cambridge mathematicians such as George Peacock (1791–1858) was motivated by his perception of the socially subversive influence of Peacock's formalistic doctrines on the nature of algebra. In another contemporary mathematical dispute, according to Barnes, Bloor, and Henry, advocates of non-Euclidean geometry in the nineteenth century used it as a political weapon against the Church:

> As long as theologians could point to geometry as a source of a priori knowledge about reality, it proved that such knowledge was possible, and hence supported their claim to be able to produce further instances of this commodity.[98]

The fact that Euclid's parallel axiom might be empirically false unmasked such claims and cleared the way for replacement of religious by scientific authority.[99]

Toppling the idols from their pedestals

There is often a certain (unattractive) delight shown by this kind of analysis in destroying the prestige of "Great Scientists." In the case of the Pasteur-Pouchet controversy we have seen that that Farley and Geison regard Pasteur as a politically driven opportunist and they argue that Pouchet was at least his equal, perhaps even his superior, in "scientific method"; Pasteur's victory was simply the result of his better connections. Elsewhere, in his book on Pasteur's "private science"[100] Geison accuses Pasteur of not giving rivals or associates proper credit for their work, which anticipated his own, and proceeding with criminal recklessness in his experiments on human beings. Geison claims that Pasteur's administration of a new vaccine to Joseph Meister, a boy who was bitten by a possibly rabid dog, was unethical. In the first place, the vaccine was a "killed" virus version due to Emile Roux, a physician colleague of Pasteur's, and

---

[97] Bloor (1981a).
[98] Barnes, Bloor, and Henry (1996), p. 191.
[99] Additional examples of the influence of politics (usually of a conservative kind) on the content of science or mathematics are discussed by Shapin in Shapin (1982).
[100] Geison (1995)

not Pasteur's "attenuated" virus vaccine. Roux feared that his vaccine was too dangerous to use; not all the virus might be dead and it might actually *cause* rabies. Nevertheless, Pasteur gambled with young Meister[101] and gave him a potentially fatal vaccine which had only been tested on 11 dogs—even though it was unclear if the dog that bit Meister *was* rabid. Clearly, this violated all modern NIH guidelines for human experimentation and today would be a probable ground for prosecution or at least denial of tenure! Dishonesty was added to recklessness! Pasteur did not give Roux credit for the use of his vaccine and allowed the press to think that he used his own attenuated type.

A similar morally debunking treatment of Newton has been given by several writers. Frank Manuel in his book *A Portrait of Isaac Newton* (1968) gives a Freudian interpretation of Newton's personality and stresses his sadism in pursuit of counterfeiters after he was appointed Master of the Mint in 1697. Much of Newton's peculiar personality and scientific achievement according to Manuel is explained by his suppressed desire for incest with his mother. Bloor and others have also claimed that the priority dispute between Newton and Leibniz over the discovery of calculus involved *au fond* the Hanoverian succession and Newton's position as Court Philosopher.[102] Hence this dispute was not really about priority issues or serious philosophic differences. Instead it reflected Newton's ego and lust for status. For George Grinell the *Principia* was essentially Hanoverian-Whig propaganda rather than a work of science, and the form of his law of universal gravitation is a function of Newton's Whiggism and desire to destroy the intellectual foundations and cosmology of Catholicism. On this interpretation, Newton emerges as more a political propagandist than a scientist.[103] Patricia Fara, a Cambridge lecturer, admits that Newton's early work on optics influenced "the ideology of scientific research" and in the two pages she devotes to the contents of the *Principia* (out of her 275 page book) mentions the law of universal gravitation which she admits has a cultural status like $e = mc^2$. But she believes that the thesis that Newton's reputation rests on the fact that "he discovered fundamental laws of nature is too simple." In the first place, according to Fara, it is not clear that "scientific knowledge can ever be absolutely and permanently true." Secondly,

---

[101] Meister felt grateful to Pasteur throughout his life. Many years later he was the caretaker of Pasteur's mausoleum. When in 1940 German officers forced him to give them access to the mausoleum and an attached museum, Meiser later committed suicide.
[102] See Bloor (1981b), p. 200.
[103] Grinell (1973).

Newton's ideas were of little help to Einstein, and his overrated book was "riddled with inconsistencies, contradictions, speculations—and even plain errors."[104] Newton also was a thoroughly disagreeable self-promoter whose glorious reputation through the centuries was constructed by ruling elites to serve their political needs.

Although our criticisms of such debunking of scientific icons will be mostly confined to a later chapter, we cannot resist the observation that this mode of *ad hominem* attack on scientists of towering achievement by relatively minor academics puts one in mind of other *ad hominem* remarks by Hegel. In his *Lectures on the Philosophy of History* Hegel contemptuously refers to village schoolmasters who think themselves better than Alexander the Great because *they* did not suffer the irrational passions that made Alexander conquer the Persians. He also quotes Goethe to the effect that although "no man is a hero to his *valet de chambre*," this is not because the former is unheroic but because the latter is a valet.[105]

A consequence of SSK analysis sketched above is that science is not "epistemologically privileged"; in its claims to knowledge it is no better or worse than any other coherent culturally constructed system of belief, for instance political ideology, religion, a school of literary criticism, and so forth. It is a one of many possible stories which tells more about the values and obsessions of our society than about nature. According to Richard Rorty[106] there is no essential difference in kind between the deliberative process of science and "the deliberative process which occurs concerning, for example, the shift from the *ancien regime* and bourgeois democracy, or from the Augustans to the Romantics."[107] Also:

> ... what could show that the Bellarmine-Galileo issue 'differs in kind' from the issue between, say, Kerensky and Lenin, or between the Royal Academy (*circa* 1910) and Bloomsbury.[108]

---

[104] Fara (2002), pp. 2, 9, 68.
[105] Hegel (1953), p. 47.
[106] Rorty is not usually considered a member of the SSK community. His interests is not so much the epistemological nature of science as arguing that science is not (and should not be) the preeminent cultural expression of modern liberal civilization. Yet the thesis of *Philosophy the Mirror of Nature* (Rorty (1979) regarding knowledge is similar to that of SSK. In the terminology of an earlier political era, it is fair to call him a "fellow traveler."
[107] Rorty (1979), p. 327.
[108] *Ibid.*, p. 321.

But the most important consequence of the SSK approach of exposing the social and political factors behind scientific theories and thus showing that they are no "better" than other belief systems is the "disintegration" of the pretensions of science, the same program Mannheim carried out for ideology. There is in much of SSK a thinly disguised motive of "knocking science off its pedestal" and deconstructing its authority in the modern world. Often this is combined with a disparagement of the Enlightenment itself where in the eyes of at least part of SSK the historic wrong step of relying on reason began. Michel Callon and Bruno Latour are quite direct. Their core purpose is to wage a "moral struggle to strip science of its extravagant authority,"[109] because:

> We no longer have to fight against microbes but against the misfortunes of reason... Because we have other interests and follow other ways, the myth of reason and science unacceptable, intolerable, even immoral. We are no longer, alas, at the end of the nineteenth century, the most beautiful of centuries, but at the end of the twentieth, and a major source of pathology is reason itself—its works, its pomps, and its armaments.[110]

Feminism and Afrocentrism

Even more extreme conclusions follow if what passes for scientific truth does not mirror reality but only the ideology (often the reflection of values of a political elite) of a local culture or dominant group of investigators. For if science is merely socially constructed in this manner, and if as Stanley Aronowitz maintains both logic and mathematics are contaminated by "the social,"[111] it follows that we can change the ambient political/social environment by changing the content and methodology of science and vice-versa. Feminists and advocates of other "deprived" groups have grasped at this idea to:

---

[109] The quotation continues by criticizing those in the SSK camp who are guilty of ideological heresy on this issue: "Any move that waffles on this issue appears unethical, since it could also help scientists and engineers to reclaim this special authority which science studies has had so much trouble undermining." (Callon and Latour (1992), p. 346.) The article is part of a complicated dispute between Callon and Latour and Collins and Yearly, on basically how to avoid the problem of reflexivity. The "waffling" Callon and Latour speak of is "treason" to the STS project of debunking science of which they accuse Collins and Yearly. See the "epistemological chicken debate" in the volume *Science as Practice and Culture* and the remarks in Chapter 11.
[110] Latour (1988a), p. 149.
[111] See proposition (ix) above.

> ... expose science as 'hegemonic', as 'masculinist', as 'dehumanizing', as 'mystifying' [and] to open up alternative visions of what science might be and how its social relations ought to be constituted...[112]

in order to benefit the particular group.

Sandra Harding, for example, believes that Western science—even mathematics—is infected by racist, classist, androcentric, and Eurocentric values. These evil aspects can be traced back to the very beginnings of modern science in the Renaissance. She feels that it can be perfectly characterized by the misogynistic metaphors of torture and rape found in the writings of Francis Bacon since these were as important as the mechanistic view of nature in the scientific revolution wrought by Newton. We may be told by "traditional historians and philosophers" that the former played no role in the development of science while the latter provide "the interpretations of Newton's mathematical laws: [since they direct] inquirers to fruitful ways to apply his theory and [suggest] methods of inquiry."[113] But this is mistaken, since:

> ... if we are to believe that mechanistic metaphors were a fundamental component of the explanations that the new science provided, why should we believe that the gender metaphors were not. A consistent analysis would lead to the conclusion that understanding nature as a woman indifferent to and even welcoming rape was equally fundamental to the interpretations of these new conceptions of nature and inquiry. Presumably these metaphors, too, had fruitful pragmatic, methodological, and metaphysical consequences for science. In that case, why is it not more honest to refer to Newton's laws as "Newtons rape manual" as to call them "Newton's mechanics"?[114]

Other early misogynistic guilty parties include Copernicus who "replaced a woman-centered universe with a man-centered one."[115] Because of him the "womanly earth which had been God's special creation for man's nurturance (sic!), became just one tiny, externally moved planet circling in an insignificant orbit about the masculine sun."[116] Since these tainted beginnings, Western science has become dominant not though its superior rationality, but because of the superior military and political power of the

---

[112] Shapin (1982), p. 293.
[113] Harding (1986), p. 113.
[114] *Ibid.*
[115] *Ibid.*, p. 114.
[116] *Ibid.*

West. It has clearly been and continues "to be complicit with racist, colonial, and imperial projects."[117]

Because of these characteristics, science has never been truly "objective," and since its inception has offered a distorted account of nature. It pretends to be free of social or political factors. However, this is a hypocritical stance given the footprints of social injustice it bears. Science is and has been dominated by white middle class males whom it has benefited along with the comfortable of the First World. On the other hand, it has excluded and disempowered people of color, the poor, and women especially in the Third World. The failure to realize this on the part of the scientific elite constitutes for Harding "Eurocentric scientific illiteracy."[118] To correct this, Harding proposes a frank incorporation of political values into the sciences. She wants a "feminist standpoint approach" or "feminist epistemology" which would privilege the perspective of women and expose masculine distortions. This should be allied with anti-racism and anti-classism. The result would be a better "successor science" characterized by "strong objectivity" and which would produce a more "useful knowledge."[119] Even better:

> ... because science's social hierarchy so closely mirrors the social order 'outside' any progressive changes ... in the social structure of science should have rapidly escalating consequences for the larger social order."[120]

While affirmative action may be worthwhile in slightly improving the status of women in science, Harding feels that radical, revolutionary political changes in both science and society are necessary, the resources for which may be created by "mere reforms."[121] Although she does not mention the need for a violent purge of white male scientists, it is hard to see how this can be avoided in order to achieve her goals.

From an even more radical feminist perspective, the French philosopher Luce Irigaray, according to her American interpreter Katherine Hayes, has argued that modern physics privileges the study of rigid body over fluid mechanics because of the rigidity of the male organ in contrast to feminine menstrual flow.[122]

---

[117] Harding (1993), p. 3.
[118] *Ibid.*, p. 1.
[119] See Harding (1991), pp. 105–63.
[120] Harding (1986), p. 80.
[121] *Ibid.* p. 247.
[122] Sokal and Bricmont (1998), p. 101.

Somewhat more restrained than Harding and Irigarway are Evelyn Fox Keller and Helen Longino.[123] Like Harding, Keller suggests that masculine values such as: (i) the urge to dominate and control nature, (ii) the vision going back to Francis Bacon of nature as a woman to be subdued, (iii) fondness for mathematical abstraction, (iv) explanations in terms of hierarchical structures, (v) the separation of the observer from nature (or "objectivity"), and (vi) the belief that nature is an essentially lifeless mechanical system, have affected the development of modern science since the seventeenth century. She contrasts these attitudes with an older hermetic tradition, mainly located in alchemy and the followers of Paracelsus, which incorporated feminine values and thought of nature as a hermaphrodite union of male and female principles. She speculates that the survival of this tradition might have led to the creation of a "different science" than the androcentric version we are afflicted with now.[124] For Helen Longino, one can distinguish between values essential to the proper conduct of science upon which all scientists agree. These she calls *constitutive* values. In her view the practice of science is a blend of these and what she calls *contextual* values. The latter derive from the social position, ideology, politics, gender, or culture in which the scientist is embedded. The two classes cannot be separated, which means that the concept of a value free science is an oxymoron. Examples of contextual values would certainly include the masculine metaphors embedded in science described by Keller or Harding; but they also might include reactionary politics as in the case of sociobiology. For both Keller and Longino, the ideological bias they describe operates mostly at an unconscious level in science. They determine the kind of problems investigated and the kind of explanations accepted by a white middle class male scientific establishment. This is not to say that the resulting science is "wrong"; unlike Harding, both admit that in several respects traditional science has been strikingly successful.[125] It is more the case that existing science is "incomplete." For Keller, we could get alternative and better theories if the holistic, psychologically softer attitudes of women—their willingness to "listen to" nature rather than trying to dominate it—were more represented in the scientific establishment. To Longino, since contextual and constitutive values are intimately entwined, a feminist

---

[123] Good expositions of the ideas of Keller and Longino may be found in Keller (1984) and Longino (1990). See also the references in Gross and Levitt (1994).
[124] See Keller (1984), pp. 43–65.
[125] Longino in particular rejects the idea of an explicitly feminist science in method and content, but speaks of exposing androcentric bias by practicing science as a feminist.

has a certain freedom in theory choice; if she detects "politically noxious" values in one theory she should not "try to avoid ideology by sticking to the data" since this is a "recipe for replicating the mainstream values and ideology that feminists and radical scientists reject."[126] Instead, she should suggest alternative theories that can also work and have better political consequences, a view of the construction of scientific knowledge Longino calls "contextual empiricism."[127]

Unlike some feminists, Afro-centrists do not yet commonly argue for a Black science or mathematics in content. Instead they try to give Black scientists greater credit for the development of modern science. In so far as Black contributions may have been ignored or discounted in earlier accounts, this is beneficial. However, there is a tendency on the part of some extremists to make up history, for instance, by claiming that ancient Egyptians could fly or knew about quantum physics.[128] On a more scholarly level, Martin Bernal has maintained that much of Greek learning and scientific culture was borrowed (as the Greeks themselves believed) from Egypt, a partially black civilization.[129] Bernal also believes that Greek cultural borrowings from Egypt may have started with a possible colonization of part of the Greek peninsula by the Egyptians in the Second Millennium B.C.E.[130]

The extremism of the more radical feminists and Afrocentrists has been devastatingly exposed by authors such as Gross and Levitt, Koertge, and Sokal and Bricmont, mentioned previously, and has unfortunately been an essential factor in launching the Science Wars between defenders of a traditional realistic epistemology of science and many of the more radical sects and offshoots of PIS.

---

[126] Longino (1990), p. 218.
[127] *Ibid.*, p. 219.
[128] See e.g. Adams (1990).
[129] Bernal (1987).
[130] There is little hard evidence for either Bernal's thesis or an early Egyptian colonization of Greece. Even if such a colonization actually happened, to use it to explain Greek scientific achievements in the age of Pericles and Plato would be like explaining Newton by Caesar's invasion of Britain c. 50 B.C.E. For convincing arguments against Bernal, see Lefkowitz (1996).

Chapter 10

The Deconstruction of Mathematics

The analysis of mathematics has from the very beginning represented a more difficult challenge to SSK than mere science. Here is a subject that is universal and which seemingly deals with a transcendental world of mathematical objects and necessary truths. To our minds, according to Wittgenstein, "Logic is a kind of ultra-physics, the description of the 'logical structure' of the world."[1] And in the words of G. H. Hardy:

> I believe that mathematical reality lies outside us, that our function is to discover or *observe it* ... the theorems which we prove, and which we describe grandiloquently as our 'creations', are simply the notes of our observations.[2]

And because mathematics is timeless:

> Archimedes will be remembered when Aeschylus is forgotten because languages die and mathematical ideas do not. 'Immortality' may be a silly word, but probably a mathematician has the best chance of whatever it may mean.[3]

If these things are true, it follows that mathematics, even more than science, must be an example of Mannheim's "knowledge as such" whose credibility and content—unlike that of a religion, philosophical system, or ideology—cannot be explained by social or historical factors.

---

[1] Wittgenstein (1956), §I-8.
[2] Hardy (1992), p. 123.
[3] *Ibid.*, p. 81.

This view is, of course, anathema to SSK theorists. Having "unmasked" the pretensions of science, they think of such blatant Platonism as the final barrier to the success of their program. They very much want to show that mathematical ideas are not reflections of a world of Platonic Forms perceived by the "eye of the mind," but are historically and culturally contingent structures *invented* by man and not *discovered* by him, and in this respect no different from any other belief system men have invented. The same kinds of sociological questions can be asked explain belief in both. To Steve Woolgar, for instance:

> It should be clear ... that mathematical statements such as $2+2 = 4$ are as much a legitimate target of sociological questioning as any other item of knowledge (some sociologists use the term 'knowledge claim' rather than 'knowledge' to emphasize their impartiality). What kinds of historical conditions gave this expression currency and, in particular, what established (and now sustains) it as a belief?[4]

As in the case of the SSK analysis of science, the truth or falsity of mathematical statements apparently plays no role in the institutional belief in them! The arguments on behalf of this sociological approach have been broadly similar to those advanced by SSK regarding science, but there are interesting differences since in mathematics one does not have to worry about the relation of physical theory to the world it is supposed to represent.

For Sal Restivo[5] the idea that mathematics is a purely intellectual or cognitive phenomenon must be rejected; every aspect of mathematics and mathematicians is a social construction. This includes the "mind" and even the "self" of a mathematician which are implements through which the surrounding culture thinks: "... individuals are *vehicles* for expressing the thoughts of communities or 'thought collectives'. Or to put it another way, *minds are social structures* ..."[6] Pure mathematics especially is not "some type of unmediated cognitive process," rather it has a firm social, material, and empirical base. The link between the "thinking individual, social life and the material world" may be difficult to discover but it is:

> ... being slowly constructed on the foundations of the works of Durkheim, Spengler, Wittgenstein, and others ... [and] ... involves, in part, recognizing that symbols and notations are actually 'material,'

---

[4]Woolgar (1988a), p. 43.
[5]Restivo (1993).
[6]*Ibid.*, p. 249.

and that they are worked in the same ways and with the same kinds of rules that govern the way we work with pebbles, bricks, and other 'hard' objects.[7]

In fact, there is no fundamental distinction between the material world, the social world, and our mental world:

> Mental facts are social facts; the ideas, concepts, and knowledge systems in their notational and symbolic materiality are higher order material objects. A mathematical object, then, like a hammer or a screwdriver, is conceived, constructed, and put together through a social process of collective representation and collective elaboration.[8]

Restivo criticizes mathematicians such as George Boole for failing to recognize these insights. Boole, for instance, thinks wrongly that the Aristotelian logical principle *dictum de omni et nullo* that whatever is affirmed or denied of a genus is necessarily affirmed or denied of a species of that genus is self-evident from the "laws of thought"; in reality according to Restivo this is an inductive inference.[9] But then like some member of the upper classes, Boole is "always busy disconnecting himself from common life and common language."[10] Poor Boole's mistake "arises from his failure to see himself as a product and agent of culture—or, more radically, as a vehicle for culture."[11] Restivo also maintains that pure mathematics (as well as pure science) has an unacknowledged political function:

> Purism is an intellectual strategy that has multiple roots and functions. As a *political* strategy ... it is a way of keeping tabs on and control over creative and innovative thinkers by giving them 'academic freedom' so long as what they do keeps them from becoming active critics of the government or actively interfering with efforts by ruling elites to try to put their discoveries or inventions to use in the interest of military, economic, or political 'advances'.[12]

---

[7] *Ibid.*, pp. 255–256.
[8] Restivo (1992), p. 137.
[9] *Ibid.*, pp. 157–164.
[10] *Ibid.*, p. 163. Boole, however, was not an aristocratic elitist. He was the son of a tradesman of limited means. He could not afford to attend a university and was essentially self-educated in mathematics.
[11] *Ibid.*, p. 158.
[12] *Ibid.*, p. 156.

Because ruling elites control educational institutions and the mathematics curriculum, they can use them as a tool to maintain existing patterns of domination and subordination. It follows that "modern mathematics is itself a social problem in modern society";[13] and in the service of ruling class interests:

> ... it can be a resource that allows a professional and elite group of mathematicians to pursue material rewards independently of concerns for social, personal, and environmental growth, development and wellbeing; aesthetic goals in mathematics can be a sign of alienation or of false consciousness ...[14]

It will take radical social changes therefore to bring about the needed changes in the organization, perception, and content of mathematics. Restivo, however, seems more a traditional Marxist than an SSK relativist and so seeks a middle way between "naive realism" and "radical relativism." Relativism is "suggested by variations in mathematical ideas, and some sort of realism is suggested by recalcitrant reality."[15] Adopting a belief in "accurate representations" is allowed on pragmatic grounds provided we "think of them as jokes so they won't turn into authoritarian dogmatic demons."[16] On the other hand, the unsettling results of Gödel, Church, Turing, Tarski, and others make accurate representation impossible in the sense of naive realism. Most satisfactory, perhaps, is the pragmatism of Richard Rorty.[17]

For David Bloor and the proponents of the Strong Program mathematics is also a social construction, but they offer a more detailed deconstruction of its content and apparent necessity and objectivity than Restivo. To explain mathematical knowledge they rely on twin foundations: (i) the empiricism of J. S. Mill and (ii) the conventionalism of the late Wittgenstein. Mill is most prominent in Bloor's early book *Knowledge and Social Imagery*. Just as for Restivo the basic intuitions of arithmetic come from counting and grouping physical objects such as pebbles. Mathematics—even the deductive part—is at bottom really just inductive reasoning. We forget this when we carry out algebraic manipulations only because the process has become automatic and mechanical. But taking equals from equals on both sides of an equation, for example, is analogous to removing the same object

---

[13] Restivo (1988), p. 18.
[14] *Ibid.*, p. 19.
[15] Restivo (1992), p. 127.
[16] *Ibid.*
[17] *Ibid.*, pp. 126–128.

from two equal piles of objects. However, while regarding mathematics like Mill as basically inductive Bloor is aware of Frege's devastating critique[18] of Mill or of any theory that identifies mathematical entities either with physical objects or with ideas in individual minds:

> Mill's theory does not do justice to the objectivity of mathematical knowledge. It does not account for the obligatory nature of its steps. It does not explain why mathematical conclusions seem as if they could not possibly be other than they are.[19]

To overcome Frege's objections Bloor in several articles and books has recourse to an essentially sociological approach which he finds in Wittgenstein's works on the foundations of mathematics.[20] Mathematics is a certain kind of "language game" which one learns by social participation in the practice or "form of life" of mathematicians.[21] There is no *a priori* necessity to it. Mathematics consists of an intricate network of conventions and norms which we assent to not because they are "true" in some Platonic sense; rather, they are considered true because mathematicians assent to them. Even the rules of arithmetic do not reflect some mathematical reality embedded in the structure of the world. We posit them because they are pragmatically justified. In other cultures or situations we might chose different rules. The only reason, for example, that $2 + 2 = 4$ is social practice motivated by the fact that we find it *useful* to add like this. This usefulness causes the construction of all mathematical rules. If it was useful to have $12 \times 12 = 145$ and everybody did the multiplication in this manner, then

---

[18] See Frege (1959) and a summary of Frege's objections in Hersh (1997), pp. 142–150. With withering sarcasm Frege points out that a collection of three pebbles is not the same as three ideas and there is no way inductively to connect the two. Also if, say, "two" was an idea in the human mind they my "two" would be different from your "two" and the "twos" of past generations and the unborn. It will not do to say that these "twos" are all the *same* idea, for one can ask "the same in respect to what? What to they have in common?" Frege would answer by saying that they are all reflections of the "Abstract Object" or "Concept Two."

[19] Bloor (1976), p. 105.

[20] Principally his *Remarks* and *Lectures* on the foundations of mathematics. (Wittgenstein (1956) and Wittgenstein (1976)) Wittgenstein's writings on mathematics in the period 1930–1950 are notoriously obscure and fragmentary. There is a vast secondary literature with differing interpretations; what follows is a rough synthesis of arguments that may found in Bloor's articles Bloor (1981a), Bloor (1982a), and his books Bloor (1983), Bloor (1997).

[21] For Wittgenstein a "form of life" corresponds to Kuhn's extended definition of "paradigm."

this would be the new rule and *correct* and there *could* not be a mistake "just as there can be no mistake in saying that the rules of chess are what they are collectively agreed to be."[22]. According to Wittgenstein (as interpreted by Bloor), the reason such an analysis may make us uncomfortable is because we wrongly

> ... think mathematical truth involves some sort of correspondence between a mathematical statement and an independent mathematical reality. The correct analysis is one where it does not matter if everything is 'wrong', because then nothing would be wrong. This must be because whatever it is that 'everything' comes to, is constitutive of the definition of truth.[23]

But what accounts for the illusion of objectivity and necessity in mathematical reasoning? The mathematician may say that he accepts a theorem because its proof "compels" agreement that it is true, i.e., that its conclusion follows from the hypotheses. He may even say that he has "observed" that it expresses a true relationship among the mathematical objects it describes. But Barnes, Bloor, and Henry in their book *Scientific Knowledge: A Sociological Analysis* (1996) argue that both assertions are misleading references to a form of psychological compulsion in modern Western society masquerading as a special "insight" into mathematical truth. Beginning with arithmetic with which we have been imprinted by the constant drill of strict elementary teachers, we have been subject to a kind of educational brainwashing which instills in us a feeling that the rules of mathematics *must* be obeyed; our compulsion here is similar to our compulsion to obey moral laws.

> Mathematical necessity is just a species of moral necessity that frequently attaches to the more important social conventions. Thus the inexorability of mathematics is just the inexorability of the social demand that we employ our techniques like this rather than like that, e.g. so that $2+2 = 4$ and not $2 + 2 + 2 = 4$.[24]

---

[22] Bloor (1997), p. 37.
[23] *Ibid.*
[24] Barnes, Bloor, and Henry (1996), p. 183.

Does this mean that there might be an arithmetic in which $2 + 2 = 5$?

> ... this seems so outrageous, and so redolent of Orwell's *Nineteen Eighty-Four*, that it shows the absurdity of any approach which gives convention a significant role in arithmetic.[25]

Nevertheless, that $2+2$ can on occasion be 5 does follow from their approach the authors cheerfully admit. But this is not outrageous. They quote an example from Lakatos in which a pair of two pound items are delivered in a one pound box. A convention of the Post Office always implicitly includes the weight of the box so that $2 + 2 = 5$. Just as easily we could say that $2 + 2 = 7$ or an indefinite number of other alternatives. Why don't we? The only reason is that the convention $2 + 2 = 4$ is "easier to organize than the others." Also we need *one* system that gives a "salient solution... automatically visible to everyone."[26] But what about "proof"? Using the logical techniques of *Principia Mathematica* one can prove (after 362 pages!) that $1 + 1 = 2$ and presumably also that $2 + 2 = 4$. The authors answer that the "proof" despite its seeming rigor implicitly assumes knowledge of the result as it depends on manipulating two groups of two logical symbols into a group of four symbols.[27] At bottom this, therefore, is an inductive procedure, not that different from Mill's manipulation of apples.

The rules of formal logic are given a similar explanation. Consider, for instance, *modus ponens*, that is, the rule $p \wedge (p \Rightarrow q) \Rightarrow q$. This again must be a convention. It cannot be shown to be valid; any such attempted proof would be circular, probably relying on *modus ponens* itself. The authors admit that the *modus ponens* convention is convincing[28] and suggests that one reason among other possibilities may be biological. The human race has evolved to think in this manner; we are hard-wired for logic. But even if we have a natural tendency to accept this and other rules they are not "absolute." *Modus ponens* can on occasion yield a false conclusion from true premises. The "sorites paradox" demonstrates this. Let S be a soritical predicate, for instance "is a heap", or "is rich" and let p_n be "a pile of sand

---

[25] *Ibid.*, p. 18.
[26] *Ibid.*, p. 185.
[27] *Ibid.*, pp. 169–179.
[28] In Bloor (1976), pp. 138–142, however, it is maintained that *modus ponens* is not universal; he gives an example of the Azande tribe who from the premises "all the male descendants of a male witch are witches" and "this male is a descendant of a witch" will (sometimes) not draw the conclusion that the descendant is a witch.

with n grains" or "a man with n dollars." We would accept as true the premises $S(p_{1,000,000})$ and $S(p_{1,000,000}) \Rightarrow S(p_{999,999})$. Therefore, by *modus ponens* we conclude $S(p_{999,999})$. The difficulty comes with repeated application of this argument which yields the false conclusion, say, $S(p_{10})$. The conclusion is that *modus ponens* is a convention capable of going wrong.[29]

If mathematics is a network of pragmatically justified social conventions, then its universality and stability should be only an illusion. Over historical time and among various cultures there should have developed alternative even contradictory systems of mathematics or logic. Restivo points out that Oswald Spengler's *Decline of the West* argued that such has been the case.[30] In the same spirit, David Bloor finds what he considers several examples of radically different historically generated types of mathematics. For instance, the Greek conception of "number" seems to have been different from ours. Thus "one" was not a number according to Aristotle and other Greek philosophers, but rather the "generator" of the other numbers. Diophantus in his *Arithmetica* had a much more primitive notion of an algebraic process than moderns; his approach is always subordinated to the individual numerical problem. He has no general method; even if the reader has read dozens of his solutions, they are of no help in solving the next problem. The Pythagoreans invested the integers with properties like "Male," "Female," "Good" and "Bad." Their famous proof of the incommensurability of the diagonal of a square with the side, essentially demonstrated that $\sqrt{2}$ is not a number rather than, as with us, an irrational number. The founders of calculus such as Cavalieri and Wallis accepted the reality of infinitesimals; Euler's arguments by no means meet the standards of modern rigor, etc., etc. All this according to Bloor demonstrates the historical contingency and conventionality of mathematics.[31]

Paralleling the view of mathematics as a social construct created by man is the thesis that it does *not* represent certain knowledge. Rather it is "fallible" and "corrigible." The strongest proponent of this opinion is Paul Ernest, a prominent authority in mathematics education. An American by birth, Ernest holds a BSc (second class honors) in Mathematics from the University of Sussex (1973), a MSc (with distinction) (1974) and a Ph.D. in Mathematics Education from the University of London (1985). He is

---

[29] cf. *Ibid.*, p. 182 and Barnes, Bloor, and Henry (1996).

[30] Spengler claimed that Babylonian-Egyptian, Indian, Arabian, Chinese, Classical, and Western civilization each had a different conception of "number" characteristic of the incommensurable "world view" of each civilization. See the discussion in Restivo (1982), pp. 133–134.

[31] Bloor (1976), Chapter 6.

presently a Professor Emeritus in the School of Education of the University of Exeter, UK.

The dominating theme of Ernest's two books and many articles is an attack on a philosophy he calls "absolutism" and which he thinks dominates mathematics. What is absolutism? It is the belief that:

> ... mathematical truth is absolutely valid and thus infallible, and that mathematics (with logic) is the one and perhaps the only realm of incorrigible, indubitable, and objective knowledge.[32]

Ernest gives several reasons why absolutism is incorrect. First of all, there is no "absolute certainty" in mathematics. Any mathematical theory depends on initial assumptions, and the attempt to justify these assumptions in terms of other assumptions leads to infinite regress.

> Thus we cannot establish the certainty of mathematics without assumptions, which therefore is conditional, not absolute certainty. Only from an assumed basis do the theorems of mathematics follow.[33]

Now any mathematician would agree that an axiomatic theory is built on axioms which are not themselves subject to proof and primitive terms which are undefined. But he probably believes that theorems in the system are conditionally true; in that if they are correctly proven they are certain consequences of the axioms. But even that is incorrect, according to Ernest, since our inferences depend on the rules of logic which are arbitrary unproven assumptions. Obviously it is a vicious circle to try to justify logic by logic.

Additional evidence against the claims of absolutism is the well-known fact that the mathematical literature is full of mistakes. Every working mathematician is aware that theorems in published refereed papers, even in "good" journals, can be "wrong," or if not wrong can have mistaken proofs.[34] Standards of rigor can and do change. For example, the informal manipulation of "infinitesimals" by Leibniz, Newton, and Euler were unacceptable to Cauchy and Weierstrass in the nineteenth century.

The intellectual fragility of mathematics is further shown by the various set-theoretical paradoxes discovered in the decade after 1895 by Russell, Burali-Forti, Cantor, and Richard, as well as two devastating theorems con-

---

[32] Ibid., p. 9.
[33] Ibid.
[34] As is the case with at least two of the writer's own papers!

cerning the incompleteness and consistency of certain formal systems due to the logician Kurt Gödel in 1931. The paradoxes have only been removed by *ad hoc* devices like the Theory of Types or the Zermelo-Fraenkel axioms of set theory.[35] The results of Gödel are even more damaging according to Ernest. In his First Incompleteness Theorem Gödel showed that any formal system $\mathfrak{F}$ with finitely many axioms, symbols, rules of inference containing certain amount of elementary arithmetic is, if it is consistent, "incomplete"; i.e., a formula $\mathfrak{G}$ can be found in $\mathfrak{F}$ such that neither it nor its negation is provable in the system. The Second Theorem (as improved by J. Barkley Rosser) states that the system $\mathfrak{F}$ cannot be shown consistent by proofs formalizable within the system.[36] Ernest argues that the Second Incompleteness Theorem, in particular, demonstrates that mathematics cannot be shown to be free from error:

> We cannot therefore know that any, but the most trivial mathematics are secure, and the possibility of error and inconsistency always must remain. Belief in the safety of mathematics must be based either on empirical grounds (no contradictions have yet been found in our current mathematical systems) or on faith, neither providing the certain basis that absolutism requires.[37]

The upheaval in mathematics caused by the paradoxes and Gödel's theorems leads Ernest to sympathize with Morris Kline's bewailing "the loss

---

[35] Barnes, Bloor, and Henry, however, suggest that this foundational crisis was not inherent in the mathematics of the time, but has a sociological explanation. It was stimulated by the general crisis of European civilization after 1918. Its origins antedate World War I:

> ... but the fact that it reached such a pitch of intensity after 1918 may be no accident. Forman (1971) has documented what almost amounted to a desire or hunger on the part of many in the mathematical and scientific culture of central Europe in those years to participate in a cultural crisis. The intriguing possibility is that once there is a motive to produce a crisis there are plenty of ways in which this can be done, even in a field such as mathematics. (Barnes, Bloor, and Henry (1996), p. 193.)

[36] Since anything including consistency may be proven in an inconsistent system, an equivalent way of expressing the Second Theorem is the statement that $\mathfrak{F}$ may be proven consistent within the system if and only if it is actually inconsistent. Two accessible accounts of these two theorems may be found in Franzén (2005) and Hersh (1997). For a more technical but still reasonably straightforward treatment see Griffiths and Hilton (1970), pp. 608–615.

[37] Ernest (1991), p. 16.

of certainty" of mathematics and claim that "it has been shorn of its truth; it is not an independent, secure, solidly grounded body of knowledge."[38]

Ernest also appeals to historical evidence in his campaign against absolutism. Like Restivo, Bloor, and Spengler he believes that mutually incompatible mathematical systems have developed in history which contradicts the mathematical universality posited by absolutism. For Ernest, the intuitionism of Brouwer and his followers and classical mathematics are examples of this incompatibility. Intuitionism rejects actually infinite sets, the law of the excluded middle in nonconstructive situations, the axiom of choice, the proof of the existence of a mathematical object by *reductio ad absurdum* when no explicit construction of the object can be given, and so on. All these forbidden things (including Cantor's arguments) are accepted without reservation by standard mathematics.[39]

Having demolished the intellectual claims of absolutism, Ernest also believes it to be evil since absolutism is correlated with anti-democratic goals in education. Absolutists believe that:

> ... the aim of education is to produce the liberally educated person, with an appreciation of culture for its own sake ... Only a minority will achieve this, those fit to lead and govern society.[40]

Even worse, under a philosophy of absolutism the mathematics curriculum:

> ... prepares a tiny minority of students to be mathematicians whilst teaching the rest to stand in awe of the subject. To let one group distort the aims of education like this, to serve its own interests, is wrong and anti-educational. It results in more persons being disadvantaged than are advantaged, meaning on utilitarian grounds alone, the system is insupportable.[41]

---

[38] Kline (1950), p. 352. It is also been fashionable to draw a vast number of philosophic consequences from the Gödel's theorems. Kadvany, for example, suggests that the First Incompleteness Theorem implies a postmodern interpretation of mathematics in the sense that our mathematics is a contingent creation which is in so sense unique. He argues that since adding $\mathfrak{G}$ to $\mathfrak{F}$ and $\sim \mathfrak{G}$ to $\mathfrak{F}$ produces two consistent but distinct systems $\mathfrak{F}_1$ and $\mathfrak{F}_2$ which also are incomplete and have unprovable Gödel sentences $\mathfrak{G}_1$ and $\mathfrak{G}_2$, we can repeat the process and produce sequences of incomplete systems $\{\mathfrak{F}_n\},\{\mathfrak{F}_{n+1}\}$ *ad infinitum*. (As reported in Franzén (2005), p. 50 f.) Various French poststructuralists such as Julia Kristeva, Paul Virilo, Régis Debray, and Michel Serres have also attempted to derive (usually) political conclusions from the theorems. See Sokal and Bricmont (1998), especially Chapter 11. Thomas (1995) has even related Gödel's theorems to postmodern literary theory.

[39] Ernest (1998).
[40] Ernest (1991), p. 170.
[41] *Ibid.*, p. 180.

Throughout Ernest's writings is a passionate feeling that the aim of mathematical instruction is to increase social justice. Since certain minorities and the lower classes usually do not do well in school mathematics, they are barred from advancement in society and higher education. The teachers emphasis therefore should be on teaching the majority, not a small elite class of students.[42]

Another writer whose ideas in part are similar to Ernest's is Reuben Hersh. Unlike Ernest, Hersh has been a research mathematician. Having aquired a BA in English from Harvard in 1946, he earned a Ph.D. in mathematics under Peter Lax at New York University in 1962. He is now Professor Emeritus at the University of New Mexico and has done research on partial differential equations, random evolution, stochastic processes, nonstandard analysis, and linear operator equations. He has produced two popular and best selling books and many articles on the nature of mathematics. His first book *The Mathematical Experience* (1981) coauthored with Philip J. Davis won the National Book Award in 1983. More recently (1997) he has written *What is Mathematics, Really*. A third book *Loving and Hating Mathematics* coauthored with Vera John-Steiner as of this writing has been submitted for publication.

Hersh rejects Platonism, formalism, and logicism, i.e., Bertrand Russell's project to reduce all mathematics to logic. His objections to these three philosophies are the usual ones. There is no reason to believe in a mystical non-physical Platonic world of timeless mathematical objects. Formalism, the assertion that mathematics is a collection of axioms, definitions, theorems, and formulae which are meaningless, neither true or false, and manipulated by rules we prescribe does not account for the actual practice and creation of mathematics. Moreover, the formalist project as envisaged by David Hilbert of putting mathematics on an absolutely secure foundation was destroyed by Gödel's theorems. Finally, the bankruptcy of logicism was shown by Russell's and other set theoretic paradoxes. These necessitated the repair of set theory by the addition of axioms such as the axiom of infinity and Zermelo's axiom which cannot be reduced to logic. All these approaches have induced a preoccupation with building a firm conceptual *foundation* for mathematics that would show that it was absolutely certain and free of contradiction. The pursuit of this goal Hersh

---

[42]This point of view which is now absolutely dominant in Mathematics Education has produced in the writer's opinion a situation where in the name of egalitarianism almost *no one* learns *any* mathematics! However, detailed criticism of the educational effects of constructivism is not the purpose of this essay.

(following Lakatos who invented the term) calls "foundationalism," and those who attempt it (the majority of twentieth century philosophers of mathematics) he calls the "Mainstream." Hersh believes that this obsession with the foundations of mathematics (1) cannot succeed and (2) more seriously, diverts attention from the way mathematics is actually practiced. It leads, moreover, to absolutist beliefs: "For the Mainstream, mathematics is superhuman—abstract, ideal, infallible."[43]

To replace these three inadequate interpretations of mathematics Hersh advocates a philosophy he calls Humanism. Mathematics is an entirely human project; its concepts and objects are invented by man in a social process. It develops in history, not in an arbitrary way but in conformity to various human needs, including the needs of science.

> Mathematical objects are a distinct variety of social-historic objects. They are a special part of culture. Literature, religion, and banking are also special parts of culture. Each is radically different from the others.[44]

Although in a fundamental sense mathematics is made and not discovered, its concepts (made by us) often have hidden consequences which mathematicians can "discover." The number system, for instance, was developed by humans but after several centuries of effort has been shown to imply Fermat's Last Theorem. Other examples which have recently been shown after enormous effort to be consequences of other invented mathematical systems include the Poincaré Conjecture[45] and the Four Color Theorem. Mathematics is thus a mixture of discovery and human creation. We call the revealing of non-obvious implications of pre-existing theories "discoveries."[46] But new theories and areas of mathematics such as Cantor's theory of infinite sets, Robinson's nonstandard analysis, Schwartz's theory of generalized functions are usually viewed as "creations."

Like Ernest, Hersh opposes "absolutism" or the view that mathematics consists of timeless truths divorced from human beings. He feels that such a philosophy has a disastrous effect on education.

---

[43] *Ibid.*, p. 92
[44] Hersh (1997), p. 22.
[45] This is the writer's example. The Poincaré Conjecture was not established when Hersh's book was published.
[46] "But then, after you *invent* a new theory you must *discover* its properties, by discovering its properties, by solving precisely formulated mathematical questions." (*Ibid.*, p. 74.)

If mathematics is a system of absolute truths independent of human construction or knowledge—then mathematical proofs are external and eternal. There're there to admire.[47]

A teacher holding such a philosophy would stress axiomatics and the shortest, most elegant, and most general proofs even if they conceal motivation and real understanding of the theorem. Such proofs are often "tricky" like "pulling a rabbit out of a hat" and conceal the motivation behind the proof. The humanist teacher, by contrast, tries to give an informal perhaps non-rigorous proof that will give understanding of what is really going on in the theorem.[48] The absolutist view that mathematics is superhuman is also correlated with an antidemocratic elitism, as the majority of students become convinced of their intellectual inferiority and inability to understand such other worldly stuff.

The public and some philosophers of science in Hersh's view think of mathematics as a collection of absolute truth because they see only the "front" of mathematics in the form of elegantly polished published papers or axiomatized theories in textbooks. The "front" of mathematics resembles a dining room in an elegant restaurant. Non-mathematicians who look at mathematics in its "served" form are like the diners who "know what's supposed to go into the ragout. They don't know for sure what *does* go into it."[49] In reality, however, mathematics is created in the "back" which is like the kitchen of the restaurant. Here before the final dish is served: "mathematics in back is fragmentary, informal, intuitive, tentative. We try this or that. We say 'maybe' or 'it looks like'."[50] This unawareness of the "back" of mathematics is what is wrong with Mainstream philosophy of mathematics. The diner cannot really understand the restaurant meal while being unaware of the kitchen. "Yet you can present yourself as a philosopher of mathematics, and be aware only of publications washed and ironed for public consumption."[51]

Concerning proof, Hersh argues that even in its polished and formal form it is more like a haphazard dish cooked up in the kitchen, than a strict formal sequence of deductions expressed in first order predicate logic, using *modus ponens*, etc., imagined by philosophers of mathematics, formalists, and logicists. Real, practical proofs found in the literature skip

---
[47] Ibid., p. 60.
[48] Ibid.
[49] Ibid., p. 36 [emphasis in original].
[50] Ibid.
[51] Ibid., p. 39.

obvious steps, have phrases in them like "it is easy to see" or "a routine calculation shows." They are simply *"what we do to make each other believe our theorems."*[52] Furthermore, if we argue that proofs *in principle* can be made certain by being translated into first order predicate logic, almost no proof of a real theorem other than trivial homework exercises can in practice be dealt with in this way. The result would be both boring and so lengthy that it would be impossible to read. And, as Hersh points out, the effort would introduce a myriad of new formal assumptions whose consistency is unknown.[53] Nor can mathematicians appeal to the infallibility of computer assisted proofs such as Appel and Haken's 1976 proof of the Four Color Theorem. Since no human mind can possibly verify such results, they are dependent on "fallible" hardware and software. In any case, trying to justify the correctness of mathematics by appeal to some infallible "method" is an illusion; just as in other cultural areas the essential criterion of "correctness" is "social." A lay person without professional training can not judge the validity of a theorem or proof. This task is reserved for a elite of qualified mathematicians. They function as as kind of an ideological politburo or Holy Office and define what passes for correct mathematics. "What mathematicians at large sanction and accept *is* correct mathematics."[54] In any event, all this means that "proofs" cannot have the absolute certainty that absolutists and formalists claim.

This real but rather unorganized and messy process of mathematical creation and proof leads Hersh to agree with Paul Ernest that mathematics is "fallible." It does not consist of a majestically unfolding, rigorously proved, sequence of eternal truths but: "Like science [it] can advance by making mistakes, correcting and recorrecting them."[55] This fallibility also means that some widely accepted platitudes concerning mathematics are myths, at best only partially correct. These include:[56]

1. *Unity*: There is only one mathematics, indivisible now and forever. Mathematics is a single inseparable whole.
2. *Universality*: If little green creatures from Quasar X9 showed us their textbooks, we'd find again $A = \pi r^2$.

---

[52] *Ibid.*, p. 48 [emphasis in original].
[53] Hersh (1997), pp. 49–50.
[54] *Ibid.*, p. 50. Hersh seems to have a "consensus theory of truth" for mathematics which parallels a common SSK view for science.
[55] *Ibid.*, p. 22.
[56] *Ibid.*, pp. 37–38.

3. *Certainty*: Mathematics has a method 'rigorous proof,' which yields absolutely certain conclusions, given truth of premises.
4. *Objectivity*: Mathematical truth is the same for everyone. It doesn't matter who discovers it. Its true whether or not anybody discovers it.

Since mathematics is a social product, it is not divorced from other aspects of culture such as politics. While mathematics can not directly express a political ideology, Hersh feels that certain standard interpretations of mathematics have historically been associated with either the political Right or Left. Plato was on an extreme right-winger in the context of the politics of his time. His dialogue *The Republic* celebrates a totalitarian politics and as Karl Popper has argued served as a blueprint for fascism. A similar orientation was present in Plato's predecessors the Pythagoreans and Eleatics who established secret societies of aristocratic youth in the Greek colonies of Southern Italy. Hersh gives a sizable list of mathematicians and philosophers, dividing them politically according to their views on mathematics. Thus Decartes, Leibniz, Berkeley, Frege (an early Nazi sympathizer), Brouwer, Thomas Acquinas, Quine, Augustine of Hippo, and Lakatos are right-wingers. Those somewhere on a leftist scale include Spinoza, Locke, Kant, Russell, Hume, d'Alembert, and Mill. A large majority of the rightists are in the Mainstream, often infected with Platonism, while those who are left-wing are what Hersh calls "mavericks" or precursors of Humanism. The correlation is a natural one. Those who believe in eternal nonphysical mathematical objects also tend to believe in an eternal, hierarchical social order. Mavericks, in contrast, who accept the social, empirical, and corrigible nature of mathematics often have similar views concerning society. However, there are exceptions. Imre Lakatos, for example, is a Humanist but became very anti-Communist after the 1956 uprising, and so Hersh puts him on the Right. Spinoza, on the other hand, with his attempt to reduce philosophy to Euclidean style deductive proofs is in the Mainstream, but left-wing in his politics.[57]

In opposing "absolutism" and concluding that mathematics is a fallible social construction whose certainty cannot be definitely established, both Paul Ernest and Reuben Hersh are influenced not only by the conventionalism of Wittgenstein, and SSK,[58] but perhaps more strongly by the work of Imre Lakatos (1922–1974) who offered a deeper and more penetrating critique of mathematics than any of the authors we have discussed so far.

---

[57] *Ibid.*, pp. 238–245.
[58] These influences are more apparent in Ernest's writings than Hersh's, however.

Born into a Hungarian Jewish family, Lakatos had studied mathematics and philosophy in Debrecen, Budapest and Moscow. After his education he became an important official in the ministry of education in the communist government before being imprisoned under harsh conditions as a revisionist from 1950 to 1952. After the 1956 Hungarian uprising he sought refuge in Britain. He studied with Karl Popper and obtained a Cambridge Ph.D. in philosophy in 1961. From 1960 until his premature death of a heart attack in 1974 he held a position at the London School of Economics. As a young man he had an unwavering faith in Communism (which was a factor helping him survive imprisonment), but in Britain he became more conservative and opposed student unrest over Vietnam which he felt threatened academic freedom.[59]

Lakatos' views on mathematics were expressed in his book *Proofs and Refutations* (1976) which was based on his doctoral thesis and in a few journal articles. Lakatos believed that all attempts to demonstrate the absolute validity of mathematics as, for example, by grounding it in logic as in Whitehead and Russell's *Principia Mathematica* or by Hilbert's program of formal axiomatization had ended in bankruptcy. Such efforts ran into the problem of infinite regress, set theoretical paradoxes, or the limitations entailed by Gödel's theorems.[60] For Lakatos mathematics is "quasi-empirical" and "conjectural," in these respects similar to physics or any other high level science. This view allowed Lakatos to extend several of Popper's ideas concerning science to mathematics. Recall that Popper believed that a scientific theory can never be considered as certain knowledge. A theory is a conjecture (the bolder the better) and the goal of scientific practice is its refutation. Lakotos considers that the same is true of mathematical theorems. Just as conjectures yield theories in science they yield possible theorems in mathematics. Negative results of critical experiments can refute conjectured theories and counterexamples can refute conjectured theorems. It follows that progress in mathematics does not consist of a sequence of deductively proven of eternally true results but is a dialectical process of conjectured theorems followed by counterexamples. Counterexamples give insight into "hidden lemmas" or unacknowledged assumptions in its proof. When these are exposed a new and better conjecture is advanced which may lead to new counterexamples or even new fields of study.

---

[59] A more detailed account of Lakatos' life quoted directly from a London Times obituary is given in Hersh (1997).
[60] See in particular the papers Lakatos (1978a) and Lakatos (1978b) for Lakatos' views on foundations and mathematical epistemology.

Lakatos illustrates this process though a "rational reconstruction of the history of the Euler-Descartes formula which relates the the number of edges (E) to the number of faces (F) and vertices (V) of a polyhedron:

$$V - E + F = 2. \qquad (1)$$

A simple intuitive argument due to Cauchy in 1813 for this formula is easy to obtain. We first remove one face of the polyhedron. Since this does does not change the number of vertices and edges, $V - E + F = 1$ for the remaining structure if and only (1) holds for the polyhedron. Then we flatten what is left into a polygonal figure and triangulate it so that each face is the union of (possibly curvilinear) triangles. Next it is possible to remove remove one triangle at a time so that $V - E + F$ does not change. We end up with a triangle where $V - E + F = 1$. Since each step is reversible, (1) follows. The trouble with this informal proof is that the formula works only for so-called "regular" polyhedra. If a polyhedron is merely defined as a solid with polygon faces, then two tetrahedrons joined at a vertex is a counterexample since

$$V - E + F = 7 - 12 + 8 = 3.$$

There are other counterexamples as well: Consider a cube with another smaller cube removed from its interior. Here $V - E + F = 4$. It took decades of effort by mathematicians to characterize which polyhedra satisfied Euler's formula and which had different "Euler characteristics."

Lakatos gives a similar historical account in the development of analysis. It is now an elementary theorem that a uniformly convergent sequence of continuous functions converges to a continuous limit function. However, in 1821 Cauchy apparently proved that pointwise convergence of continuous functions guaranteed the continuity of the limit. This agreed with the intuition of all previous and contemporary mathematicians. But the difficulty of Cauchy's result was that it contradicted Fourier's earlier demonstration in his *Mémoire sur la Propagation de la Chaleur* (1812) that a sequence of trigonometric functions could converge to a discontinuous limit. There was a confused discussion of the problem by several mathematicians lasting more than a generation. Only in 1847 Seidel showed that Cauchy unconsciously assumed a property that what we now call "uniform convergence."

Lakatos believed that these and other illustrations show that the history of mathematics, at least as rationally reconstructed, follows a pattern of conjectures followed by refutations which means that mathematics as it develops is always "fallible" and "corrigible." But this is concealed by the dead hand of "deductivism" (clearly the analogue of Ernest's absolutism and Hersh's formalism) where mathematics is presented as an array of eternal static truths deduced in the Euclidian manner from definitions and axioms. Of course, once a theory is axiomatized, a proved theorem *will* follow from the axioms. In this sense the theorem is "true," and there will be no counterexamples. But the axiomatization is the least interesting part of mathematics and plays no significant role in mathematical discovery. Also, it may be premature and cause interesting results and new directions to be missed. Moreover, deductivisn is educationally unsound because the real motivation *behind* the theorems is concealed. Much more interesting and vital is the process of conjecture and counterexample which is the real essence of mathematics. Thus Lakatos has the same negative opinion of the role of deduction as Popper has of induction. For Popper induction plays no valid role in the logic of scientific discovery, and for Lakatos neither does deduction in the logic of mathematical discovery. By failing to understand this historians of mathematics miss the complicated and fallible development of their subject. Just as scientists do, they write "Whig" history where past mathematics is viewed as an inevitable deductive progression to contemporary results, only temporarily hindered by the mistakes of past mathematicians.[61]

---

[61] These opinions and Lakatos' historical examples are to be found in Lakatos (1976).

Chapter 11

Epistemic Issues

Since the dialogue *Theaetetus* philosophers have argued about what knowledge is, how we know we have it, and whether or not knowledge is possible. Since they have never agreed on any of these issues, it may be too ambitious a goal to demolish all the theses shared in common by SSK and PIS—at least to the satisfaction of all reasonable minds—as decisively as mathematicians have demolished circle squarers. Insofar as epistemological ideas which deny science access to objective knowledge of reality are based on some form of skepticism they cannot be definitively refuted. After all, the skeptic might argue as follows:

> Since we might really be brains in a vat, or be perpetually dreaming, or be systematic distorters of what enters through our senses, the only correct conclusion is that we do not know what we unthinkingly take ourselves to know. The concept of knowledge must therefore be eliminated, and some less demanding description of our epistemic powers devised.[1]

Proponents of the SSK/PIS position would certainly accept the conclusions of this argument, but—so far as we know—they have not yet claimed that scientists are "brains in a vat." Some, such as Steve Woolgar, take an idealist position—only representations exist. But the strategy of others, probably the majority, is, as we have seen, to appeal repeatedly to a set of variations on a metaphysical idea due to Kant. Since they are not by and large trained philosophers, the underlying metaphysics is not developed in any rigorous systematic way. But the essential idea is as follows. Our world—what we call "reality"—has two components: There is first of all the world of what is variously called sense data, appearances, phenomena, or

---

[1] McGinn (1993), p. 113.

representations. To the naive realist (which includes most scientists) this *is* the "real world" of sticks and stones, cabbages and kings, cats and dogs, advocates of PIS, and so forth. This world is mind-independent. Many of the objects in it exist independently of human perceptions. Dinosaurs, for example, existed long before human beings discovered them; Mt. Everest and (probably) cockroaches will continue to exist long after our species has vanished. To Kant, however, this phenomenal world is *not* mind-independent. In essential respects it is constituted by the human mind via the imposition of *a priori* concepts such as space, time, number, and causation. Kant, as Michael Devitt observes, was not a relativist since he held that these concepts were common to the human race.[2] On the other hand, we do have an intuition that there is something underlying the phenomenal world which *is* mind-independent. We can make a distinction between the world of our perceptions and the world as it is "in itself," which Kant calls the noumenal world and assumes to exist in order to avoid the idealism of Berkeley. An element of this world is a "thing in itself" (*Ding an sich* or *res in se*). By definition the contents of this world are unknowable. But they supply the "raw material" which is somehow molded by human *a priori* concepts into the familiar world of appearances. Various strands of SSK and PIS have modified this Kantian doctrine in different ways. But they all abandon Kant's belief in universally human *a priori* concepts which give meaning and structure to the world. Instead, this role is taken over by "concepts" appropriate to a local group. Thus, for Ludwik Fleck the phenomenal world is essentially the product of a thought collective; for Kuhn, it is shaped by the shared paradigm of a group of normal scientists, and is "caused" by the unknowable "stimuli" which correspond to Kant's noumenal world. To Feyerabend, the same shaping function is performed by the world view of some "tradition" which may differ from that of another tradition; for example, the tradition of a modern atomic physicist produces a different phenomenal world than that of a shaman or priest of the Olympian gods. For SSK and its allies "nature" is the noumenal world and the perspectives due to the scientist's gender, race, social class, skill at laboratory politics, or the classifications and methodologies permitted by his society or paradigm generally determine the scientist's "concepts," (or "theories") and thus his phenomenal world of laboratory results and "facts." And in more general nonscientific realms studied by anthropologists or sociologists this kind of Kantian world construction may be due to the belief system of a culture or

---

[2]Devitt (2001), pp. 144–145. I am indebted to Devitt's essay for several ideas presented in this section.

the ideology of a social group. Let us denote this family of doctrines having the common feature that science or other belief systems are somehow invented by us, and do not reflect what is actually given in the world by the term "Constructivism."

Epistemic relativism and the possibility of incommensurability are immediate consequences of Constructivism. Since the scientist does not have contact with things as they actually are but only with a phenomenal world which he or she wholly or partially "constructs," "truth" is relative to this "world" and may differ from group to group or belief system to belief system. To express this idea more precisely, the nouns and verbs of some statement S have meaning only in respect to a phenomenal world which is constituted by some paradigm, world view, etc. The same words in the negation $\neg S$ may have different meanings in a "different world" with the consequence that both S and $\neg S$ may be "true" without contradiction in these worlds.[3] For similar reasons, according to Kuhn's semantic notion of incommensurability, scientific theories formulated within different paradigms may not be comparable with each other since they may refer to differing concepts or events in different, incompatible worlds. Key descriptive words in both theories may be the same, but have different meanings and imply different world structures. Even if there is some mathematical relationship between the two paradigms—such as, for instance, Newton's mechanics being a limiting case of Einstein's—the terms in the formulas appropriate to Newton denote entirely different entities than the formulas from which they are derived in relativity theory. This kind of incommensurability, as the constructivists happily note, is present in every aspect of human thought as well as in science. It is especially common in political argument. For example, consider the discourse of a Salafist Muslim and an ardent Zionist or the political analysis of a Michael Moore and an Ann Coulter. Just as in the case of rival Kuhnian paradigms, the opposing sides may be said to live in different worlds; it follows that there is little *epistemic* difference between science and any other belief system; all are purely human constructions imposed on an unknowable world which is in Frederick Jameson's words a "formless chaos of which one cannot even speak..."[4]

There are several other consequences of Constructivism concerning science. The most obvious is scientific anti-realism. If the ordinary "representations" that make up our world are in part constituted by the contents of our minds, this is certainly also true for the world of non-observable

---

[3] See Chapter 6 for Ludwik Fleck's concept of truth which is similar.
[4] Quoted in Devitt (2001), p. 147.

scientific entities such as muons, quarks, and electrons. These are the pure creations of theory. Furthermore, since a scientific theory has nothing to do with the noumenal world, it cannot be "true" in the usual sense of correspondence. To what state of affairs do the claims of a theory—especially one involving unobservable entities—correspond? How would we recognize such correspondence if it existed? We can only compare one theory with another, not a theory with "reality." In such a situation how should scientific claims be evaluated? If one wants to appeal to more than the consensus of scientists making the claim (perhaps induced by social or political factors)[5] the only "objective" answer is to invoke pragmatic criteria. The noumenon may, if it exists, be unknowable, but we personally[6] or socially "construct" statements about it. If these statements "work" or have good consequences for us, they are accepted. In the case of science this leads to some form of instrumentalism or "empirical adequacy." Acceptable scientific theories allow us to organize and control our phenomenal world. Or from the more sophisticated point of view of Kuhn they allow scientists to "solve" ever more elaborate puzzles.

What can we say about these various neo-Kantian ideas which underly the Constructivism common to SSK and PIS? Let us admit we cannot show that they are false. Even if we reject these specific later modifications of Kant, it is possible that the world is wholly or in part determined by the human mind. Martians, if they exist, may live in a totally incommensurable world with respect to ours; their minds and sensory organs might be so different from ours that nothing in their phenomenal world would be understandable to us. Perhaps, for instance, they might "see" only gamma rays and not the spectrum of colors available to us. Nevertheless, we can identify weaknesses which will demonstrate that Constructivism is at least as problematic as the scientific realism it attempts to replace. First of all, what reason is there to believe in the existence of this mysterious noumenal world? By its definition, we know *nothing* about it and have *no* contact with it. It is outside time and space and all the *a priori* concepts by which humans construct the phenomenal world. What can its connection to this

---

[5]This alone, however, is a sufficient reason for acceptance for much of SSK. Going beyond "consensus" would be an illicit appeal to non-sociological reasons for belief.

[6]The "personal construction" of knowledge is the position taken by the "radical constructivism" of Ernst von Glasersfeld (see e.g., von Glasersfeld (1995)). Like SSK von Glasersfeld rejects a correspondence theory of truth on the grounds that reality is inaccessible to us. Rather individuals construct their own "truth" to help manipulate their environment to achieve their goals. Such constructions may be judged not by their truth, but by whether or not they are "viable."

world possibly be? It will not do to say that it is a "cause" of the phenomenal one, since as Devitt has noted "causation" according to Kant is an *a priori* concept imposed by us.[7] Thus the noumenon is a far more vacuous and empty concept than those of quarks, neutrons, or electrons—which in fact explain in a way we can understand many things in the phenomenal world— and yet are said by SSK to be social constructions. Why then should we not also view the idea of an unknowable world-in-itself distinguished from a mind-dependent world of appearances as a social construction? It seems to merit such a label more than a quark or electron. One motive to postulate its existence is probably to provide a constraint to scientific theorizing and to explain the "resistance" that normal science encounters in making a theory "fit" the phenomenal world.[8] But how can an object with no known properties at all exercise such a constraint? We have to admit that the noumenon is a completely redundant idea with no content or role to play. We should therefore abandon it. All that is left (as Woolgar maintains) are representations. Since they are in fact *my* representations, it is hard to see how solipsism (not just idealism) can be avoided. As we have previously pointed out it is no solution to claim—as SSK does—that representations are *socially* constructed. Other individuals, minds, and society in general are part of my phenomenal world and therefore just as constructed as anything else in it. Solipsism, of course, is irrefutable and the existence of a mind-independent world is unprovable. It is possible that Berkeley's logic is correct: all that exists are my representations. But if this is the case for whom am I writing this essay and for whom is Woolgar writing his book? As Searle has observed, asking for a proof of a mind-independent reality is rather like some one asking for a proof that logical reasoning is valid. The very demand presupposes the thing in question. As soon as we try to

---

[7] Devitt (2001), p. 147.

[8] Another may be psychological and based on a fundamental world view of science itself which has been dominant at least since the seventeenth century and can be traced back to the ancient atomists. For Aristotle and his scholastic followers qualities such as color, taste, odor, etc were *real* attributes of an object which we directly perceive. For both seventeenth century corpuscular philosophers and modern scientists they were "secondary qualities" or sensations processed in the brain from signals transmitted by the nervous system and caused by light waves, various molecular motions or vibrations, and so forth. These causes alone were "real" in the sense of being mind-independent and quite different from the sensations they produce. This conception may be the psychological origin of the distinction between the phenomenon and noumenon. Paradoxically, however, these unobservable physical causes of our sensations have been demoted into mere "constructions" by scientists and their role has been taken over by a concept empty of all content.

communicate with anyone–for instance, to ask for a proof of an external reality we are presupposing that that person is part of an external reality which is not just a collection of representations.[9]

While certain forms of Constructivism may imply incommensurability, we should note that the reverse implication does not hold. Even if we reject Constructivism it is possible that theories may in some sense be incomparable. Even if there is a a mind-independent world, perceptions of it by scientists may differ radically. Sketchy and ambiguous evidence can lead to incommensurable interpretations. This is not necessarily the result of "theory-laden" observations caused by different paradigms. If the evidence was clear and complete then all scientific observers might agree on its meaning and significance. But suppose it is fragmentary and seems to have contradictory aspects; in this case the human mind will have difficulty making sense of it in a uniform way. One scientist may stress one part, and "explain away" another part; whilst his colleague may do the reverse. To give an analogy suppose we find an incomplete papyrus in the Egyptian desert seemingly written by a minor Greek author in very bad condition and composed in a poorly understood Ionian dialect. A beginning fragment, moreover, appears to contradict something at the end. Are there actually two works by the same author in the papyrus? Two authors? Would a more complete version remove the contradiction? Or perhaps we just have a mistranslation? Such problems may generate bitter disputes among classicists, producing the intractable disagreement characteristic of incommensurability, but without being the result of any form of Constructivism. The evidence is simply ambiguous and reasonable people may disagree; also unfortunately, psychological factors common to the primate species *homo academicus* may then turn these disagreements into all out warfare. We suspect that many scientific disputes, especially in the early days of theory construction can be of this type and have nothing to do with incommensurable paradigms.

One radical type of incommensurability introduced by Kuhn which is more specific than than the rather vague concept expounded in his early work is "semantic."[10] Kuhn, as we have seen, argued in his later writings that the basic terms and concepts of one paradigm can not be translated into the terms and concepts of another. Even if some of the words are the same, they have different meanings and are associated with different world views. "Mass," for example, is different for Newton and Einstein. The use of

---

[9]Searle (1995), p. 194
[10]See also the discussion in Chapter 7.

the word in both paradigms as if it has common meaning creates misunderstandings. Similarly it is almost impossible to translate Aristotle's *Physics* so that it is intelligible to a modern physicist.[11] Kuhn's conclusion, almost certainly influenced by the Sapir-Whorf thesis, is that the language appropriate to different paradigms organizes the phenomenal world in different incommensurable ways just as Navaho and English do. Although Kuhn was a Constructivist in our generalized sense, this form of incommensurability is evidently consistent with scientific realism since it is possible to argue that the semantic aspects merely distort nature instead of "constructing" it.

But does semantic incommensurability actually exist and if so is it significant? The difficulty with Kuhn's thesis is that while it may be true for a high-energy physicist—untrained in Greek philosophy—trying to read Aristotle, the linguistic barriers may be overcome. One may become bilingual. Historians of Science, including Kuhn himself achieve this every day when they explain archaic theories and world views in a way that the modern reader can understand. Also, by necessity, the originators of a new paradigm are always thoroughly familiar with the old one. Galileo, since he followed the standard university course at Pisa, knew his Aristotle thoroughly; Kepler, Descartes, and Newton certainly did as well. It is not that they misunderstood Aristotle—they just thought he was wrong.

In a purely *practical* sense, however, situations mimicking incommensurability (that is, the impossibility of settling quarrels by appealing to common standards) *do* seem to exist. We have already touched on one kind arising from attempts to make sense of fragmentary incomplete evidence which may be consistent with different interpretations. There is also another form which we have all experienced in arguments with others. For example, consider a dispute between a member of Hamas and some ultra-orthodox settler on the West Bank concerning Palestine. Now it is tempting to argue that that they live in semantically "incommensurable" intellectual and emotional worlds, and that their situation is analogous to proponents of different paradigms in a period of scientific revolution, so that not even in theory does there exist a way to translate one paradigm intelligibly into another. To say this *sounds* profound and seems to support epistemological relativism. But the real situation is more pedestrian. The premises or interests of the contending parties may just be too far apart or different at too many points for their quarrel to be rationally settled in a finite time even if the task is in principle logically possible. As an

---

[11] As indeed Kuhn discovered for himself when he tried to read Aristotle in preparation for the course he was organizing for Conant at Harvard.

(admittedly unreal) thought experiment we suppose that both share the same respect for logic, the same basic ethical system, and a willingness to look impartially at evidence. Consequently there seem to be no deep philosophic reasons why they may not ultimately be brought to agreement, and indeed become good friends. This indeed is the hope of UN mediators and Middle Eastern specialists. But as a practical, common sense matter this is not going to happen. Even if they can be prevented (with difficulty) from immediately killing each other, the discussion would involve a near infinite number of differences over the interpretation of historical events, theological claims, values, psychological disposition, material interests, the existence and role of the firms of Allah, Yahweh, or Jehovah and Son in the real estate business. Even if we suppose that each point can be individually settled, there are just too many of them; every point that is resolved would also be immediately replaced by another larger point of contention *ad infinitum*. To take another example, suppose we have an argument between a Young Earth Creationist and a Darwinian and the former unexpectedly admits that the conclusions that the latter draws from the fossil record are convincing, but goes on to claim that this is only because the Devil created the fossil record so that men who believe in it should be damned. If now the two parties attempt to investigate and resolve the argument, they will be drawn into a bottomless swamp of theology, beginning with a proof of the existence or nonexistence of the Devil.[12] Especially in political and religious matters such deep differences in premises which breed endless regress in argument probably reflect rival and contradictory interests compounded by emotion and egoism. (One cannot imagine, for instance, the Hamas member or the settler, each regarding the West Bank as "home," agreeing with the other's claim.) Again we repeat that in each of these cases the contention is not a case of some kind of semantic incommensurability between mutually unintelligible paradigms. Both parties understand each other's language and concepts *perfectly* (indeed, with respect to the Hamas member and the Israeli settler that is the main part of the problem).

---

[12]Exactly the same kind of impasse was evident in sixteenth century Germany from scholastic disputes between Catholics and Lutherans which were sometimes held under the auspices of the local prince. Each side would be represented by well trained theologians who shared a common university training in philosophy and logic. Their syllogisms would be perfect. But the opponent would deny the premises. In this way there would be what amounted to an infinite regress, but not just in a simple linear way. To clarify the differences in premises, distinctions and other premises would be introduced. The result would be a constantly growing nest of conflicting premises which could not be evaluated in a finite time.

Fortunately, the disagreements between opposing scientific theories are usually not as profound as in politics or religion. Rival interests concerning status, recognition, and tenure may be present, of course; but they do not usually breed the all consuming fury of disputes over land or salvation. Also, scientists on both sides of the issue may share many of the same technical premises. But, as we have pointed out, the evidence may at least initially be obscure and subject to multiple interpretations; consequently there may be good and rational arguments on both sides. The "wrong" side may at first even have the better of the argument. As Feyerabend and others have claimed, such may have been the case in the dispute between Copernicans and traditional Ptolemaic astronomers around 1600. The entire body of accepted physics and philosophy (not to mention the tower argument and the paradoxical consequences of Copernicanism concerning the size and distance of the stars) seemed to support the Ptolemaic view. Nor was the Copernican system superior in accuracy. By way of reply, all the Copernicans could initially offer were Pythagorean speculations and mathematical elegance. Similarly, the remarks of Fleeming Jenkin, a British engineer, on heredity and the arguments of Lord Kelvin on the age of the Sun seemed to be fatal objections to both Darwinism and uniformitarian geology.[13] Many similar examples may be found in the historical record. In general, this kind of incommensurability comes at the early stages of a theory when the evidence is sketchy and poorly understood. In such circumstances appeals to philosophical considerations and values may be made. It may be agreed by all concerned that a theory should be accurate, fruitful, have a wide scope, and be simple. But *which* value is the most important in the specific case may be unclear. It would, of course, be naive to deny that rival interests may not also play a role in the dispute—no one likes for his own interpretation be proven wrong and suffer professional demotion as a result. The natural human response to a challenge to one's own interpretation— especially by those one considers academic inferiors—is to circle the wagons and break out the Winchesters. In these circumstances it is not surprising that agreement in the short run may be impossible.

In the light of this common sense analysis, the dream of Leibniz that fundamental disputes in science, religion, and politics can be settled rationally

---

[13] Jenkin argued mathematically that the hereditary effects of individual variations would be so diluted in subsequent generations that they could not possibly modify populations. Kelvin "proved" on the basis of accepted physical principles that the Sun was only a few tens of millions of years old, which did not allow enough time for Darwinian evolution or for uniformitarian explanations of geological change.

by a process akin to calculation seems naive.[14] But the "practical incommensurability" that we observe every day is a contingent phenomenon, not necessarily a consequence of some relativistic epistemic doctrine or deep semantic principles. In the special case of science disputes are sometimes settled fairly quickly; at other times rival groups of scientists seem to be speaking past each other and no real communication between them can be achieved, perhaps for many years (and woe to the untenured assistant professor in a department dominated by the other faction!). The causes in such cases may, as we have already hinted, may range from the ambiguity of evidence that is subject to conflicting interpretations to personal factors such as ego involvement, a struggle for scientific power, pure pigheadedness, or even who is sleeping with whom. In these circumstances we admit that victory by one side or another may indeed be won by charisma, laboratory politics, homicide, or other sociological factors as SSK maintains; but the victory will not be permanent unless the theory really *does* work and continues to be supported by the evidence. Unlike some strands of SSK we do not believe that the victors in a scientific dispute can indefinitely control and shape the flow and interpretation of evidence to their liking so that their theory is forever victorious. Scientists lack the talent of a Karl Rove or Lee Atwater in this respect. But to see what is going on in these situations, we need a case by case analysis, not a philosophical doctrine.

We should also realize that the "paradigm shifts" and "different intellectual worlds" or impossibly esoteric experiments that so fascinate SSK/PIS are fairly rare. Much of the time science is "normal," and simple experiment, observation, or straightforward analysis is decisive. Well-designed clinical trials in medicine reveal that certain therapies are effective and others are not. Wind tunnels and the complex mathematics of aerodynamic theory weed out bad aircraft designs. Doubtlessly a scientifically literate reader can think of many other examples where scientific facts are arrived at in a fairly direct, noncontroversial way.

We pass now from general philosophical issues to a detailed criticism of the Strong Program. At first glance, there is nothing objectionable about

---

[14]Leibniz believed it possible to resolve intractable quarrels in every field in a purely algorithmic way. The theses of both parties would be translated into a quasi-mathematical universal philosophical language which he called the *Ars Characteristica* or *Lingua Philosophica*. This language would be very precise and not subject to the ambiguities of the natural languages; in addition consequences could be exhibited by a mechanical computational process. An error in religion or politics would then be like an error in arithmetic. Catholics and Lutherans could sit down together and settle their differences by saying: "Let us calculate."

items (i) and (ii) of the Strong Program. Past or present beliefs about science (or anything else) certainly have causes, and it is the function of psychologists, sociologists, and historians to find them. Sensitive historians of science, are indeed "impartial" about past scientific theories, whether or not they are now considered correct or mistaken. In order to understand Aristotelian cosmology, Cartesian vortices, the concept of the aether, or Robert Koch's discoveries in bacteriology it is important to try to understand the theory in question "on its own terms," and not to evaluate it using our present knowledge. This is standard historical practice in other areas as well; for example one does not *understand* Bismark's policies between Olmütz and Sedan by arguing that their ruthlessness prepared the way for Hitler (although this judgement may be true). Similarly, Chamberlain's policy of appeasement may have been mistaken and the cause of terrible consequences; it is certainly legitimate to point this out. But such judgement is completely separate from the task of understanding what motivated Chamberlain and what exactly he was trying to do. (iii) also as we have seen is in part a sound methodological principle. In many cases as has already been pointed out "rational arguments" support both sides of a scientific argument, especially in the initial stages. The evidence is usually incomplete and capable of contradictory interpretations. The authority of established scientific principles (perhaps later shown to be false) may support one side or another. Defenders and opponents of a new theory use complicated mixtures of fact and theory, some of which turns out to be correct and some mistaken. An excellent example of this, as pointed out above, was the controversy over Copernicanism from 1543 through the mid-seventeenth century. If arguments are reasons for the affirmation or denial of a belief, it is certainly the same type of "cause" regardless of the correctness of the belief according to present standards. Sometimes, however, the "causes for belief" can be different. If A appeals to logic and observational evidence to support a theory on the nature of sun spots and B appeals to a personal revelation from the sun god Elagabalus in opposition to A's theory—and these actions are the causes for their respective beliefs, then the causes are obviously different in kind.[15] To give a simpler example consider two persons, each of whom believes that there is a chair in front him. In the first case, the person is normal and the reason for his belief is

---

[15] Note that we are not necessarily assuming that A is correct and B mistaken; it is possible that Elagabalus, being a sun god, has an intimate knowledge of the Sun's chromosphere which is far more reliable than mere human logic and observation. All we are claiming is that the causes for belief may differ.

that the chair is really there. The second person, however, is having a drug induced hallucination which is the reason for his belief.[16]

The difficulty in the Strong Program lies not so much with the mere statement of directives (i)—(iii) but with what SSK does with them. David Bloor, in particular, while admitting that other than social causes (e.g., biological causes such as mental hard wiring for certain types of argument) may play a role in belief is unwilling to admit that logic, rational argument, evidence, or perception of truth may be the *real* causes of belief. At best, these factors amount to self-deception; at worst, appeal to them is pure propaganda. In all cases they reflect more profound social causes. Moreover, since the symmetry postulate (iii) forbids that we explain belief in accepted scientific theories by appealing to their rationality while seeking sociological or psychological causes for theories that are considered "wrong," the same general type of social causes must explain *all* beliefs whether "true" or "false" or "rational" or "irrational." Only this thesis Bloor argues permits a truly scientific approach to the genesis of belief. Up to a point Bloor is perfectly correct. Social causes of various sorts, along with human brain structure and neural chemistry are in some sense general causes of belief. If Newton had attended a madrassa in Afghanistan instead of Cambridge he might have become a great but heretical Muslim theologian, but he would not have written the *Principia*. Nim Chimpsky[17] because of the differing organization of his brain has only a few of the thoughts of Noam Chomsky, and probably none of the politically radical ones unless they involve bananas. But as in the case of SSK and the notion of "social construction," Bloor and other advocates of the Strong Program want to go far beyond such trivial observations. Their work never seriously considers the actual *arguments* advanced by scientists past or present on behalf of theories as a reason for acceptance.[18] In Marxist terminology these are just a "superstructure", consciously or unconsciously masking more fundamental social phenomena which in nearly every case turns out to be *political*, a thesis that Bloor and company support by numerous "case studies." Thus Newton is

---

[16] This example is due to Newton-Smith and is given in Bird (2000), p. 215.

[17] A chimpanzee employed by some specialists in primate cognition at MIT.

[18] This, however, is not always the case among others of the SSK school who are not associated with the Strong Program. Two notable exceptions which do not neglect scientific arguments are Collins and Pinch (1988) and Pickering (1984). Both books find it necessary to explain the ideas they consider in technically competent detail if only to illustrate "experimental regress" in the case of Collins and Pinch or as Pickering tries to show that the success of an advanced nuclear theory may be explained by the institutional interest of the researchers involved rather than by the theory's "merits."

variously "explained" in terms of the needs of rising British capitalism or Whig ideology. Both Newton's theory of matter and Boyle's corpuscular philosophy are said to originate in their fear and opposition to the social upheavals of the Puritan revolution and especially to the leveler concept of an *anima mundi* informing matter. Likewise, it is claimed that Hamilton's dislike of Peacock's algebraic formalism was motivated not by mathematical values, but by Hamilton's political and religious conservatism. Pasteur's opposition to spontaneous generation is supposed to have been a consequence of his desire to accommodate the ideology of political and religious conservatives during the Second Empire. And the Weimar physicists created quantum mechanics because they wished to pander to a post World War I intellectual climate hostile to causality and determinism, thereby recouping the social prestige they had lost after Germany's defeat, and so forth and so forth—such examples may be multiplied nearly *ad nauseam* in the Strong Program/SSK literature.

What can we say in response to all this? It would be foolish to deny a linkage of social or political factors to all human thought, including science. In the case of some scientific questions which directly involve political values, this linkage can be both obvious and profound. (We will discuss this situation in some detail in Chapter 14.) Yet in most cases, scientists correctly think of themselves as rational observers of nature, not as interest driven propagandists. They are certain that they believe the scientific theories they do because of good arguments and evidence. Is this merely a rationalization? To be sure, the extra-scientific factors, if there are any, may supply metaphors and general concepts perhaps on an unconscious level. Darwin did not "invent" Natural Selection to justify the excesses of British Capitalism or to protect his status as an independently wealthy rentier. Darwin was genuinely interested in the problem of the origin of the species. But Malthus and general observations of the British economy may have supplied him with a vague framework of ideas which he could sharpen and apply to his problems. Similarly, the Weimar physicists were faced with several longstanding puzzles involving the atom which they could not explain satisfactorily by the methods of nineteenth century classical physics. The approach they chose may have been made more palatable via the fashionable emphasis on acausality in post-World War I German culture. But there is no evidence that the physicists deliberately thought: "we will concoct an acausal physics—who cares if it is correct—in order to improve our social status."[19] Pasteur's opinion on spontaneous generation may have been naturally correlated with his conservative Catholicism. But he sincerely believed that it was impossible and gave rational reasons for his position.[20] As Roll-Hansen has shown, in the actual controversy with Pouchet between 1859 and 1864 Pasteur (unlike Pouchet) never mentioned political or religious values, and in a number of public speeches was careful to separate religion from science. He argued that science might indeed support religious tenets (which was a good thing) but religion could in no way establish scientific facts.[21] Whatever his political and religious beliefs, the many intricate and ingenious experiments Pasteur designed and carried out to test the hypothesis of spontaneous generation were far superior in acuteness and mastery of scientific method than Pouchet's experiments. This was so obvious that the official committee set up by the Paris Academy of Sciences had no choice in their verdict for Pasteur.[22]

We are dealing here several levels of explanation. Consider the following statements:

(i) Newton created the *Principia* on the basis of elaborate mathematical arguments and empirical evidence.
(ii) Newton created the *Principia* as a means of sublimating feelings of aggression, sexual frustration, and abandonment because of his mother's remarriage after his father's death to a hated stepfather.[23]
(iii) Newton created the *Principia* because (like Boyle) he feared the lower class rabble, wished to destroy the foundations of Catholic theology, support the Whigs, and to become England's court philosopher.
(iv) Newton created the *Principia* because of a certain biochemical state in every neuron of his brain.
(v) Newton created the *Principia* because of the total state of the universe (or at least the solar system) just before he began to write it.

---

[19] For a detailed critique of Forman's thesis emphasizing professional and internal factors in the development of the new quantum mechanics after 1924 see Kraft and Kroes (1984).
[20] Farley and Geison strongly hint that Pasteur's conscious motivation in opposing Pouchet was entirely political and motivated by a desire to "suck up" to the Emperor. This would make him no better than a cynical propagandist. While Pasteur's opposition to spontaneous generation was consistent with his explanation of fermentation, it was, they claim, inconsistent with his earlier work on crystallography which may have favored the possibility of spontaneous generation. I am not sure that this conclusion follows. But it may be the case that Pasteur simply changed his mind on the issue, something which happens frequently in science.
[21] Roll-Hansen (1979), p. 275.
[22] *Ibid.*, pp. 281–89.
[23] This is close to the thesis of Manuel (1968).

Suppose further that each of the above statements is true and can be justified on the basis of an elaborate investigation. Which cause of belief should we prefer? The Strong Program would accept "causes"(ii)–(v) but would probably emphasize (iii). These explanations all have different virtues. Certainly, (ii) through (v) (assuming a correct psychoanalytic theory behind (ii)) are listed in order of increasing generality. Clearly, (iv) and (v) are the most "scientific." On a trivial level they are immediate consequences of the determinism which Bloor would accept. In the present state of our knowledge they are unfortunately rather uninformative; but if it could be shown exactly *how* they imply Newton's construction of the *Principia*, say, by the year 2700, then this would be an incredible triumph of science. (ii) would be fascinating to a psychoanalyst. (iii) would reinforce a cynical attitude that science is just another power hungry ideology. Note also that as generality increases we become ever more distant from the actual content of the *Principia*—although in theory, perhaps, an ingenious psychoanalyst could make some specific contact between (ii) and Newton's treatment of certain forces, rigid bodies, and fluids. If David Bloor or George Grinell are correct, (iii) gives an explanation of Newton's theory of matter and his Copernicanism, although of none of the details associated with these theories. (iv) and (v) tell us nothing about the content of the *Principia* although it might be possible as a macabre scientific exercise in the future to make a (fragmentary) copy of it by an analysis of the state of every neuron in what remains of Newton's brain based on an exceedingly sophisticated postmortem examination conducted in Westminster Abby. But at present, only by investigating (i) and going through Newton's arguments can we reach any true understanding of the intricate mathematical physics in the *Principia*.

To generalize this example, we admit that explanations similar to (ii) or (iii) (but alas not yet (iv) and (v)) may in a certain abstract sense "explain" a scientific belief. Background sexual, ideological, or social factors may suggest, motivate, or reinforce a scientific hypothesis and some detailed arguments for that hypothesis. But while it may be interesting to know (ii) or (iii), they really tell us nothing about the *science*. There may be a correlation between sexual neurosis, political ideology, social influences, etc., and the actual content of scientific theories. But by not seriously considering the arguments given on both sides of scientific issues historians or sociologists of science miss what is really important, an actual *understanding* of the science. Newton was one of the world's greatest mathematicians, in the same league as Archimedes or Gauss. His geometric arguments based on the classical tradition of Euclid and Apollonius have a transcendental beauty. Likewise, his analysis of the tides, demolition of Descartes' theory of vortices, determination of the motion of bodies in resisting fluids, study of the isochronous pendulum, or even his erroneous derivation of the speed of sound in air,[24] all of which are found in the *Principia*—whatever the political or psychoanalytic motivations which may possibly lie behind them—are monuments to the human intellect, comparable in their way to the greatest literary or artistic achievements. But to do justice to *these* aspects of Newton would require some genuine knowledge of mathematics or physics which is often lacking in a sociologist of science and certainly not evident in the writings of Patricia Fara or George Grinell. Therefore, it is easier to concentrate on Newton's politics and to argue that his view, like Boyle's, of the nature of matter was somehow conditioned by a fear of of the lower classes or entwined in Restoration politics. Understanding Newton's actual achievements is a prerequisite for any "deeper" investigation, and trying to deal honestly with it should not be dismissed as Whiggish hagiography.

To give analogies in a different field, the reductionism favored by the Strong Program if applied to music would lead to an evaluation of Beethoven's symphonies solely in terms of the composer's attitude to Napoleon or his antisocial and unhygienic personal habits, while ignoring the technical structure of the music. Or suppose that a sociologist could show convincingly that upper class chess players favor strategies involving bishops and rooks, while lesser breeds favor pawns and knights. But we do not "understand" a game in any normal sense of the word if we just observe that White is able to mate in three moves using his rook and bishop because he is upper class. It would be ridiculous to insist on this sociological discovery as the decisive explanation of the content of a chess game. We want to know if White will indeed "mate in three," and that will take an analysis of the chess board. To give a more realistic example, it is clear from Plato and other sources that ancient Greek geometry was an aristocratic pastime, a way of losing oneself in "impractical" abstract thought and distancing oneself from the tumult and vulgarity of the *Demos*.[25] We can also note that whatever mathematics (if any) Pythagoras

---

[24] It was 20% too low and the derivation was only finally corrected by Lagrange in 1759.
[25] As well as for Plato training for future Philosopher Kings.

discovered,[26] he was also a leader of a secretive aristocratic political cult involved in various *coups* and conspiracies in the Greek colonies of Southern Italy. Yet even if this is so, this level of explanation does not give any insight into Greek geometry. We still want to know *how* the Pythagorean Theorem is proved or *how* Euclid demonstrated the infinity of primes. If an advocate of the Strong Program told us that the proofs the above theorems are not the "cause" of our belief in them, but instead is a consequence of our aristocratic antidemocratic sympathies, we would question his sanity. The point is that the sociological and "rational" levels of explanation fulfill different functions. One *may* explain motivation, by the sociological characteristics of (some) people on one side of a scientific issue or involved in some scientific or mathematical practice, but not the detailed content. For that we need to understand the concepts and arguments and realize (with rare exceptions) that we cannot merely dismiss them the way SK does with ideological argument.[27]

By ignoring the actual arguments and reasons scientists have given for their beliefs and concentrating on their "extra theoretic" function SSK is essentially following Mannheim's recipe for the "disintegration" of an ideological position by revealing the political purposes it serves. But scientific arguments are usually much more than ideological rationalizations. If Pasteur had refuted Pouchet by appealing purely to metaphysical speculations or to Catholic theology, or Newton had disproved Descartes vortices by attacking the Jesuits (who supported Descartes) the SSK strategy would be justified. We would be tempted to regard Pasteur or Newton's arguments as *post hoc* rationalizations. But Newton showed on the basis of fluid mechanics that Descartes' vortices were unstable. Pasteur proposed many detailed ingenious experiments which strongly supported his case. Any opponent of Newton or Pasteur would have to come to grips with their arguments on their own ground in a way that is unnecessary for the Divine Right of Kings or for a theory supporting slavery. Of course, we do not wish to claim that science *always* differs from ideology in this way, only that it often

---

[26] Almost nothing is known about Pythagoras' specific contributions to mathematics. The sources identifying him or his followers with musical ratios, the Pythagorean Theorem, or the irrationality of $\sqrt{2}$ were all written hundreds of years later and are considered unreliable. The only thing we have good evidence for (in Aristotle) is the Pythagorean fondness for numerology.

[27] Of course there are exceptions such as anthropological arguments on behalf of slavery, German racial "science" in the 1930s, etc., where explicit political content or interest is built in to the theory. But most science is not of this type.

does. And as we have already pointed out and will consider further below,[28] some "sciences" are either (1) so loaded with political consequences that the arguments pro and con on certain issues necessarily have an ideological character (2) so vague and undeveloped that "metaphysical" speculation is all that is possible. Often characteristics (1) and (2) go together, but in any case either (1) or (2) make disputes difficult to settle. But no general rule can be given. Whether scientific claims are more like the claims of Newton and Pasteur or like those of James I on the Divine Right of Kings has to be investigated on a case by case basis.

The most obvious (and perhaps the strongest) criticisms of the Strong Program (and SSK generally) are a cluster of arguments following from the reflexive directive (iv). They all involve the observation that the Strong Program, taken seriously, damages itself. At the shallowest level, we have seen that the Strong Program denies that rational argument, logic, etc., is a cause of belief. Yet, the advocates of the Program wish to change *our* beliefs and they attempt to do this by writing books full of rational argument, logic, etc. Is there not some inconsistency here? It is as if a solipsist wrote books for others to read proving solipsism! More fundamentally, if one applies SSK to its own claims as (iv) commands, it is hard to see how SSK can avoid "disintegration" in the Mannheim sense. SSK believers use logic, observation, etc. to support their claims. They are proud of this and believe that their method of analyzing the sociological causes of scientific or mathematical belief is truly scientific. In fact, according to them it is the only method that avoids Platonic mysticism or some kind of teleology (i.e., an inner drive towards "truth" which is "out there" independent of the individual scientist). Consequently, they are "social realists" just as scientists are "physical realists" and they claim that their methods resemble those of science. But SSK also believes that it has demolished scientific realism—the belief of scientists that they have access to objective knowledge about the world, that scientific progress is cumulative, etc. Hence, SSK applied to itself would, in the same way that it has done for science, demolish its own social realism and claim of objective knowledge. Its own ideology, like any other, reflects "existential" factors and cannot be "knowledge as such." In particular, if it is useful to explain the content of a scientific theory by searching for the social and political interests behind it, we can do the same for SSK. Why not simply bypass the arguments of SSK, as it does in the case of Newton and Boyle, and "explain" their "extra-theoretical function"

---

[28] In Chapter 14.

by "unmasking" the hidden sociological and or political motives for which they are a mere ideological superstructure? Perhaps the authors of these arguments are those on the Left who wish to discredit science because of the vital role it plays in the modern capitalistic state. Or perhaps these arguments reflect the class interests of sociologists who wish to seize (at least academic) power from scientists.[29] Licensed by the reflexivity of Strong Program, we can dismiss as a confidence trick any claim that it represents an honest effort to discover "truth" about science, the way it really *is*.

To express the same point differently, if the Strong Program is true and the decisive element of a scientific theory is not the rational arguments given in its behalf but some concealed ideology, does not the theory lose credibility? Do we have good reasons to believe it? For example, why should anyone have ever believed Pasteur's arguments against spontaneous generation if the interpretation of Farley and Geison is correct and had existed in 1860.[30] Since by reflexivity a similar analysis can be employed for the Strong Program, why should it remain undamaged in credibility?[31]

Secondly, SSK in general as well as the Strong Program tend to believe that the so-called "facts" about nature are the *result* of a purely sociological process of "negotiation" among scientists and not somehow the "cause" of their agreement.[32] It follows, as we have frequently noted, that something is called "true" or a "fact" in science simply *because* there is a consensus in its favor. By reflexivity this means that the judgements of the Strong Program are true only *because* there is a consensus in *their* favor by sociologists of science, presumably also the result of "negotiation." Thus we have a universal consensus theory of truth.[33] Hence in particular, there can be no atemporal difference in truth status between statements by scientists or Strong Program proponents and statements by priests of Amun in 1500 B.C.E., statements by the Holy Office, or statements by the Politburo, provided that they all reflect the consensus of the appropriate local group at a certain time.

---

[29] We will revisit these possibilities in Chapters 13 and 14.
[30] This point has been made in Brown (2001), p. 143 who observes that it seems inconsistent with Bloor's professed admiration of science.
[31] We also note that the reflexivity postulate (iv) of the Strong Program implies an infinite regress. The proposition that SSK applies to itself is itself a statement that can be analyzed by SSK which in turn is a statement to be analyzed by SSK and so on *ad infinitum*.
[32] This is not stressed by the Strong Program but seems to be implicit in the cognitive relativism of the Program and the fact that science and mathematics are viewed as entirely "conventional."
[33] This was noticed by Fine in Fine (1996), p. 134. See also Heilbron (2005).

Note that the previous paragraph would be uninteresting and trivial if we are allowed to interpolate "believed true" or "thought to be true by consensus" where needed. We can certainly do this and take no position on the "truth" in the "ordinary" sense of the term (i.e., a statement about something that is really the case) of beliefs we are reporting. Historians, anthropologists, and sociologists commonly adopt this technique, and it is called "methodological relativism." Sometimes the Strong Program innocently claims to be only doing this as well; but elsewhere they write that "knowledge" or "truth" *is* whatever is accepted by the group or culture. Clearly, the latter position which the Strong Program apparently believes to be universal goes far beyond methodological relativism. But we suspect that no one (including the Strong Program) would *actually* accept such a theory when the chips are down. For example, the writer believes that the present essay is not just a report of a consensus (or partial consensus) with Sokal, Gross, and Levitt. Nor do Bloor, Barnes, Latour, et al, seem to think of themselves as simple journalists reporting on a consensus. All parties believe that what they are saying *is* the case, and give *evidence* and *argument* for their positions, which they do not cynically believe to be mere propaganda or rhetoric. Nor is a whole range of statements David Bloor might make about his activities at the Edinburgh Science Studies Unit last week true because of a consensus with colleagues in his office.[34]

Still a third argument closely related to the previous two is that the relativism inherent in the Strong Program or in any version of SSK is self-refuting. The form given by Mary Hesse (b. 1924), who is sympathetic to SSK, is as follows:

> Let P be the proposition "All criteria of truth are relative to a local culture; hence nothing can be known to be true except in senses of 'knowledge' and 'truth' that are also relative to that culture." Now if P is asserted as true, it must itself be true only in the sense of 'true

---

[34]This is an instance of simple argument that a universal consensus theory of truth must be incorrect. To give another example, my belief that I had corn flakes this morning is not the result of a social consensus. Both examples are adapted from one given by Sokal and Bricmont:

> If I regard the statement 'I drank coffee this morning' as true, I do not mean that I *prefer* to believe that I drank coffee this morning, much less that 'others in my locality' think I drank coffee this morning. (Sokal and Bricmont (1998), p. 81.)

The authors claim that this argument originates from a criticism made by Russell of the pragmatism of James and Dewey. (*Ibid.*, n. 111.)

relative' to a local culture (in this case ours). Hence there is no grounds for asserting P (or, incidentally, for asserting its contrary).[35]

Hesse refutes this argument by arguing that:

> ... it is fallacious to ask for "grounds" for P in some absolute sense: if P is asserted, it is asserted relative to the truth criteria of a local culture, and if that culture is one in which the strong thesis is accepted, then P is true relative to that culture. We cannot consistently ask for absolute grounds for accepting either P or the strong thesis.[36]

As far as the logic goes, Hesse's argument is correct and we have apparently arrived at a stronger assertion than the claim that scientific statements are true only because of the consensus of scientists. Now the consensus is the product of the fact that a statement can be shown to meet the truth criteria of the wider culture. But this just pushes the difficulty a little farther back. It is essentially the same perspectivist position as Feyerabend's with the "truth criteria of a local culture" playing the role of his "tradition." Assuming that a standard position of SSK is that (at best) modern science is "true" only relative to the standards of the modern West, it follows that it is not "truer" in any absolute sense to "other ways of knowing" that conform to the standards of other cultures. Therefore, the theories of archeologists are not superior in any absolute sense to the creation myths of American Indians concerning the Kennewick man. Orthodox medicine is not superior to Chinese traditional medicine, Philippine psychic surgery, or to the incantations of shamans. Darwinian evolution is not superior to faith in Genesis by North African orthodox Jews. The belief of astronomers in the approximate sphericity of the earth is not superior to the theses of the Islamic Flat Earth Society. And finally SSK itself cannot be superior either to the science it studies or to a "way of knowing" of a culture of rational absolutists, for instance that of Torquemada. All we can say (like Fleck or Feyerabend) is that all these rival pairs are mutually "incommensurable" either because they reflect radically different "thought styles" or "traditions." Thus SSK would use sociology to explain the beliefs of Torquemada, and Torquemada would use concepts of heresy or demonic possession to explain belief in SSK. Torquemada would not get a contract at the Edinburgh Science Studies Unit, and David Bloor would be the object of an *auto de fé*

---

[35] Hesse (1980), p. 42.
[36] *Ibid.*

at Toledo. We may prefer Bloor's arguments because we are "modern," but no absolutely justified trans-cultural preference is possible.

Again, there seems some (probably unconscious) intellectual dishonesty in the above argument. Casual observation shows that the articles and books advocating SSK vehemently argue that SSK theory is "true"; from the passion of the argument and the severe criticism of the epistemological pretensions of science the reader might be pardoned for believing that the truth of SSK is meant in an absolute sense; this is the way science really *is*. We certainly *do* get that impression from certain remarks by Sheila Jasanoff, a professor at Harvard's John F. Kennedy School of Government and an important figure in SSK, who has argued that standard assumptions about the objectivity of science are false since "each of these assumptions are at odds with well-established findings in the sociology of science."[37] Nowhere is there a warning label saying that "we merely demonstrate (using the relative standards of our culture) that SSK is 'true' in the sense that it merely conforms to these standards." This would perhaps be embarrassing because Western standards are usually criticized by certain SSK or PIS sects as "hegemonic," "sexist," and "racist." A possible reply, however, would be to simply declare SSK and not science "objective." That is, one could declare that SSK offers an accurate description of the way science is, while science does not accurately describe how the world is. How this thesis could be defended, however, is unclear given that SSK declares its approach to science "scientific," the reliability of whose method and general objectivity it elsewhere enthusiastically undermines.[38] This would also have the counterintuitive consequence that SSK alone produces the pure milk of "knowledge as such" while the sciences (including mathematics) are contaminated by various social, racial, sexist, or political factors.[39]

On a philosophic level still another manifestation of the reflexivity problem is seen in the universal rejection by SSK and postmodern theorists of the correspondence theory of truth at least in connection to scientific practice. Their views are founded they claim on a careful study of the history

---

[37] Quoted in Levitt (1999), p. 363.

[38] Some relativists have attempted to eliminate the self-contradiction inherent in the claim that "there are no absolute truths" by amending it to "there are so absolute truths except this one." One could, of course, take such a statement as an axiom, but there is no reason to believe it as it stands. It would have to be defended by *other* propositions which could be deconstructed by sociological or historical investigation, so the reflexivity problem is still present.

[39] This point has been made by Norman Levitt who in replying to Sheila Jasanoff's remarks given above that "apparently, sociology of science is allowed to have well-established findings, but not high-energy physics." (Levitt (1999), p. 363.)

of science. The perpetual change and falsification of previous theories and influence of social and political factors they believe rules out the naive view that science is "true" because it accurately mirrors the way the world is. But this judgement suffers from a similar flaw as the "social construction" argument. If true, it is true because it "corresponds" to the way science actually *is*. More abstractly, philosophers who reject the correspondence theory are arguing that their rejection is true because the world *is* not of a kind to admit the correspondence theory. Nelson Goodman, for instance, believes that instead of one unique world with which we can compare our statements, there are many worlds, all "made" by humans and not "found," so that a proposition may be true in one world but not in another. Such judgements are of the form X (denial of the correspondence theory) is true because it "corresponds" to an actual situation Y (the nature of the world or its inaccessibility to human reason) which implies X. But this is an instance of the correspondence theory of truth as direct as the judgement 'snow is white' is true because it corresponds to the fact that snow is white.

We might also point out that the extreme anti-realism favored by SSK theorists like Steve Woolgar (and implicitly perhaps by Latour and the Strong Program) suffers in addition to its other problems from its own form of a self-contradictory reflexive paradox. All versions of realism imply that there is a way the world is which is independent of our opinions, desires, or mental states. Anti-realism denies this. In its strong form it argues that *no* judgement reflects the way the world is. All judgements are representations which in turn are functions of our minds rather than "reality." Then if this is true, it is *also* describing the nature of the world. This is the case even of the most extreme anti-realism, for example solipsism. Hence if reality consists only of representations or if we are mutually isolated Leibnizian monads (or brains-in-a-vat) whose "world" is a preprogrammed movie on some kind of interior screen, then this is the way things *really* are; in other words, *whatever* the world is like—even if it consists only of mental states—constitutes a fact which is independent of our mental states and therefore as "real" as we commonly consider Mount Everest to be. But on the other hand, the anti-realist thesis is also a "representation" and thus cannot be a true statement about the way the world really is.

To be fair, most of the self-destructive problems associated with reflexivity including the infinite regress (cf. note 31) were recognized by SSK itself "within the first few nanoseconds of the relativist big bang."[40] This

---

[40] Collins, Pinch, and Yearly (1992), p. 304.

recognition soon generated a large article literature, as well as various sects and heresies within SSK. For instance, Steve Woogar and Michael Mulkay created a method called "discourse analysis" which would analyze SSK as SSK analyzed science, but was itself immune to such treatment. But this maneuver was soon criticized by the leaders of the Bath School of SSK, H. M. Collins and Steven Yearly, since it:

> ... paved the way for more radical deconstruction which goes under the title 'reflexivity'. Those of the reflexive persuasion have noted the omnipervasiveness of the social conception of truth... a conceptual rather than an empirical point—and have set out to sidestep all truth-making conventions.[41]

As a result of the more spectacular exercises in reflexivity: "Subtle reflexivists realize that their work leads nowhere. For example, Woolgar has said that getting nowhere should be seen as an accomplishment."[42] The reflexive methodology, moreover, destroys the possibility of knowledge:

> Typically, as we have remarked, the sociologist knows less than the natural scientist, while the sociologist of science knows still less. Those engaged from day to day with the problem of reflexivity would, if they could achieve their goals, know nothing at all. We might say that SSK has opened up new ways of knowing nothing.[43]

Hence, it is not surprising that Collins and Yearly demand that reflexivity be ignored since it has become a futile game of "epistemological chicken" which accomplishes nothing except the weakening of SSK. Instead, they recommend "meta-alteration": scientists and sociologists of science should be respectively realists regarding nature and social realists regarding science since these positions help each party get its job done. Naturally this does not answer the epistemological problem raised by reflexivity; but to Collins and Yearly efforts to take it seriously are equivalent to the adolescent male game of "chicken," i.e., foolhardily running across a road on a dare in the face of oncoming traffic or in the case of full-blown reflexivists "jumping into a hole in the road from which there is no escape."[44]

---

[41] Ibid., p. 305.
[42] Ibid.
[43] Ibid., p. 302.
[44] Ibid., p. 323.

As early as their 1979 "field study" of the Salk Institute *Laboratory Life* Bruno Latour and Steve Woolgar had also noticed the ease with which SSK self-disintegrates due to reflexivity. By the mid-1980s they together with Michel Callon, believing that reflexivity made purely *social* explanations bankrupt, decided to push SSK in a different—in some ways more radical—direction. In so doing they created a potent heretical challenge to orthodox SSK doctrine—"actor-actant network theory." Recall that item (iii) of the strong programme was the requirement of "symmetry": the same types of causes should be sought to explain "true" and "false" beliefs. In the Anglo-American philosophical tradition this was a sufficiently radical step; but Latour now proposed that *every* type of explanation should be symmetric. In particular all attributes including those such as "agency" granted to humans should also be granted to non-humans. The motive of Latour and his coauthor Michel Callon was to: "disprivilege a prevailing asymmetry... the relationship between humans and nonhumans. Their (especially Callon's) writing grants a voice to nonhumans as well as humans."[45] Both human and nonhumans participate as "actants" in scientific situations on an equal plane; they all form complicated networks, all parts of which interact. The new theory serves an emancipating goal—freeing nonhumans "from the double domination of society and science" which if successful would be "the finest result of that perhaps clumsily begun 'anthropology of the sciences'":[46]

> However, in order to reach that aim, we have to abandon many intermediate beliefs: belief in the existence of the modern world, in the existence of logic, in the power of reason, even in belief itself and its distinction from knowledge.[47]

Some fruits of this approach were briefly sketched in Chapter 9 in our discussion of *Pandora's Hope*. There the microbes and Pasteur have equal status. Both act on each other to their mutual profit. The ferments, for example, become known and respected as living beings which is a promotion from their previous status as disgusting agents of chemical putrefaction while Pasteur becomes famous. There is also a paper[48] by Michel Callon on research by three French scientists to see if the reproduction of scallops

---

[45] Woolgar (1992), p. 335.
[46] Latour (1988a), p. 150.
[47] *Ibid.*
[48] Callon (1992).

in St. Brieuc Bay can be enhanced using a Japanese method of cultivating scallop larvae in special collectors. The larvae were viewed as "actors" coequal to the scientists, ocean currents, and parasites. Complex "negotiations" were necessary involving all parties in order that the scallops should be "willing" to cooperate with the researchers by anchoring to the collectors. The diplomacy was difficult and resembled the "tripartite Vietnam conferences held in Paris."[49] In another article Latour explores the idea of granting agency to automatic door closers as well as to the people who walk through them. As for "great" scientists like Pasteur in Latour's book *The Pasteurization of France*, they occupy a role in respect to the bacteriological revolution like Napoleon or General Kutuzov in respect to the battle of Tarutino as portrayed by Tolstoy in *War and Peace*, issuing orders but actually in command of nothing and swept along in the chaotic flow. Except for his political skills in securing allies among hygienists, Pasteur is hardly more significant than the microbes he fights.

At this point we can only agree with Bloor:

> Latour never manages to make this metaphysical vision even remotely clear. Metaphors are introduced, monads and entelechies are invoked, a new terminology of 'quasi-objects' and 'actants' is suggested, and mental exercises recommended, such as using purposive language to talk about inanimate things and mechanistic language to talk about people—but all to no avail. Embarrassingly, the project is mired in impenetrable obscurity.[50]

Paul Gross and Norman Levitt or Alan Sokal and Jean Bricmont in their respective books *Higher Superstition* and *Intellectual Impostures* could not have said it better!

Let us next look again at Latour's thesis that Pasteur's microorganisms did not exist before Pasteur and that Rameses II did not die from tuberculosis bacilli since these did not exist before Koch. If we reexpress this by saying something like "Pasteur's microorganisms or Koch's tuberculosis bacilli did not exist *for us* before Pasteur or Koch" then we have a trivial truth. It is possible that despite the flamboyance of Latour's prose this is all he means to say. But he *does* seem to claim that these microorganisms

---

[49] Also see Collins, Pinch, and Yearly (1992), p. 313; Callon (1992), pp. 211-216.
[50] Bloor (2004), p. 941.

or bacilli had no existence in *any* sense prior to their so-called "discovery." Summarizing the arguments given in Chapter 9, he believes this because: (1) There is no "substance" "standing under" the attributes of a thing; a thing is totally defined by its collection of attributes and relations with other things. (2) Because of the "historicity of things" these attributes and relations change in time, and consequently the identity of the thing changes in time also. (3) It is Pasteur and Koch who "retrofit" the past with these microbes considered as permanent unchanging substances and help to create powerful "institutions" to maintain this belief. (4) The subject-object distinction of the "modern settlement" makes it natural for us to believe in the permanent existence of substances such as microbes.

We confess that Latour's thesis is formally irrefutable. There is no way to *prove* that the microorganisms were "out there" all along in their present form before Pasteur "discovered" or possibly "fabricated" them. This thesis, moreover, has extraordinary scope. Using it we can argue that North America, for instance, did not exist before the Vikings or between them and Columbus. Or even if it *did* exist it changed in time. In 1492 it was obviously Japan or part of the Chinese empire. Four hundred years earlier it was Vinland, a green and pleasant island west of Greenland suitable for growing grapes. For Amerigo Vespucci and other explorers after 1497 it transformed itself into a new continent between Europe and Asia. Powerful social institutions have retrofitted this belief into the past and continue to maintain it, so that it *appears* to be a timeless truth (which ought to have profound implications for the geological sciences—especially plate tectonics!). We now live in Vespucci's world rather than that of Columbus or the Vikings. The reader may amuse himself by performing a similar analysis of other "discoveries." Did planets travel around the sun in elliptical orbits before Kepler or because of the "historicity of things" were they governed by epicycles? Was the inverse square law of gravitation functioning before Newton? Was the the moon made of an ultra-pure crystalline substance before Galileo's observations of 1608? Did witches really exist in the sixteenth century, but not in the nineteenth? What about the flatness of the earth before and after the monk Cosmas of Alexandria in 535 C.E? We could go on *ad infinitum* with such examples. And if we find all this less than convincing, is it only because the distinction between subject and object allows historical change only for human subjects and rejects "the historicity of things"? Yet, the thesis is highly original and like belief in Pasteur's microbes clearly merits maintenance by an institution; but whether it should be the Institute for Advanced Study or one of a different kind is unclear.[51]

The contextual explanations of SSK, if they work at all, work best analyzing science at the frontier, especially when the effects predicted or measured are obscure and the apparatus and experimental design is complicated and delicate. Then things tend to go wrong and the hermeneutic resources of the paradigm have to be used, often in contradictory ways, to explain the phenomena. As a result, there is sometimes a jungle of quasi-political infighting by rival groups of scientists for SSK to gleefully analyze. The studies of H. Collins in *The Golem* of controversies over the existence of gravity waves or chemical memory transfer in planarian worms are good examples of this type of SSK. On the other hand, SSK has trouble with simple statements of fact or straightforward experimental success. While the calendar is certainly a social construction, it is hard to see how the statement "today is Friday" is. Similarly, what is the sociological status of my colleague's report that "my cat Messallina pooped on the rug this morning"? To take another example, Anthony Flew has argued:

> The cause of our belief that the ferry canoe is where it is on the Zaire River does not lie in the social structure of our tribe. It is to be found, instead, in certain intrusive non-social facts: that when we turn our eyes towards the right bit of river the canoe causes appropriate sensory impressions; and that those heedlessly placing themselves in the water rather than the canoe are incontinently eaten by crocodiles.[52]

---

[51] As far as we know Latour has not yet generalized the thesis of *Pandora's Hope* to mathematics although, as we have seen, he has attempted to expose the reactionary motives behind Socrates' fondness for geometric argument. An attempt at such a generalization may be found in the virtual manuscripts of one Hilbert Archimède Latour (University of Pieland), whom the author has just "discovered" using the techniques of Pasteur and "inscription devices" in order to "retrofit" him into the past. Our discovery may be justified by the doctrine of "the historicity of persons" which is far more obvious than the "historicity of things." To earlier researchers Hilbert Archimède was Alfred E. Neuman in the same sense that Pasteur's yeasts had previously been ferments. This Latour may be a distant unacknowledged cousin of Bruno and is related to a prominent family of French hair pomade manufacturers. Just as the ferments were made famous by being promoted into living yeasts, Hilbert Archimède Latour has been be promoted from his previously degraded status as an idiot and should become a celebrated philosopher of mathematics for arguing that large primes are entirely an artifact of the supercomputers used to "fabricate" them. According to him, they had no real existence before. They are now maintained by institutions such as mathematics and computer science departments whose unceasing propaganda retrofits them into the past.

[52] Quoted in Barnes and Bloor (1982), p. 31.

Barnes and Bloor attempt to deal with the status of simple factual statements such as Flew's example by saying that our ability to evaluate them depends more on biology than society; we share this ability with animals. Asking SSK to explain this is comparable to explaining "how a dog retrieves its buried bone."[53] Bloor elsewhere, however, claims that the nouns in these factual statements such as "cat," "canoe," or "poop" represent socially created classifications. They are entirely conventional "and cannot be determined by the way the world is. There is no such thing as a natural or uniquely objective classification."[54] Thus, for example, I see a "cat"; an individual from another culture might see a "furry evil spirit" or the Egyptian goddess Bastet. I see "cat poop"; someone else "food of the gods," or perhaps for Jean Fernel (1497–1558), physician to the French Royal Court, a powerful medicine—which when mixed with mummy—corrects fatal humoral imbalances involving the yellow bile. Even using less specific nouns such as "animal" or "thing" and saying "the smelly thing is on the flat thing" does not evade the argument since the meaning of all words is a social product:

> There are no privileged occasions for the use of terms—'no simple perceptual situations'—which provide the researcher with 'standard meanings' uncomplicated by cultural variations.[55]

But such arguments are vulnerable to the criticisms at the beginning of this section of the neo-Kantian Constructivism which seems the common denominator of SSK. The above quotation implies that the object "behind" the culturally determined noun signifying it must be like a Kantian noumenon. Everything said about it comes through social filters. It is difficult therefore to see how factual statements about cats, poop, or canoes *are* determined by the objects called "cats," "poop," or "canoes," or even what is really being said about them. Since they are unknowable, even their existence seems an unnecessary metaphysical conceit. All that is left is a network of socially determined linguistic conventions. But this assertion is in turn subject to the same analysis. What is the status of these conventions or the "society" that generates them? To say that our answers to these questions are also socially determined begs the question and is the

---

[53] *Ibid.*

[54] Bloor (1982b). As Ian Hacking has observed SSK entails nominalism. (Hacking (1998), pp. 82-84.) Clearly, if the classifications inherent in our common nouns represented "natural kinds" scientific realism would follow. (Bloor (1982b), p. 269.)

[55] Barnes and Bloor (1982), p. 38.

beginning of an infinite regress. On the other hand, if we argue that the social roots of our linguistic conventions are "real" in the same sense that scientists believe that their objects of study are real we face the problem of reflexivity again.[56] The remaining alternative is that, say, my conceptions of society is just another "representation," a position that as we have already seen soon terminates in complete solipsism. Since Barnes, Bloor and many other SSK advocates insist that they do not reject the "real world" or the "pragmatic" achievements of science, it appears that their doctrine has done its job *too* well.

The issue of the status of "simple factual statements" is important because a good part of science does consist of such statements, in principle no more complicated than the examples given here. We can think of the common facts of chemistry, such as boiling points, densities, reactions with acids, etc., or of medicine such as the fact that TB is contagious,[57] that there are no pores in the ventricular septum as Galen believed, or that male and female humans have (contrary to certain Biblical fundamentalists and biology students at Persepolis State) the same number of ribs. We also *know* that when a person has jumped off the top of the Empire State Building without a parachute he will hit the ground at a high velocity (about 120 mph), that the moon is not made of an incorruptible fifth element, that Mars does not contain canals as Percival Lowell thought in 1910, that the Earth is approximately a sphere and not flat; nor is it hollow with the solar system residing inside,[58] and that Titan contains a large quantity of menthane. We know these and many other facts for the simple reason that either we or our cameras have witnessed the truth of these claims. Much of science is a vast array of facts similar to these which are simply "true" and are, moreover, likely to remain invariant—even if their significance and

---

[56]That is, why should sociological realism be privileged over scientific realism? As has been pointed out in the previous section Bloor sometimes argues that it is a misconception that SSK claims that the only influences on science are social. Science is also influenced by the objects in its environment; but at the same time, he argues that the "causal impact" of these objects underdetermines our beliefs about them. The objects are necessary but not sufficient for belief. In view of his nominalism and the exclusively social analysis he gives elsewhere, this position is incoherent.

[57]As late as the mid-nineteenth century, however, this fact was contested. Many physicians believed that the disease was due to certain occupations or to a morbid predisposition in the individual. The problem was a failure to distinguish TB from other conditions such as silicosis. (Watson (1848), pp. 657-658.)

[58]A belief that Paul Feyerabend hints that some people might "rationally" hold, and if so, has an equal right to be taught in the schools with other theories. (Feyerabend (1978), p. 74.)

relation to a theory may change—under any conceivable paradigm shift. In this sense of an increasing accumulation of facts about nature science *is* cumulative. No reasonable person can doubt that we know in an absolute sense many more facts than we did in, say, 1900, and in turn scientists of 1900 knew more than those of 1700, and so forth. We do not deny, of course, that some contemporary "facts," especially those of recent origin and coming from the research frontier, may become tomorrow's "mistakes," but they are usually a minority even in a specialized field. Many facts are stable. Despite the views of the Flat Earth Society (recently headquartered in southern California) the earth is unlikely to ever become flat unless there is a cosmological catastrophe.

SSK or the Strong Program as we have argued can give no satisfactory account of this cumulative process of fact acquisition. Either by Barnes and Bloor's own admission simple observational facts are off-limits to SSK (although the interpretations of these facts for theoretical purposes may not be) or they too are mere social conventions referring either to nothing "out there" independent of society or to something superfluous because it is unknowable.

Likewise, it is difficult to see how SSK can account for experimental verification of a theory or nontrivial prediction. Consider the treatment Barnes, Bloor, and Henry give to Robert Millikan's famous oil drop experiment designed to measure the charge e of the electron. Several of Millikan's experiments did not work as he had hoped, as they seemed to find charges that were fractions of the expected value of e. Here Millikan had to use interpretive methods of the Duhem-Quine type to justify throwing out these results.[59] But some trials did work; the expected value of e was found. Why was this? The authors answer that:

> ... the routine unproblematic successes that theories sometimes enjoy (i.e. successes that are not explained by selective or interpretive work), belong to the realm of pure contingency. They are not, for the detached analyst, explained by the truth of the theory. They are not explained by anything.[60]

---

[59] Recent research, however, has shown that Millikan was far more honest and careful with his data than Barnes, Bloor, and Henry imply. Recalculating e using the data he threw out gives a value very close to what he finally accepted. Only one measurement gave a possible fractional value of e for the charge on the electron. See Franklin (1981).
[60] Barnes, Bloor, and Henry (1996), pp. 30–31.

Let me propose a much simpler experiment than Millikan's. There is a theory that my colleague's cat Messallina poops on the upstairs rug in the early evening because of her dislike of her food. Therefore tomorrow afternoon he will give her food (Science Diet) she doesn't like and then lets her upstairs. According to Barnes, Bloor, and Henry we can explain a failure to poop by invoking Duhem and Quine, but not "success." Pooping is in the realm of pure contingency.[61]

It is true of course that correct prediction does not guarantee the truth of a theory. Messallina's behavior if successfully predicted could be the result of worms, separation anxiety, or a neurosis induced by her male companion Claudius rather than a nutritional protest. Likewise, Ptolemaic astronomy could "save the appearances" essentially as well as early versions of the Copernican system. Babylonian temple priests could also make accurate predictions of planetary position based on careful extrapolation from detailed records without assuming *any* model. The situations seems different, however, when a theory or discovery leads to a qualitatively new and unexpected phenomenon. While this also cannot imply the truth of the theory, it is strong common sense evidence that we should take the theory seriously. Once the telescope was available so that the planet's disc could be seen, Copernican theory predicted the phases of Venus, an event incompatible with Ptolemaic theory. The phases were duely verified by Galileo around 1610.[62] Newtonian mechanics and irregularities in the motion of Uranus led to a prediction of Adams and Le Verrier of the existence and approximate location of a new planet which was duly discovered by observers in Berlin in 1846. Although the possibility of sources of unlimited power was hinted at by Einstein's realization of the mutual convertibility of mass and energy, the atom bomb was unimaginable before James Chadwick discovered the neutron in 1933 and Otto Hahn fission in uranium in 1938. It also could not be built until Stan Ulam successfully predicted on the basis of a mathematical analysis that the bomb could not ignite the earth's atmosphere.[63] All this would be inexplicable and in the realm of pure "contingency" (as according to SSK Millikan's correct measurement of the charge on the electron was) unless some key parts of this physical theory really mirror in some sense the "way things are" and have a differ-

---

[61] Which in the case of this particular feline is almost true!
[62] Of course this prediction is not decisive. Tycho de Brahe's theory would also predict the phases of Venus.
[63] Eddington used to joke that a supernova was merely the result of a nuclear experiment gone wrong by some alien civilization.

ent epistemological status than "ideology." Even Harry Collins who usually believes that nature has nothing to do with the resolution of scientific controversy recanted as we have seen in the case of Einstein's theory of matter and energy. This is not to say that the theories behind the Bomb are not subject to revision; future science may change them in surprising ways–so that the whole world view and tissue of meanings built on them may also radically change, just as according to Kuhn the world view of relativity replaced that of Newton; but the existing theory will suffice to make atomic bombs—just as the inverse square law allows us to calculate orbits and send a probe to Saturn's moon Titan. Unless there was some sort of correspondence or partial correspondence in the case of at least *some* theories to the way the world really is the success of repeated, novel, unexpected, and correct predictions by these theories would be in Putnam's words a "miracle." In trying to deal with this problem Richard Rorty has claimed that explaining the "success" of science in terms of the objective "truth" of its theories is vacuous and is like arguing:

> Why are we able to predict eclipses so well? Because Ptolemy's *Almagest* is an accurate representation of the heavens. Why is Islam so spectacularly successful? Because of the will of Allah. Why is a third of the world Communist? Because history really *is* a history of class struggle.[64]

This argument is less than convincing. Ptolemaic astronomy is one of many examples that, although false, provided a fairly accurate representation of the phenomena it describes and predicts. However it does not predict verifiable *unknown* phenomena; those predictions of this nature that it makes are (mostly) false. The other explanations ignore the fact that complex historical and psychological factors can explain the "success" of political or religious idea independently of the ideas "quality." Examples would include the witchcraft craze in sixteenth century Europe, Nazi ideology, or the writings of Rorty and other advocates of PIS today. Does Rorty really think that the success of science rests only on causes of this kind? Is it really vacuous to argue that Martian rocks have the chemical composition they have or that the Sun is 93,000,000 miles from the from the Earth because our chemical analysis and measurements have shown these facts to be the case? Do these claims really have the same status as the claims of Communists, Moslems, or for that matter of SSK/PIS? It is difficult to

---

[64]Rorty (1991), p. 54.

avoid the conclusion that no account of science which reduces it to a purely sociological phenomenon or which accounts for it only in terms of "empirical adequacy" or "instrumentalism" would be a sufficient explanation of its success. The latter tactic, indeed, seems a begging of the question since no explanation is given of *why* a theory can organize experimental results (instrumentalism) or make correct predictions (empirical adequacy) unless, of course, these achievements are only a collective delusion of scientists.[65]

Moreover, alternatives like instrumentalism, and empirical adequacy seem just as incompatible with SSK as scientific realism, since all three refer to an *objective* property that a theory may have which is independent of our opinion of it; in particular the degree of "empirical adequacy" of a theory is an objective measure of its "fit" with the observational data. Now for a particular theory it may take a group of scientists to decide whether it offers a "realistic" explanation of a phenomenon or is just a method to "save the appearances," so that one can speak of a sociological process of decision.[66] However, if a theory is to be a "social construction," i.e., the product of internal power struggles in the laboratory or of external ideological factors, then Latour's "Third Rule of Method" or the equivalent judgements of Collins or Woolgar must be correct. In such a case it is difficult to see *any* connection between such a product and either a mind-independent nature or the ability to account for or to predict observations.

At this point the partisans of SSK or PIS may reply that instrumentalism or empirical adequacy is built into the theory since observation is "theory-laden" and always able to be interpreted or explained away by the theory, just as some political "fact" can be interpreted or explained away by the Party. This claim is worth examining closely since it is so important to the anti-realism of SSK/PIS. The appropriate response is that on the basis of the historical record it is seldom true in the sense intended! If it were, it is hard to understand why there should ever be anomalies or experimental results that worry scientists. Consider Ptolemaic astronomy.

---

[65] We are aware that there is an enormous pro and con literature on the issue of scientific realism and associated epistemological issues, the analysis of which would require a new and very long book. The most thorough defense of realism is Psillos (1999). Also see Newton-Smith (1981), Brown (2001), Brown (1994), Brown (1989), and Devitt (1984). Versions of "empirical adequacy" have been most forcefully argued in Van Frassen (1990) and Laudan (1984). A short and well balanced account of the issues may be found in Parsons (2005). See also Klee (1997).

[66] Note that we do not take a global philosophic position on scientific realism or its alternatives. We think it best to decide this case by case. Some theories may have a strong claim to mirroring what is in nature; others may be essentially accounting devices.

This theory owed much to external religious and philosophical factors. It is true that observations by medieval astronomers were "theory-laden" in the sense that the observer of the motion of a planet could "see" the epicycles given by the theory, much as a modern physicist "sees" a certain elementary particle from a track in a cloud chamber. At the same time, Ptolemaic astronomy had empirical consequences that bothered astronomers. Most were not realists with respect to the fine details of the theory. They had an instrumental point of view. To them the test of a good astronomical model lay in its ability to "save the appearances." But by 1500 there was universal agreement that the Ptolemaic model in any of its variations did not perform this task adequately. No amount of theoretical "spin" could eliminate the problem. To repeat, the most reasonable explanation for the existence of an anomaly in a theory is that (at least in some cases) it is an indication of a misfit between the theory and a world independent of it. It is not a sociological creation by a rebellious subgroup of scientists.[67]

Another difficulty for SSK/PIS is to explain the near unanimity of scientific beliefs. One of the most obvious facts about the history of ideas is that religions, political ideologies, philosophical systems constantly fission. Every major religious system generates innumerable heresies and sects. For Islam we have Sunni orthodoxy, the Shiites, Sufis, Wahhabis, and at least a dozen smaller subgroups. Christianity has been even more fertile in its generation of sects which until recently have despised each other. The same phenomenon can be seen in Judaism and Buddhism. Marxism over the last 150 years has also seen its share of dogmatic disputes and rival sects. Philosophy has throughout its history been an arena of universal warfare. There is no enduring consensus on virtually *any* philosophical issue. This situation is also true in almost every other humanistic discipline. The most severe and violent measures have often been required to preserve the integrity of a political ideology or religious system. Academia is not immune—the denial of tenure, merit raises, or promotion being its equivalent to an *auto de fé*. Now we do not deny the reality of scientific disputes or the formation of different scientific "sects." The latter frequently happens as Kuhn has noted in the pre-paradigm stage of scientific development and is especially common in the case of the soft sciences where theories may have explicit political consequences. As SSK and all historians of science have noted, there are often also vicious disputes at the frontier—especially when the reigning paradigm is under revolutionary attack. In these situ-

---

[67] We do not want to make a universal claim here. It is easy to find counterexamples in the "soft" sciences, closely connected with politics. See the discussion in Chapter 14.

ations the defeated can suffer the loss of their career, if not their lives as may be the case for defeated religious and political heretics. Still in most cases (especially in the hard sciences), the disputes are sooner or latter settled. Unanimity emerges among almost all scientists who are in the field. The same paradigm and normal science is practiced by everyone whether they live in Saudi Arabia, North Korea, Cuba, the US or elsewhere, even if the scientists have radically different and mutually hostile political and religious beliefs. Any exceptions such has Lysenkoism in the Soviet Union are the result of grotesque political manipulation by the state and collapse when that manipulation ceases. The analog of this phenomenon in religion would be the spontaneous and sudden conversion of the vast majority of world theologians to, say, Wahhabism. What explains this? Do scientists have better means of indoctrination and coercion than the Imams of Saudi Arabia, Goebbels, the KGB, or the Holy Office? Kuhn, as we have noted, stresses the narrow, intensive, and dogmatic training of scientists as an explanation of their agreement on a paradigm. But this kind of training is also practiced by religions and political ideology. In the short run it often does guarantee unanimity, but heresies and sects always emerge. Neither Kuhn or SSK/PIS theorists have any convincing explanation for these differences. Since the latter mostly deny that nature or experimental results have anything to do with the emergence of common belief, they stress such things as the scientist's rhetorical and networking skills in the laboratory or perhaps his prestige as a factor in the victory of his point of view. Now we admit that this is possible and probably happens in certain cases. But why should a scientist's victory spread beyond his laboratory to the entire network of scientists in the research area? If science is similar to politics as SSK maintains, this seems improbable. We should expect the emergence of all kinds of rival groups with different "ideologies." Competing scientists in other laboratories have different interests; perhaps if the ideas of the first scientist prevail, their own funding might collapse. Their best strategy might be to push a different theory and support it by convincing propaganda. Every laboratory would then perpetually maintain competing and mutually contradictory theories, for instance, in high energy physics. In the case of N-rays, for example, there should have been a century long conflict, still raging, over their existence with equally prestigious laboratories taking opposite sides. But this does not happen; unanimity is achieved. To give a political analogy, it is as if after the triumph of the Jacobins in France, the Kings of Prussia and Austria, and Czar Alexander I of Russia had accepted the French paradigm (perhaps after initial resistance) because they felt it the best way to govern a state. Can we explain this odd ability of some scientific theories to prevail on any other grounds than the fact that scientists sincerely believe that the victorious theory has better supporting empirical evidence (as contaminated as this evidence may be by theory), solves long standing puzzles better than alternatives, or simply gives a much more convincing account of part of nature? By what right can SSK claim that this is an illusion?

There is also the matter of the connection of science to the vast explosion of technology in recent decades. How is this possible unless science has some connection to the way the world really is? Mostly SSK ignores this problem. A few, however, such as Michael Mulkay solve it by trying to abolish the link between science and technology.[68] Mulkay argues that pure science is now so remote from practical affairs that it almost never has any influence on technology which is a matter of engineering. In the case of medicine, if science is given credit for new cures, why should it not be held responsible for the many failures of scientifically created therapies? Besides, the improvement in health and lifespan over the past century has more to do with better hygiene, clean water, and so on, than medical advances.[69]

We admit that Mulkay's thesis concerning medicine has elements of truth. If we returned to the medicine of Galen or Paracelsus, but kept modern hygiene, there might only be a slight decrease in the average lifespan. However, who would want to be bled almost to death, treated with hot irons, or prescribed ground mummy or toxic heavy metals? Who would want to give up modern surgery because many of the conditions which it successfully treats are rather rare? Would we allow people with a burst appendix to die in agony as they did in the 1600s even though only a few people in a thousand ever have this problem? Then again there are the vaccines which have eliminated smallpox, typhus, and a host of childhood diseases. All of these are examples of medical technology based on scientific advances. There have of course also been failures. Until the late nineteenth century medicine certainly caused as much human morbidity as the diseases it was supposed to treat. Even very recently some well established therapies have turned out to be useless or quite harmful.[70] But

---

[68] Mulkay (1979b).
[69] Another diametrically opposite approach is to argue that science is "merely technology and nothing more: physics, says Rorty, is the construct of a community with a shared interest in pushing objects around." (Forman (1998)).
[70] From the 1930s until at least the 1950s children's tonsils were commonly excised. This did little damage but was useless. As late as the 1950s a standard treatment for adolescent acne, to which the author was subjected, was bombardment for ten minutes

the balance at least over the last century is certainly positive. There is real progress in medicine. As far as the writer knows, there has not been the kind of "pessimistic induction" among historians of medicine which is common among historians and philosophers of science. No one asserts that since most medicine before 1900 was harmful or futile, the same will turn out to be true for our medicine. The absence of such a thesis is tacit recognition of both medical progress and of medical scientific realism. Not even Ludwik Fleck or the Strong Program doubted the reality of bacteria or viruses and their role in infectious disease or would have denied that most assertions concerning them are "truer" than the humoral theories of Galen.

There is also a certain validity to Mulkay's first point. Modern academic science is rather remote from technological applications and engineering. In fact the complete abolition of basic scientific research might not be noticed by the industrial world for several years. But who can convincingly deny that modern technology is built on the the scientific achievements over the past two centuries. There may not have been commercial radio without Marconi or similar entrepreneurs,[71] but its very possibility depended on the work of Hertz and Heaviside, not to mention Maxwell and Faraday before them. Could X-ray technology ever been developed without the discovery of X-rays by Röntgen, or atomic bombs and nuclear reactors without Chadwick and Hahn? Likewise, it is difficult to see how PET and CAT scans would be possible without the existence of positrons or the Radon transform. We can show in fact that the very possibility of many (though not all) technological innovations depends on some scientific discovery often decades before the innovation.[72]

Returning to epistemological themes, SSK often maintains that the demonstration of social causation in accepted scientific belief does not mean that the science is "false." But what this statement really means is unclear. If the science is accepted it is "knowledge"; its claims are "facts," and it is "correct" at least at the time, for there is no other test as we have seen repeatedly for SSK *other* than acceptance by the contemporary "local cul-

---

or so with X-rays in a lead lined room. This had deplorable long term effects on the thyroid. In 1848 a treatment celebrated in Watson (1848) as the latest French advance was the application of leeches inside the rectum. The author was a leading physician of his time.

[71] No less an authority in pure science than Lord Kelvin said that radio had no future.

[72] But not in all cases. The Watt and subsequent steam engines were invented without a correct theory of heat. This example illustrates the fact that sometimes technology affects science. The very presence of steam engines created the need for a better theory of heat in order to make them more efficient. The road between science and technology is bi-directional and the precise relationships need to be analyzed on a case by case basis.

ture" of scientists; any alternative would imply the possibility of "objective" nonhistorical judgement. Later when the paradigm, politics, gender, race, politics (or whatever) of the scientists change, the previous science will be "false" and the new science now "correct." This is clearly implied by the relativism of the SSK/PIS project. We must conclude, therefore, with Kuhn that newer scientific explanations are not "truer" than older ones, but are just "incommensurable" with them. It would follow that Galen's theory of blood "circulation", the theories in Aristotle's *De Caelo* or Ptolemaic astronomy, Pouchet's views on spontaneous generation, or that the theory that the proper balance of the four humors: blood, bile, phlegm, black and yellow bile is essential to health are all just as true or false as the the theories of William Harvey, NASA, Copernicus, Pasteur, or the latest advice from the Mayo clinic.

But do the advocates SSK really believe any of this? If they do, shouldn't they follow the implications of of such belief in daily life? What reason, for instance, could they give for not choosing Philippine psychic surgery in the case of serious illness over bloody, dangerous, and invasive Western operative techniques? Neither is connected to the way the world is. Both can be explained by a purely sociological consensus of appropriate groups. Both "work" at least for their believers. But one suspects that—with the possible exception of Paul Feyerabend—few would choose the Philippine option.[73] As A. N. Wilson has put it

> In all likelihood, our post-modern habit of viewing science as only a paradigm would evaporate if we developed appendicitis. We should look for a medically trained surgeon who knew what an appendix was, where it was, and how to cut it out without killing us. Likewise, we should be happy to debate the essentially fictive nature of, let us say, Newton's Laws of Gravity unless and until someone threatened to throw us out of a top-storey window. Then the law of gravity would seem very real indeed.[74]

The logic of the SSK position would imply that the only reason that most of its advocates would make the "Western" surgical choice is cultural. We have not been brought up in a culture that accepts psychic surgery. We cannot be sure that either choice would be medically better for us, but we would be uncomfortable with the Philippine option since we have been brainwashed by our modern techno-scientific civilization. Similarly, is the universal belief

---

[73] In his final illness, however, Feyerabend accepted treatment from conventional neurologists.
[74] Wilson (1999), p. 178.

that people land with high terminal velocity if they jump from the 90th floor of the Empire State Building instead of flying to the Chrysler building only the result of being so embedded in the current scientific culture that one psychologically cannot doubt it?

Once again, this whole issue is connected with reflexivity. The Strong Program insists in many places that "knowledge" or "truth" consists only of those beliefs accepted as knowledge by the consensus of some group whether scientists, Philippine psychic surgeons, or the Azande. That a statement is true because it states what *is* the case is emphatically rejected. But very rapidly their *own* theories become "justified true beliefs." That science may also be "knowledge" of this sort is hinted at by the Program's admission that it may not be "wrong," no matter how shabby or political its origins. These unacknowledged (and probably subconscious) oscillations in the meanings of "truth" and "knowledge" pervade SSK writings and, as has been observed, make its program incoherent. A similar incoherence may be found in Kuhn's writings. Kuhn also appeals (independently of Larry Laudan) to "pessimistic induction" in order to deny that paradigm change brings scientific progress in the sense of approaching more and more closely to the truth. Almost all theories once held to be true, he says, were later shown to be false, so it is highly probable that the same will be the fate of our current theories. Supposing that this observation is true, we can ask if Kuhn is speaking merely of a change of consensus by normal scientists or does he think that past theories are "false" in the sense of not reflecting nature? The former meaning would imply that, given the right circumstances consensus might be reestablished in favor, say of Ptolemaic astronomy or Galen's theory of the heart, just as consensus might be reestablished in favor of Marxism as some socialists fervently hope. To the ordinary reader, Kuhn *seems* to intend the latter meaning which contradicts his view that what scientists call true reflects the consensus of normal scientists, not a correspondence to reality.

It is true that social conditioning and education do play a role in our belief system. In the case of a political ideology or a religion, however, it is quite common for a person educated dogmatically in some doctrine to suddenly reject it and believe something else. Thus student Trotskyites of the 1930s become prominent neoconservatives today, preserving only a common belief in the efficacy of violence.[75] The process of switching core beliefs in politics or religion is similar to a *gestalt* switch. The same

---

[75] In the one case violent revolution, in the other the violent imposition of Democracy and Free Market ideology on recalcitrant states.

can (possibly) happen in science as Kuhn argues when someone switches allegiance from one paradigm to another or, on a different level, simply abandons modern medicine altogether for homeopathy or Christian Science. But the bulk of "scientific" beliefs like the circulation of the blood or the terminal velocity of suicidal financiers are never challenged in this way. If the reasons for both cases are "sociological" and have nothing to do with the relation of the belief to "reality", it is the obligation of SSK to explain the difference; having done this, they can explain the sociological reasons why—to refer to an earlier argument—the writer believes that he had cornflakes this morning.

To see more of the odd consequences of the above views, let us, for the sake of argument admit that some key theses of SSK are correct, for instance, that politics, power struggles, ideology, and social factors play a decisive role in the creation and content of scientific theories, and also that many of the theories we now believe were initially accepted for reasons having little to do with standard conceptions of the scientific method; so that, for instance, Pasteur prevailed over Pouchet mainly because his politics were favored by Napoleon III and the scientific elite of his time more than the politics of the luckless Pouchet. But was Pasteur wrong? Is it possible that the germ theory is false, that spontaneous generation might come back on its merits? We can ask the same sort of question for Aristotelian cosmology, for Galenic physiology, or any of the other supplanted theories we have mentioned. If the answer is "no," is the only reason because the modern world is "brainwashed" or conditioned by the totalitarian methods of modern science to believe in contemporary theories and reject older ideas?[76] The answer is that they cannot return because they are wrong! Nature simply does not work in the way these theories assert. That Barnes, Bloor, Collins, et al., probably also believe this (at least in carrying out the tasks of daily life when they are not in a SSK literary mode) demonstrates the inadequacy of their approach.

---

[76]We don't deny these discredited theories might return but it would have to be as a result of a social cataclysm leading to a new Dark Age, just as knowledge of the earth's sphericity was lost to sixth century Franks. Another possibility would be foreign conquest. The writer has read of certain Madrassas in remote parts of the Islamic world teaching Ptolemaic astronomy and the medicine of Rhazes. One can imagine political or demographic events whereby these theories would be orthodoxy at the University of London in 2200. In a sense therefore the content of science could be a function of political change, "verifying" the SSK thesis. If so, was it also verified by the fact that "racial science" became orthodoxy at the University of Berlin between 1933 and 1945? The point is that it makes sense to argue that a theory is *really* right or wrong in an absolute sense, independently of social or political factors.

What finally can we say about the feminist critique of science? In its epistemological approach it is part of SSK and hence is vulnerable to the same criticisms based on reflexivity. If traditional masculine oriented science reflects "contextual values" and ideology more than nature, why should not the same be true of feminism in its vision of science? Since both science and feminism are "stories" having the same (shaky) epistemological status, why should the stories of feminism be privileged over those of existing science? Furthermore, if SSK is correct feminists have no business in claiming that masculine values have distorted science since the seventeenth century, since what "distortion" can mean here is unclear. This is especially evident in the position of Helen Longino. She feels that contextual ideological factors are always intimately intermixed with scientific practice, so that the "facts" or "data" never speak for themselves; they always require interpretation. Unfortunately Longino proves more than she probably wishes. For if scientific theories or models depend on interpretation based on contextual factors rather than nature, we have no basis for saying that one set of of contextual values produces less distortion than another since there is no unambiguously "true" science to be distorted. In other words, since decisions at bottom reflect ideology there can be no *epistemological* reason for the choice between a science reflecting feminist values and science as it is currently practiced; the choice will just reflect the interests and relative power of the parties making it. This conclusion which appears to follow from Longino's theory is close to Mannheim's "relationism." Few feminists, however, would endorse such a position since they inconsistently think of their vision of science as being "better" in some objectively nonpolitical sense than the traditional type.

There are other problems in the feminist position. Feminists like to contrast the gentler feminine virtues of subjectivity, love, spirituality, and sympathy with nature with masculine lusts for domination, control, and "objectivity." The trouble is that elsewhere they believe that these feminine virtues are constructions of patriarchal society designed to keep women in a subordinate position; they are not innate. Aside from anatomical differences feminists believe that all other apparent difference between men and women are social constructions which they designate by the term "gender." Why then should they identify feminist science with these "feminine" characteristics? In a non-sexist society women ought to be as "dominating" towards nature as any man.

Keller's contrast of the psychological properties of masculine and feminine science, moreover, is not even historically new. Gross and Levitt noticed that her critique of standard science parallels Goethe's critique of Newtonian science. Goethe opposed what he conceived as Newton's dead mechanical system, based on mathematical abstractions, and the urge to dominate nature. He contrasted this with the sympathy and closeness to nature exhibited by his own *Farbenlehre*.[77] Besides Goethe, many others have made criticisms of science similar to Keller's but for other political purposes. Phillipe Lenard (1862–1947) and Johannes Stark (1874–1957), two Nobel prize winning but Nazi oriented physicists, used the same kind of vocabulary to contrast "Jewish" with "Aryan" science. Jewish science is characterized by trying to dominate nature and force hypotheses upon it. It is also excessively theoretical and abstract, and the theories it propounds are not founded on reverent observation or loving affinity with nature, but rather on aggressive Jewish values. In contrast, according to Lenard and Stark, Aryan physics rejected "mechanistic materialism, rationalism, theory, and abstraction, objectivity" and was linked with an organic view of nature.[78] These and other characteristics of Aryan science seems very close to feminist science. Also, both as a consequence of their analysis reject the notion that science is value free or objective. For feminists the fundamental nature of science is determined by the sex of its practitioners. For Lenard and Stark it is determined by race, blood, and soil.[79] As a thought experiment one could replace in some of Evelyn Fox Keller's, Sandra Harding's, or Helen Longino's writings "masculine" with "Jewish" and "Feminine" with "Aryan." We should also substitute an appropriate phrase for "contextual values," perhaps "values of race and blood." Jews then instead of men would emerge as rapists of nature (just as in anti-Semitic fantasies they desire to rape German women). The writer has made several experiments of this kind. What emerges could have been an elegantly written submission (better so far as I can tell than the usual fare) to that esteemed (but short lived) journal *Forschungen zur Judenfrage*.[80] Conversely, the inverse transformation could be applied to discourses on Aryan physics by Lenard and Stark. We wish to emphasize that we are not claiming that feminists are Nazis; only the most radical would think of men the same way as Nazis

---

[77] Gross and Levitt (1994), pp. 137–139.
[78] See Beyerchen (1978), p. 136 and Ch. 7.
[79] While SSK rejects the romantic view of science shared by Aryans and feminists, at least some strands of SSK accept another view which both hold in common: that science depends on some characteristic of the scientist rather than nature; in the case of SSK it is the scientist's politics or class while for Aryans and feminists it is the scientists race and sex.
[80] Eight volumes, 1937–1943. Hamburg: Hanse.

did of Jews. We are only pointing out the similarities in both ideologies concerning the interpretation of science. Further we can ask the truly interesting question: why should the thinking of contemporary feminists be nearly structurally isomorphic to a German romantic tradition exemplified by the *Farbenlehr* of Goethe or the Aryan science of Lenard or Stark? The attempt to find an answer is beyond the scope of this essay, but would be a subject for a fascinating historical investigation.[81]

None of these writers, moreover, has offered any real evidence for their position that all or most science are contaminated by androcentrism. At best there are only some isolated examples connected with primatology, reproductive issues, male and female infant development, etc.[82] The writer found himself wondering how the male desire to "torture" nature is found in the analysis via Eulerian angles of the motion of a top,[83] Yang-Mills theory, or the chemistry of heavy metals. Sandra Harding in her many books, in particular, constantly repeats her views that Western science is a diseased product of imperialism, racism, male domination with only the slightest justification for her claims or evidence of concrete knowledge of *any* science.[84] Harding also makes embarrassing mistakes of fact, the most obvious being the domination of present science by white males and the exclusion of "people of color" from the Second and Third Worlds.[85] While this may have been true two or more generations ago, it has certainly not been the case for at least the last twenty-five years in the US. The emerging majority of scientists, mathematicians, and engineers, both as students and faculty, are definitely *not* white and Western. They are Indian, Chinese, Korean, Japanese, Middle-Eastern, immigrants from Eastern Europe

---

[81] In fact all criticisms of overly abstract, souless, and dominating Western science develop similar themes. Sandra Harding, for instance, looks with sympathy at the "African world view" which in its view of science stresses the need to preserve human harmony with nature, communal values, etc. which is contrasted favorably with Eurocentric rationality and objectivity. Again, if we substitute "German" or "Aryan" for "African" we would have another vision of science that Lenard or Stark would have approved. See Harding (1986), pp. 167–172.

[82] The best examples, such as they are, may be found in the writings of Keller and Longino. Also see Schiebinger (2008).

[83] The only torture here is of students studying classical mechanics.

[84] This is in contrast to Keller who was trained as a mathematical biologist and who has done some valuable research on slime molds (for which she suggests the possibility of masculine and feminine models and interpretations).

[85] We pass over her comments on mathematics, revealing complete technical ignorance, which the reader is invited to examine in Harding, *Science Question*, pp. 48–52. (p. 50 is especially clueless.)

and the former USSR together with a fair number of Africans.[86] The same is true for undergraduates in all subjects at the nation's most selective universities. The percentage of Asian undergraduates at places like Berkeley, Stanford, MIT is commonly above 30%. Asians in fact are now very similar to Jews in the 1930s and 40s in their dedication to education. The main reason for this phenomenon is that US students whether male or females, white or Afro-American, rich or poor are simply not intellectually competitive with these newer immigrants or their offspring. Their own high school and undergraduate education has been "dumbed down," doubtlessly in the quest of the "equity" and "social justice" that Harding seeks, while the postcolonial "others" at the graduate level are survivors of ferociously competitive educational systems in their own countries. The US, in fact, is in the process of developing a two track higher education system. On the one hand, there is a "soft" undergraduate option oriented towards demotic values and fit for "natives," who wear their baseball hats on backwards and complain about their foreign instructors. On the other, there is a much more rigorous graduate system heavily patronized by exceedingly hard working immigrants who are fortunate enough to have done their undergraduate work at *real* universities abroad. This means that in some high level graduate classes in technical subjects with small enrollment it is quite unnecessary even to lecture in English.[87]

---

[86] For instance, a very large component of the engineering faculty at the writer's university are Muslim. A colleague of the writer at a sister campus is from the Congo and trained in Belgium. Another is from Zimbabwe.

[87] At various universities the writer has seen board work in graduate level mathematics in Russian or Chinese in classes where both students and instructor are of the same nationality.

Chapter 12

The Fallibility of Conventionalism and Fallibilism

Having dealt with the SSK/PIS view of science, how valid is its conception of mathematics? As in the case of science, direct refutation is impossible. There is no universally accepted theory of the origins or foundations of mathematics. Every account that has been offered—whether Plato's view (as revived by Frege) that mathematical objects are in some sense "real," or the twentieth century doctrines of formalism or logicism—has been shown to have fundamental weaknesses and to offer an unsatisfactory account of mathematics.[1] Our goal will be simply to demonstrate that the same is true of SSK by examining some of its implications and to see whether or not SSK conforms to or can explain the intuitions of practicing mathematicians concerning their subject.

Mathematicians on the whole are not an especially philosophical lot. They are usually more interested in obtaining the next interesting result than in pondering the epistemological status of their subject or the reasons why they accept mathematical proofs. Without devoting much thought to epistemological issues, most are uncritical Platonists.[2] For the average mathematician (say, at Persepolis State) entities like groups, vector spaces, measure spaces, or the positive integers are as real as his 1975 Pinto, his minimal raise, the malefic behavior of Persepolis State's administrators, and student complaints about mathematics in general and his teaching of it in particular. His world consists of both mathematical and other physical and mental objects whose properties may be discovered and seem to exist

---

[1] See, for example, the discussion and criticism of these doctrines in Körner (1960), Hersh (1997), or Davis and Hersh (1981).

[2] However, he is a formalist "on Sunday" or if he is forced to answer embarrassing questions about Platonism. See Davis and Hersh (1981), pp. 321-22.

independently of human minds.[3] Also statements about mathematical objects are, if correct, *necessarily* true. They cannot be otherwise than what they are, and their truth is independent of space and time. $\sqrt{2}$ would be irrational and the number of primes would be infinite in the Andromeda galaxy or during the Precambrian era. For anyone who has felt the *force* of a mathematical proof, this feeling of timeless necessity can be overpowering. Additionally, correct theorems give information that we did not know before about the world of mathematical objects. The information is as much a discovery as anything in chemistry or high energy physics. In many cases, e.g., the recent proof of Fermat's Last Theorem by Andrew Wiles that $x^n + y^n = z^n$ admits no nontrivial solution in the non-zero integers for $n \geq 3$, mathematical discovery has been a culmination of effort extending over centuries. Finally, because of its universality and independence from politics mathematics is very different from other deductive systems such as theology or political ideology. Here is perhaps one of very few subjects where Teichmüller, a world rank mathematician and fanatical Nazi, and a suitably trained Russian commissar could have agreed on the properties of "Teichmüller spaces" before killing each other.[4]

As we have seen, SSK rejects all these claims about mathematics. In the first place, mathematics is viewed as a vast network of social conventions, "true" because they are accepted, not accepted because they are true. The grounds for acceptance are pragmatic; the practice of mathematics is a "form of life," and the subject matter consists of various "language games."

Let us take a closer look at the conventionality thesis. To say some activity or rule is a convention means that (i) it is a creation of human beings and (ii) it *could* have been different; it need not be completely arbitrary, but there is no necessity to it. This is a very large category. We can think of an enormous number of disparate conventions: rules of the road (such as driving on the right or left), the rules of a bureaucracy, table manners in polite society, the French language, games such as chess or bridge, school dress codes—we can go on endlessly; just about any human activity which

---

[3] Yet there is some ambiguity. Upon reflection, the mathematician may realize that the definitions of mathematical objects have been constructed in time by other mathematicians. Their definitions did not always exist; they were written down by mathematicians at particular times in particular publications. Did the objects come into being with the definitions, or did the definitions permit the human mind to perceive and describe structures which were previously invisible?

[4] Although Teichmüller was a vicious anti-Semite and went to his death fighting for his Führer in the Dnieper region of the USSR in September 1943, his space still survives and is respected enough to merit its own web page.

is not the result of basic biological drives can be portrayed as a convention. "Convention" is also closely related to another vague term "social construction" which we have previously discussed, but the latter has a wider more "static" application. We say that some practice is a convention, but that money, the IRS, literary criticism, or the platform of a political party is a social construction. Hence, to just insist that mathematics is socially constructed or that its rules are conventions does not distinguish it from the vast array of other very different socially constructed things. Since almost every mathematician believes that his methods generate necessary truths in the form of theorems, they seem very remote from conventions which are in no sense "necessary." Therefore, if mathematics is conventional its conventions must be radically different from other more ordinary conventions. Just what is the difference is not explained by SSK. Conventions and social constructions also vary and may contradict one another. This is certainly true with respect to ideologies, philosophical doctrines, moralities, the institutions of various states and cultures, and so on. But there seems to be only one mathematics and it is universal in the sense that each mathematical theory is accepted by all regardless of culture and political belief. We don't, for example, see different, incompatible ring theories for Islamists, North Koreans, and Europeans. SSK (and Wittgenstein) have no convincing explanation that among practically all social constructions this is uniquely the case or for the fact pointed out by Reuben Hersh that:

> Some people dare question the Holy Trinity, the American flag, whether God should save our Gracious Queen, and so on. But *nobody* questions elementary arithmetic ... we *never* get letters claiming that $3 + 5 = 9$.[5]

But is the uniqueness and universality of mathematics really true? What about intuitionism which rejects existence proofs based on the law of the excluded middle, actually infinite sets, the axiom of choice, etc., while conventional mathematics accepts all these things as well as existence proofs with no constructive procedure given to produce the object whose existence is claimed? As we have indicated earlier, intuitionism rejects a good deal of classical mathematics, and each has different standards of proof and truth. If this is not bad enough, there is the case of alternative logics such as modal and many valued logics. So we can and do have at least two and perhaps several apparently incompatible mathematical systems! To this one can reply that any mathematician can agree and verify that the theorems of an

---
[5] Hersh (1997), p. 206.

alternative system are correct deductions from its axioms. Moreover, the theorems of intuitionism do not actually *contradict* the theorems of conventional mathematics. Its goal is to rebuild mathematics from the ground up while avoiding what it considers illegitimate methods and assumptions. To the degree that it succeeds there is no inconsistency. It only rejects classical theorems if they cannot be proved according to its standards. The writer, for one, is perfectly happy to work in either system should there be interesting problems. Everything provable in an intuitionist system is provable in orthodox mathematics, but the reverse is not true. Theorems that depend on nonconstructive proofs may not be provable using intuitionist techniques. They are not contradicted by intuitionist results but are simply not regarded as part of mathematics. The disagreement, therefore, between the two systems is a disagreement in the *philosophy* of mathematics—what mathematics ought to be like, not *in* the mathematics. Turning now to the question of different logics, what we really have is different mathematical instruments built for different purposes. There is no contradiction between them any more than a field contradicts a ring or chess contradicts checkers. The law of the excluded middle $p \vee \neg p$ in particular is compatible with either two or multivalued valued logics. That is, p may be "true" or "not true," and if p is true then $\neg p$ is not true. But the notion of "not true" may be equivalent to "false" as in classical two-valued logic or it but may embrace many possibilities such as an intermediate truth value or a continuum of truth values representing increasing degrees of improbability as in fuzzy logic.[6]

According to the proponents SSK still more evidence of the conventionality of mathematics can be found in the historical record. They claim, echoing, Spengler that over time different cultures have developed alternative mathematical systems, and that in particular each has had a different concept of number. This thesis may be true as long as attention is confined to issues of interpretation, philosophy, and method relating to mathemat-

---

[6] At this point we make a concession to the "conventionality" thesis of SSK. Logic is a system constructed by man for the purpose of sorting and keeping track of statements about mathematical objects. In part it reflects our way of thinking, but in part it is artificial. The truth functional behavior of $p \Rightarrow q$ has no counterpart in ordinary language since we do not normally view a false statement p implying any true statement q, and we usually demand some connection in meaning or relevance between p and q. In a similar departure from normal usage, it is useful to think of the the connective $\vee$ in $p \vee q$ as equivalent to "and/or." Neither these or any other logical principles are part of the deep structure of the universe. They are invented by man. For purposes other than the study of pure mathematics, logicians have constructed a large number of alternative logics, such as modal, deontic, temporal, conditional, continuous, and fuzzy logics.

ics. Pythagorean number mysticism may still be found in the "New Age" section of modern bookstores, but no longer taken seriously by modern mathematics departments. Similarly, the Greeks viewed what we call $\sqrt{2}$ as the hypotenuse of a isosceles right triangle with unit sides. Their proof showed that the hypotenuse was incommensurable with respect to the side, in other words there existed no line segment such that the lengths of hypothenuse and side were both integral multiples of its length; we, on the other hand, think of $\sqrt{2}$ as an irrational *number*. Since the Greeks did not have an arithmetical outlook and had no concept of an irrational number or of common fractions, this result for them meant that a "magnitude" as represented by a line segment (such as the hypotenuse) was a more basic concept than number. This meant in turn that their mathematics was almost entirely geometric and not computational or algebraic as was the case for the Babylonians or the Indians.[7] It is also true that mathematics is in a constant state of change. Old theories are rejected and forgotten. We no longer use Eudoxus' theory of ratio as recorded in Book V of the *Elements* or prosthaphaeresis as a tool for the multiplication of large numbers.[8] As can be the case with archaic languages (or even the mathematics article published twenty years ago) old mathematical techniques and results can be essentially inaccessible to contemporary mathematicians. The classical geometric approach to conic sections due to Apollonius, for example, which Newton used to derive and present his results in the *Principia* is certainly not understood by modern Ph.D.s in mathematics unless they are specialists in the technical aspects of the history of mathematics. Likewise, standards of proof and argument are not invariant. Since Cauchy and Weierstrass mathematicians no longer are content with the non-rigorous, intuitive methods of Euler. The history of mathematics is full of many similar instances. SSK in its enthusiasm for pointing out this historical diversity in mathematics forgets, however, that there is *no* contradiction between mathematical systems developed in different eras or by different cultures. Chinese or Babylonian mathematics, or the mathematics of the Rhind papyrus gets the same answers as we do, however differently expressed or discovered. The Pythagorean hypotenuse *is* incommensurable with the unit side of the triangle. Eudoxus's theory of ratio is correct. The

---

[7] For an analysis of the peculiar view of the Greeks to arithmetic and in particular fractions and ratio, see Fowler (1987).

[8] This method converts multiplication to addition using trigonometric tables and identities like $\cos A \cos B = (1/2)[\cos(A + B) + \cos(A - B)]$ to aid in the multiplication of large numbers.

answers to the problems in Diophantus' *Arithmetica* are correct, however oddly explained. 823 has a stubborn way remaining prime in any culture or period that has a concept of a prime integer. It is this essential universality that conventionalist theories of mathematics cannot explain.[9]

What would a real "alternative mathematics" whose existence has been claimed by SSK be like? The difference from our own mathematics should be more than a matter of style, philosophy, or a few mistakes (like, for example, some of Euler's assertions concerning infinite series). If mathematics is solely a matter of social convention resembling other cultural discourses it should be possible to exhibit a system or body of theorems that we now regard as totally false, like Aristotle's physics or Ptolemaic astronomy. No mathematics of this kind has been shown by SSK to have ever existed, and it is difficult to see how it could arise other than by some kind of *political* transformation that would allow mathematical cranks to gain power. As a thought experiment we could imagine a dictator (call him O'Brian) of an imaginary third world state. Having come to power in the Revolution of 2084, O'Brian rules with an iron fist and an intense personality cult. A quondam Ph.D. candidate in mathematics whose thesis was rejected, he has had his professors shot and his old university razed. O'Brian had tried to prove in his thesis that the number of primes is finite. The belief that they are infinite he claimed is a nonconstructive "reactionary mystification" based on "bourgeois delusions of contradiction." Now O'Brian's mathematical doctrines (which also include his more recent discoveries that the square roots of non-square integers are rational and that the rational numbers are uncountable[10]) are taught in every school and university. Countless books are written by Party ideologues and by O'Brian himself defending his ideas; they fill the libraries of all state mathematical institutes and universities.[11] All aspiring Party members must memorize abridged versions of these arguments presented in O'Brian's little *Blue Book*. Those who do not accept what O'Brian calls "The People's Mathematics" are executed by being buried alive in black board chalk dust.

---

[9] We certainly would require an explanation if we made the surprising discovery that all political ideologies and religions said exactly the same thing underneath the cultural and historical variations in their expression.

[10] O'Brian proves this by a Cantor diagonal argument similar to that used to prove the same result by an ex-physician, working as a laborer in a warehouse, who submitted it to my Department in the 1980s with a cover letter denouncing Cantor's "lies."

[11] The writer found something like this situation existing in 1978 Romania. The works of the dictator Ceaușescu filled the library of the Mathematical Institute in Bucharest, almost to the exclusion of orthodox mathematical texts.

What could various proponents of SSK say about this obvious nonsense? We can only speculate, but our speculations are grounded in their writings. Certainly the conventionalism of SSK or the constructivism of Paul Ernest and other critics of "absolutism" would permit the observation that O'Brian's mathematics contradicts our own. But there seem no grounds in the SSK approach to say that it is "wrong," if we mean by "wrong" more than "it conflicts with our mathematical conventions." Indeed, if mathematics is a matter of social convention, at best justified pragmatically, and his version of it and ours are equally "credible" according to the standards and norms of his and our respective cultures, how can we condemn it? The difference between the theorems of O'Brian and those of, say, Gauss would be similar to the difference between two political or religious systems founded on different conventions, like a republic or monarchy or Islam and Christianity. More precisely, according to Wittgenstein we are just dealing with two different "language games" with different but completely contingent rules. O'Brian's language game is analogous to everyone writing in a certain culture that $12 \times 12 = 162$. Relative to that culture 162 would be the correct answer, and the person who wrote "144" would be mistaken. SSK might also try to say that the two systems have equal epistemological status because they are incommensurable in the sense of Kuhn or Feyerabend. But what about the well-known "proofs" of the existence of infinitely many primes or the irrationality of $\sqrt{2}$? According to SSK what counts as a proof is also a matter of convention. O'Brian's conventions just signify that both proofs which we accept are deficient for him. That is all that can be said, since there are no absolute standards of mathematical validity over and above whatever conventions O'Brian adopts.[12]

---

[12] It is tempting to continue such speculations in different directions. We can imagine, for instance, that the more politically oriented among SSK believers might go beyond a Wittgenstein-like analysis and reason that since "everything is political" opposition to O'Brian's mathematics derives from elitist opposition to his Revolution. They compare his leading mathematical critic, Professor Goldstein, to Robert Boyle or John Wallis who likewise opposed Hobbes' brilliant theses on physics and circle squaring for reactionary reasons. Still others admit that Goldstein may be "technically correct." But to dwell on "minor mathematical questions" now may "compromise the real achievements of the Revolution and play into the hands of its enemies." Only the power of the Restoration political elite ensured the victory of Boyle over Hobbes with the deplorable consequence that science and mathematics have been on the wrong path for the last three centuries. Likewise, without the vigilance of true humanists and progressives, the machinations of the CIA may cause "absolutists" like Goldstein to defeat O'Brian, sabotaging the emergence of a truly liberatory anti-elitist mathematics fit for the masses.

How does the conventionality thesis of SSK explain the mathematician's feeling of mathematical necessity, given no convention can be said to be "necessary"? The answer is that SSK maintains that such feelings are purely psychological. Neither "truth" of a theorem or the fact that we are convinced that the "proof" is correct explains our belief in it. Rather, they mask the fact that we are psychologically imprinted with a form of *moral* compulsion; mathematical conventions and rules *have* to be obeyed—just as we must obey the prohibition against murder. We have been conditioned to have such feelings beginning with the authoritarian instruction in arithmetic in elementary school. Acceptance of mathematics, then, is the result of a form of "brainwashing" and does not derive from anything inherent in the subject.

There is some truth in this analysis. Certainly O'Brian's subjects are conditioned to feel that his mathematical pronouncements *must* be obeyed. Elsewhere, elementary arithmetic and algebra is (or used to be) communicated to students in the early grades by drill and punishment. To many of them the multiplication table or the method of adding fractions is a mysterious convention proclaimed by the teacher, Miss Wormwood—"she who must be obeyed" like H. Rider Haggard's Ayesha (only much less attractive). The students memorize the rules without understanding them.[13] We observe the same phenomenon in higher level mathematics classes such as calculus. To many students at Persepolis State the chain rule, for example, is a meaningless formula to be memorized and blindly reproduced on tests. Psychologically the SSK account seems like a philosophical abstraction of this kind of "blind" learning. It is tempting, but probably unfair (and certainly *ad hominem*), to wonder if it results from SSK adherents (or their

---

[13]The following anecdote reinforces this impression. The father of a mathematical colleague of the writer, being Jewish, could not get a standard university job as a research mathematician in Moscow; instead, he was working in a pedagogic institute. One of his duties was to evaluate mathematics teaching in nearby collective farms. In one of them the truck driver was assigned to teach arithmetic to the students. The lesson was the addition of fractions. His procedure, which he required the students to master, was to write

$$\frac{a}{b} + \frac{c}{d} = \frac{a+c}{b+d}.$$

When my colleague's father pointed out the error privately and showed him the correct method. The truck driver announced to the class "Children, today we have an announcement from the Party in Moscow. We now should add fractions in a new way." Remarkably, the truck driver was unfamiliar with Wittgenstein.

guru Wittgenstein's) own experience with mathematics.[14] But it is hard to see how a feeling that the rules must be followed can be transmuted into the feeling of mathematical necessity that a person who *understands* the material possesses. At age 6 one learns by rote (and fear) that $8 \times 6 = 48$ rather than 37 which would have seemed equally likely had not Miss Wormwood commanded otherwise. Many such operations—even including standard calculus techniques—were learned long ago in this manner, and are now automatic. Like driving, one does them without conscious thought.[15] But if one continues to study mathematics, knowledge of the subject gradually becomes much more than an automatic response like memorization of the alphabet. (The automatic response is still present certainly; one usually does not want to think about routine parts of a calculation.) Increasingly there is the ability to *explain* a result, to see why it must be true, to "play" with the ideas behind it, and if one is lucky to generalize them and perhaps develop a new proof. This "understanding" or "mathematical maturity" is the real goal of mathematical education. There is no similarity in any of this to "moral compulsion." Hopefully, one does feel a moral obligation not to lie, steal, or murder, but my feelings towards the identity

$$e^{ix} = \cos x + i \sin x \tag{1}$$

have no moral aspect at all. What they consist of depends on the situation. If I were a sociologist and did not understand (1), but just happened to glance at it in a textbook, I might not not care one way or the other about it. If pressed for an explanation I would declare it a "convention" and a typical ideological manifestation of decaying Late Capitalism in the West. If I was taking a course where the formula played a role, the test was tomorrow, and I did not understand it I would be worried.[16] If I

---

[14]Probably, however, not in the case of Wittgenstein since he was a trained engineer before becoming a philosopher. It is tempting, however, to speculate if Wittgenstein's peculiar deconstruction of mathematics might have been semiconsciously motivated by envy of his Cambridge colleague, G. H. Hardy, one of the most accomplished mathematicians of the time.

[15]This is not a necessarily bad thing. One would not want to have to meditate on the theory of the brake and accelerator before using them when driving on the Pennsylvania Turnpike.

[16]However, this is a rather old fashioned response. Today the preferred reaction, at least at Persepolis State, would to complain to the Vice-President for Academic Affairs about the instructor's pathological unfairness in inflicting such a difficult formula on the students and demand that he be fired.

understood the formula and could derive it, I would simply observe that it was true. Where does SSK come in? Naturally to learn (1) I may have had to go to college and be taught the proof or at least read a book on my own. These possibilities *do* involve a great many sociological factors like the existence of colleges and libraries. Perhaps initially, I just memorized the proof for the test (facing the competing demand of a 1,500 word English paper on Catherine Morland's reaction to Mrs. Radcliffe's *The Mysteries of Udolpho* the same night). But after further study, I can *see* that each step was correct; still later I can devise my own proof, perhaps based on power series expansions of both sides or on the fact that an existence and uniqueness theorem implies that $\{\cos x, \sin x\}$ is a basis for the solution space of the differential equation $y'' + y = 0$. In fact, the more I think about it the more I can break apart the components of the steps of the proof into *their* components and analyze them also. At no point as far down as I choose to go am I aware of blindly following a convention or of any feelings similar to "moral obligation."

There is certainly a moral aspect to mathematics, but it does not account for beliefs; rather it prohibits certain behavior. It is the obligation of honesty. Most mathematicians when they write research papers feel the obligation not to cheat or deliberately fake proofs. This would be very easy to do. The material is so technical, so unlikely to be read *closely* by anyone—even the referee or the rare person who actually uses the result— that an argument gliding over a gap would be easy to fabricate and probably would not be detected; or if detected, could be said by the perpetrator to have been an accidental mistake. Honesty usually prevails even if the paper is badly needed for next year's tenure application, and the author may have have wasted a year on the argument before discovering a tiny, fatal, but easily disguised flaw. This seems to the writer one of the few situations where mathematical behavior has something to do with moral obligation. If, however, one regards mathematics as a mere network of social conventions, having more to do with "power" and "interest" than with truth, it is difficult to see why Alfred E. Neuman's attitude of "what, me worry" should not be an option in this situation. I could reason that the objective situation is just the same as if I were unaware of the gap. The institution of mathematics would go on, people would cite the paper; best of all, I would get tenure and therefore avoid a career as a bag boy at the A&P like other unemployed algebraic geometers. If the flaw is detected by somebody (in spite of my best efforts to disguise it), only then do I need to write a retraction, try to fix it, lead a mob of radical students to attack the Tenure Committee, etc.[17]

To see how peculiar the SSK philosophy of mathematics is, let us compare mathematics with other deductively oriented belief systems. As far as SSK is concerned there is no *essential* epistemic distinction between them. For example, mathematicians believe unconditionally in such things as $2 + 2 = 4$, p and $p \Rightarrow q$ implies q, or the Radon-Nikodym Theorem. Followers of Charles I or Louis XIV believed unconditionally in the "Divine Right of Kings." Like mathematics this doctrine has complicated arguments, "proofs," and an axiom system. Both are surprisingly alike in formal deductive structure.[18] It is true that SSK, agreeing with Mill, claims an empirical origin for mathematics. The partitioning of stones or gingerbread cakes in certain ways suggests useful conventions. This is probably not true in the same sense for the Divine Right of Kings.[19] But in either case the believers have mistaken sets of certain socially sanctioned conventions for "objective" truths. Also, there is nothing *intrinsic* to either mathematics or Divine Right that forces credibility; both networks of conventions are believed true and followed because of a mixture of psychological conditioning and pragmatic considerations. The severe conditioning and savage punishments inflicted by Miss Wormwood or by seventeenth century grammar schools run by priests inculcate belief in algebra or Divine Right which is preserved into adulthood, doubtlessly by a process akin to the Stockholm syndrome. Likewise, it is *useful* to believe in mathematics, at least until the revolutionary triumph at Persepolis State of Calculus Liberation Front or, failing that, the takeover of the Mathematics Department by radical von Glasersfeld-style constructivists from the College of Education; analogously it was dangerous *not* to believe in the Divine Right of Kings until the execution of Charles I or the triumph of the Jacobins. Thus the compulsion to

---

[17] The same observations, of course, apply to science. Assume that science is merely a matter of propaganda and political success which is either independent of or actually creates "nature." Then it is difficult to make sense of the notion of "cheating." The possibility of cheating seems to imply the existence of "facts" which are independent of the scientist's volition. If science is *only* a social construction, why not "cheat" if it helps in the victory of our construction? We would not be falsifying anything for there is nothing to falsify, but only indulging in a clever tactic in Latour's "agonistic field."

[18] This is clear if one examines the writings of, say, Bishop Bossuet justifying the absolutism of Louis XIV based on an "axiom system" drawn from Scripture.

[19] However adherents of Divine Right would probably admit that the existence of Kings is needed to prepare men to receive this Divine Idea, just as the slave boy in the house of Meno needed a diagram plus the questions of Socrates in order to "remember" mathematics.

believe in these systems does not derive from their "rationality" or "truth," but should be investigated by sociological methods which are neutral in respect to the two belief systems. The same kind of analysis could be used to demonstrate the epistemological equivalence of mathematics to other deductive systems of conventions like Catholic dogmatic theology, the "science" of witchcraft as revealed by the *Malleus Maleficarum* in the sixteenth century, the extremely technical late scholasticism[20] of the fourteenth and fifteenth centuries, or systematic Marxism taught at a good Russian university in 1952. In some ways the comparison is unflattering to mathematics, since according to Wittgenstein mathematics has *no* subject matter; there is no reality to which it refers. It consists solely of rules and algorithms for calculations; numbers have no existence apart from calculation. The other systems at least are *about* things of intrinsic importance.

Even if we assume that mathematics is made up entirely of rules, conventions, and social constructions which are not grounded in any inner necessity, other features of the SSK approach are unconvincing even when it is applied to something which everyone agrees is conventional. Consider the game of chess, for example, which men have obviously invented rather than discovered, and whose rules are historically contingent conventions that could have been different. Suppose a player acts on a belief that he should move a knight in a particular correct way during a game, and a SSK critic of chess wants to explain his "belief." Plainly although sociological and historical factors may explain the origins of chess, such explanations are besides the point in the case of *this* move of the knight; even if we could prove that the origins of chess somehow involved power politics in the court of an Indian or Persian ruler 2,500 years ago, this fact would have no relevance at all. Nor can any political or social characteristic of the player have any explanatory significance; at most, as we remarked in Chapter 11, they might help to explain *why* he or she became a chess player or plays in a certain style—if it could be shown, for example, that people in certain social strata, or from a certain culture, or with a certain ideology tend to like chess and prefer particular strategies. Beyond this, a sociological explanation could not go. Also, in chess as in mathematics there is no moral feeling that the rules *must* be obeyed. There is just the observation that if one decides to play chess, then this entails that the rules be followed. As in the following of mathematical rules, there is no moral obligation to do

---

[20] Anyone looking closely at the intricate arguments of Duns Scotus or William of Ockham cannot fail to be impressed by their formal similarity to mathematical argument (especially point-set topology).

so since one can write

$$\int \frac{dx}{x+1} = \int \frac{dx}{x} + \int \frac{dx}{1} = \frac{x^0}{0} + \ln 1 + C$$

as beginning Persepolis State calculus students commonly do, move the knight like a rook, or upset the table—but then we would not be playing either mathematics or chess. The only moral aspects present are (i) the obligation—as in mathematical research—not to cheat and (ii) the obligation to keep promises if one has agreed to play. If we seek an explanation for the actual move, the only possible answer is the player's account of what he is doing. He will say perhaps "I think I can mate in three moves from this position for the following reasons ..." or "this move is the best at this stage of the King's Indian defense because ..." Just as in the case of mathematics or science we would have to listen to the actor's arguments to find the "cause" of his belief. Of course we could go deeper, pointing out that since the player has a 1300 cc. brain with a certain well developed area in his cortex that controls chess he is capable of planning a chess strategy. But that observation by itself would not be very informative. Unless we can correlate biochemical neurological states exactly with the particular move and further explain the "cause" of these states we would have to be satisfied with the player's explanation of his action. There are also in chess analogues of mathematical theorems, such as a claim that a certain position entails checkmate by White in three moves. The "proof" of the claim would have the same kind of certainty and independence of space and time as proofs in mathematics. To repeat, it would be nonsensical to ask SSK to explain either the *content* of this "theorem" or the reasons we believe in it. The explanation for both is just the argument demonstrating the checkmate. The same applies to mathematics. The answer to "why do you believe the theorem?" is to give the proof.

We should note that the above analysis applied to either chess or mathematics is compatible with (i)—(iii) of the Strong Program. (i) allows for non-sociological causes. Suppose the chess player has made a mistake in his move either in the game or the solution of the "mate in three" problem or that the mathematician has made a mistake in his proof. The explanation for the error would also be of almost the same "type" as the correct analysis. We would give the person's analysis or proof and just point out the logical gap, observing that it is quite easy to make mistakes in complex deductive arguments. This would satisfy "impartiality" and (nearly) "symmetry" ((ii) and (iii)). The only weakness of the Strong Program is, as we have seen, the unwillingness in most cases of its supporters to accept any explanation other than sociological or political ones.

We now look briefly at the claim that mathematics is inductive. There is a (small) kernel of truth here. If humans were pure minds living in a nonphysical realm with no discrete objects at all, it is doubtful if any mathematics would have been discovered or invented. The primordial need to keep track of animals, loot, women, or slaves must in part explain the development of arithmetic.[21] However, to see that mathematical statements cannot be purely inductive generalizations, we need only note that any such generalization can be *imagined* to fail given additional examples. A person would be surprised if tomorrow the sun exploded instead of rising, or if the pen slipping out of his hand flew to the heavens rather than falling to the ground, or if the crow in the next room was not black but pink; but these possibilities are *conceivable*—we know what it would be like for these things to happen. But we cannot imagine what it would be like if in the next room two groups of each of two distinct objects did not make four objects, any more than we can imagine what it would be like for the left rear corner of the room to be the origin of a *physical* five dimensional coordinate system.[22] A case can be made following Kant that both arithmetical statements and our descriptions of perceptual space are synthetic *a priori*, but we don't want to get bogged down in this issue.[23] However, we can conclude that they are not inductive under any normal interpretation of the term.

What about the assertion that mathematics is fallible and corrigible? Recall that Paul Ernest in knocking down what he calls "absolutism" argues that trying to justify the certainty of mathematics leads to infinite regress, since trying to justify the axioms would lead to additional justifications without end. On the other hand, if we view theorems as hypothetical statements and regard them as true if they are correctly deduced, then this, according to Ernest, presupposes the correctness of our logical rules which themselves cannot be shown to be true. The trouble with this argument is that it would also apply to chess. We could axiomatize chess by carefully stating the properties of the pieces and the structure of the board. The rules of inference would amount to recipes for getting from one position

---

[21] A Platonist however would argue that the presence of individual objects to count is at most only an stimulus and neither a necessary or a sufficient condition for mathematical insight.

[22] This is not to say that we cannot develop the logic of such a system. It is a special case of a finite dimensional vector space that mathematicians deal with every day. Nevertheless, we cannot *physically* conceive of such a system in our perceptual space.

[23] For additional discussion see note 30 below.

on the board to another via a "move." As we have previously noted we can imagine the solution of a chess problem as a proof of a "theorem"; in the latter case the conclusion of the theorem would be the final position on the board. Then to be consistent Ernest should regard chess as he does mathematics as "fallible" and "corrigible" because there there is no "justification" for the way the rook moves or for the number of squares on the board!

When one contemplates the "fallibist" thesis in mathematics, one is reminded of Dr. Johnson's refutation of Berkeley's idealism by kicking a stone. It is perhaps *ad hominem* but effective to ask: "what mathematical theorem in the corpus of accepted mathematics (not a poorly refereed paper from Yemen in a backwater journal) might be incorrect"? Is it possible that we may discover that the set of prime integers is *finite* in number, that there really is a largest prime integer, and that the belief that there is no largest prime could go the way of Ptolemaic astronomy. Or perhaps this result, while true now, was false 200,000,000 years ago or is false now on Mars. The same questions can be asked for the irrationality of $\sqrt{2}$ (or if you reject the existence of irrational numbers, its incommensurability with the unit side of the right triangle). Could π really be 3 as the Indiana legislature is reputed to have once legally maintained based on measurements of the columns of Solomon's temple given in the Old Testament (I Kings vii.2.3)? Same question for the Radon-Nikodym Theorem, the Lagrange Four Square Theorem,[24] or the Banach Fixed Point Theorem. Can we really believe that there was a time 70,000,000 years ago when two tyrannosauruses met two plant eating sauropods and the result (at least for a short time) was not four dinosaurs.[25] In what sense are such statements "fallible"?

It is true that proponents of fallibism are aware of the perceived certainty and universality of mathematical knowledge. Therefore, they sometimes straddle the issue by slipping in a nonpsychological justification for these perceptions while still trying to claim that mathematical knowledge is "contingent." However, the success of this strategy is not clear. Ernest, for example, takes a semi-formalist position. Mathematicians have invented "the game of mathematics"; they agree on the fundamental assumptions and the rules to be followed so that:

---

[24] That is, the fact that every positive integer is the sum of at most four integer squares. For example, $39 = 6^2 + 1^2 + 1^2 + 1^2 = 5^2 + 3^2 + 2^2 + 1^2$.

[25] This example is due to Martin Gardiner. Note also that the example due to Lakatos that $2 + 2 = 5$ presupposes normal addition. The weight of the postal box is just not written down in the sum. The same is true for modular arithmetic, e.g. $2+3 = 1 \mod 4$ which is also sometimes given as an example of an alternative addition convention.

> ... many of the consequences of mathematical theories may follow inevitably from them, just as checkmate for white may follow inevitably in three moves from a certain chess configuration.[26]

But elsewhere he agrees with the late Wittgenstein's philosophy of mathematics:

> In Wittenstein's philosophy, necessity is founded in following socially accepted rules (i.e., those that are standard practice in mathematical practice). It follows that neither the individual's personal interpretation of rule following, nor the logical necessity of mathematical proof, in particular, has the absolute and extrahuman basis of certainty presumed by absolutism. [27]

Or again:

> Once certain assumptions, definitions, and rules are accepted the greater part of mathematics *does* follow by ... necessity. But that necessity rests on a set of assumptions that themselves are not necessary. Therefore the fallibist claim is that ... the body of mathematical knowledge as a whole is contingent truth.[28]

Ernest, however, fails to recognize that the second passage contradicts the first. To interpret of mathematics as a "game" or as a set of Wittgensteinian "socially accepted conventions" and then assert its fallibility is inconsistent. It is claimed that necessity in mathematics is like necessity in chess. Is chess fallible? Is the checkmate by white in three moves uncertain or a contingent truth because the rules of chess aren't justified? As we have already observed such a question is pointless. From a game-theoretic or formalist perspective the mathematical apparatus behind a theorem consists of axioms and rules of inference. Until there is some interpretation of these axioms and rules they are not held to be either true or false. They are just there.[29] A "proof" is just a finite list of steps where each follows from one or

---

[26] Ernest (1998), p. 259.
[27] Ibid., p. 82.
[28] Ibid., p. 192.
[29] However, if we wish we could think of "truth" is an *undefined* property which the axioms are assumed to possess (we might if we want call the property "blue" or "constipated"). This property is assumed in designing the system to be preserved by the rules of inference; i.e., a rule of inference applied to an axiom or previous step(s) having the property preserves the property. Saying that a result is "true" therefore is just a shorthand for saying that the proof is correct.

more previous steps by a rule of inference, just as a game of chess is a finite list of correct moves on the part of the players. The question of fallibility or corrigiblity does not arise in either chess or mathematics unless there is a mistake in the proof or an illegal move has been made. Under such an interpretation a mathematical result can have the "extrahuman" certainty that the absolutists claim, just as the solution to the chess problem does. In fact, the demand that the axioms, the rules of inference of a system or logic itself be shown to be "true" or "necessary," and then to claim that the system is "fallible" or "contingent" if this cannot be done is nonsensical. In the case of logic, giving a proof that logic is valid would seem to demand the use of logic and is thus begging the question (rather like proving for someone *else* that the external world is real). More fundamentally, what can "true" or "necessary" mean in this context? If we agree with some kind of correspondence theory of truth, e.g., "Snow is white" is true if and only if snow is white, to what situation in the world does the Law of the Excluded Middle or the axioms of group theory "correspond"? From a formalist point of view these are *our* inventions; the axioms in particular as we have said are held by *fiat*. If we wish as in the third passage we can say that the theorems we prove are "contingent" or not "necessary," but only because we could have been proving theorems in ring rather than group theory, just as is the case for the "mate in three" procedure in chess because the rules *could* have been different—we could be playing checkers or poker instead of chess. But this trivial observation does not invalidate the fact that both the proof and the solution to the chess problem may be "absolutely" certain and necessary once we have agreed on the rules. Ernest's criticism would only be legitimate if, say, Euclidean geometry claimed to be a set of absolute truths about "physical space." Then we could demand justification for this claim.[30] Oddly enough, however, Ernest's thesis concerning the fallibility of mathematics is consistent with Platonism. With Plato and G. H. Hardy we could claim that mathematical objects like groups and rings

---

[30] We have to be careful here. Human "perceptual space" (a Kantian sense is intended here) may very well be Euclidean. Like residents of Abbott's "flatland" (Abbott (1884)) who cannot grasp the third dimension or picture spheres, we cannot "picture" any other kind. In this case Euclidean geometry may be a set of "absolute truths" about *our* space. But actual space whose properties may be revealed to us by refined scientific tests might be non-Euclidean; or for that matter, we may be embedded on the surface of a ball in an a separable infinite dimensional Hilbert space. In the same way flatlanders may discover that they live on the surface of an three dimensional sphere whose properties they can intellectually explore without being able to visualize. For them the geometry of the plane would consist of synthetic *a priori* truths.

actually exist independently of human minds, and that our mathematical theories about them record our observations with the "eye of the mind" of their properties. But vision can be mistaken. Ernest could ask how we know that the structures we erect to aid in this vision are correct. Perhaps, the axioms we think of mirroring a group do not accurately capture the structure of this object.

Another context for which we could speak of fallibility would be in applied mathematics. Suppose we set up a probability model to calculate life insurance premiums, or some axiomatic game theoretic scheme to help the Pentagon predict the body count in counterinsurgency warfare in some fly-blown third world country or in the Mifflin district of Madison. If our mathematical model fails to model reality closely enough then we could say it is corrigible—and try to improve it. But this is not a deep philosophical insight produced by the wisdom of social constructivist ideologues; mathematically oriented scientists have been aware for centuries that their models are idealizations of reality and thus only approximate.

We come now to the question of "mistakes" in the mathematical literature which for Ernest is additional evidence of the fallibility of mathematics. Everyone—especially a working mathematician—will admit that the literature (including at least two of this writer's own papers) contains mistakes. In many cases the theorems are still "correct" in that the error may be repaired and a correct proof given. If this is the case, if and when the author becomes aware of the mistake he will often publish a short note in the journal where the paper appeared entitled something like "A correction to Theorem X of my paper Y. Sometimes, unfortunately, the error is more serious. Either the repair is unknown so that theorem now descends to the status of a mere conjecture, or there is a counterexample showing that the theorem is false. If the result is of sufficient interest this may generate a note with a title "Concerning an error of Theorem X in Z's paper Y." Probably, in most cases the error is harmless because nobody has any interest in the result, so that no one cites it. On the other hand, if the result is of some interest, the mistake will almost certainly be eventually uncovered. Either one has to give a talk on the result at a conference, or the method is needed for one's own research, or the result must be presented in a graduate seminar. In these situations it is important to understand *every* detail of the argument. (Psychologically many mathematicians, including this author, have the nightmare being caught out during a colloquium talk by a comment like "Isn't the identity map a counterexample to your theorem?") If after intense study one cannot understand a step in the proof, something is probably wrong.³¹ "Mistakes" however are practically always at or fairly near the frontier. The mathematics in standard graduate courses that merely furnishes the basic concepts, language, or results to be used later as tools to eventually reach the frontier has been so digested, worked over, streamlined, processed, puréed, and taught to innumerable students, that the presence of a serious mistake (other than one caused by a misprint) is almost impossible.

The strongest support for the fallibility thesis is Gödel's Second Incompleteness Theorem which states that any formal system containing Peano arithmetic cannot prove its own consistency. If indeed the axioms of a mathematical theory are inconsistent, a consequence is that any proposition whatever which is expressible in the theory can be derived. In the case of calculus, for example, we can show that both

$$\int_0^1 x^2 = \frac{1}{3}$$

$$\int_0^1 x^2 = i,$$

although the "proof" of the second statement would probably be difficult without directly invoking the contradiction.³² This objection is formally irrefutable. However, it is also true that (i) no contradiction anywhere in mathematics has ever been discovered since the paradoxes in naive set theory were patched up early in the twentieth century (ii) any new contradiction that may be discovered probably will be in remote realms of set theory and can be "removed" as the earlier contradictions were. We can also remark that to emphasize the seriousness of a possible contradiction in a mathematical system contradicts the view of Wittgenstein, SSK's favorite philosopher, who in some much criticized passages maintained the mathematicians should not get alarmed about a contradiction in the system. If the contradiction is undiscovered, it causes no harm. If it is discovered, it is—to use a modern analogy—rather like a flaw in the software. Most of the time the program runs fine. If it starts an infinite loop we fix the

---

³¹Exactly this happened to one of my colleagues when he was trying to master a topological result from the 1960s which had won its author an endowed chair at an important southern university. It turned out that the result was false; a counterexample existed. Fortunately, however, all of the many consequences of the mistaken theorem cited in several papers were *true* and could be proven by other means–the demonstration of which resulting in a new well-received paper by my colleague! This of course is an example of a process much like Lakatos' "logic of mathematical discovery."

³²For a proof that $p \wedge \neg p \Rightarrow q$ for any well formed formula q, see Shanker (1987), p. 241. $p \wedge \neg p$ and *modus ponens* then allows us to conclude q.

problem then. In the case of mathematics all we need do to keep out of trouble according to Wittgenstein is to forbid, by a new rule if necessary, any deduction from a contradiction.[33] Wittgenstein in short:

> ... set out to demonstrate not that there are no contradictions in mathematics, but rather, that their existence is both familiar and harmless, and further, that it makes no sense to speak of a *hidden* contradiction.[34]

Wittgenstein's position is controversial, but it is closer to the actual position of working mathematicians (despite their official rhetoric) in the sense that they lose little sleep over the remote possibility of inconsistency within an established part of mathematics—ring theory, for example.[35] We should also remark that the very improbable possibility of inconsistency in mathematics has nothing to do with the certainty of a particular theorem T. If the proof of T is correct it is an "absolutely" certain consequence of the axioms whether or not the theory is consistent.[36] The fearmongering, therefore, concerning mathematics of supporters[37] of SSK seems an exaggeration, and considering the *real* contradictions[37] within their system, calling mathematics "fallible" on the basis of Gödel's theorems is especially ill-conceived.

According to my dictionary "fallible" means: (1) "liable to err", (2) "liable to be misled or deceived", or (3) "liable to be erroneous or false." Hence, knowing that something is fallible licenses one to doubt it. As we have admitted, this word certainly can apply to the process of research and so the fallibility thesis of Ernest, Hersh, Lakatos, and others is most persuasive when applied at the mathematical frontier. Every day mathematicians

---

[33] See the discussion of this issue in *Ibid.*, Chapter 6 which supports Wittgenstein's position.

[34] *Ibid.*, p. 227. This reasoning is unsound however. In trying to prove p in an inconsistent system we might prove $\neg p$ and conclude that p is false. But if the system were "repaired" and the axiom(s) causing inconsistency removed p might be a theorem of the system as desired.

[35] One can imagine, however, the Calculus Liberation Front at Persepolis State bringing a lawsuit demanding that their time not be wasted by being compelled to learn a possibly inconsistent subject. An alternative strategy might be a demand for a universal grade of "A" since unless mathematics can be shown to be free of contradictions any formula written on a test might be as good as any other in the sense that it can be deduced from the axioms.

[36] Of course if the theory from which T is derived is inconsistent, the same is true of $\neg T$. This does not make the theory "fallible"; rather, it is useless.

[37] Connected with reflexivity, for example, or their dual view of truth as sometimes relative or sometimes apparently absolute.

(including the writer) conjecture and even "prove" false theorems; usually, however, the mistakes are discovered by the researcher himself in the cold light of analysis when a few days later the initial euphoria has evaporated. This psychological phenomenon appears to be quite common and probably occurs at the research frontier in many other fields besides mathematics. If "fallible" were limited to such situations there would be little to quarrel with the use of this adjective. However, given its definition and enthusiastic application to all areas of mathematics by Ernest and others, a bystander unfamiliar with mathematics and indoctrinated with the idea of mathematical fallibility might reasonably conclude that the subject is riddled with errors, and wonder why it has any prestige or is required in school.[38] Is this an accurate description of mathematics? If mathematics is said to be fallible, we may legitimately demand examples of serious errors, not mere mistakes in proofs similar to mistakes made in the solution of a chess problem or oversights in the research process. Ernest gives none.

Whatever the philosophical problems about the status of mathematical "truth," settled areas of mathematics—the puréed kind found in undergraduate or graduate level textbooks as distinct from the frontier—appear more certain than the claims of almost any other human discipline, including the hard sciences such as physics or chemistry. In particular *all* nontrivial philosophical, political or sociological statements (including Ernest's theses on the fallibility of mathematics) can be and are contested, while anyone of normal intelligence can be brought to agreement with simple mathematical theorems. The dialogue *Meno* may not, as Plato intends, prove that the soul remembers mathematics before birth. But it does prove the possibility of this agreement; teachers use similar methods every day.[39] We can call mathematics "fallible" if we want. It is a free country, but it makes the word almost an empty epithet. Like calling a political opponent of whatever strip "fascist" in the 1930s, it degrades meaning and covers over what should be important distinctions between mathematics and other belief systems such as Hermeticism, Mesmerism, flat earth cosmology, neoconservatism, and even SSK/PIS itself. The charge that one must be an "absolutist" if one rejects fallibism as a useful adjective to describe

---

[38] This attitude reminds one of an old "Calvin and Hobbes" cartoon. Calvin, faced with with another unpleasant mathematical lesson from Miss Wormwood, demands that a lawyer be summoned since mathematics must be a form of religious indoctrination forbidden in public schools.

[39] But it is fair to point out that had Socrates dealt with a freshman fraternity member at Persepolis State instead of the Greek slave boy in the house of Meno, the whole history of Western philosophy would have proceeded on different lines.

mathematics seems ill-founded. In the first place, the epithet describes a politically convenient straw man. No mathematician known to the writer is an absolutist in Ernest's sense;[40] he suspects that very few exist outside of philosophy departments. We all believe that mathematics changes and evolves; some theories lose interest, are forgotten, and are replaced. We all know that mathematical discovery is initially uncertain and prone to error. These are hardly profound observations, and to use the term "fallibility" referring to them seems overkill; it would be more accurate to substitute a less loaded term like "revisable." However, the parts of mathematics that are supplanted, unlike Galenic physiology or Ptolemaic astronomy remain valid in that their results are (almost always) still correct. This fact seems not to have been noticed by the fallibilist school of PIS.

To put the inconsequence of "fallibism" in still another way, it is like saying that Zeno's still unrefuted demonstrations that motion is self-contradictory[41] and that failure to prove the existence of other minds means that the belief that automobile travel and psychiatry are possible is "fallible". Of course, in an extended sense *any* statement is fallible and can be doubted including the statement that I am now at my computer writing this, or that I had BBQ last night for dinner. These could be hallucinations. But if we are warranted to believe such statements, we certainly have equal or greater warrant to believe the main theorems in Royden's *Real Analysis* or Lang's *Algebra*.

So far we have concentrated our criticism on the general SSK approach to mathematics as exhibited by Sal Restivo, David Bloor, and Paul Ernest. What can be said of Reuben Hersh and Imre Lakatos?

Reuben Hersh has given us a valuable account of the way mathematics really works and develops. His distinction between the "front" and "back" of mathematics is especially apt. His book *What is Mathematics, Really* is a refreshing change from the arid productions of philosophers who actually have had no experience doing real mathematics. But like Ernest with whose educational goals he sympathizes, he may over emphasize "fallibility." and "absolutism." Some philosophers of mathematics are certainly absolutists, but at the risk of repetition we can only say that they seem to be a rare species among practicing mathematician.

For Hersh mathematics is a huge network of concepts originally created by men but soon becoming independent of any individual. Once created

---

[40]However, G. H. Hardy comes close.
[41]To see that Zeno's paradoxes are still taken seriously, and that their standard refutation by limit arguments in calculus books misses the point, see Salmon (2001)

"mathematical objects are there. They detach from their originator and become part of human culture."[42] Their status is similar to other aspects of human culture that seem like mathematics to be neither mental or physical such as money, the Supreme Court, sonatas, the rules of the road, the game of chess, etc. These are social institutions which are neither physical objects or ideas in someone's mind.[43] What makes mathematics interesting and accounts for the naive Platonism of many mathematicians according to Hersh is that its concepts, although socially constructed, have consequences which are hard to discover. But once discovered, these consequences seem timeless and to have always been "there." Certainly, there is some validity to Hersh's thesis. The practice of mathematics, involving journals, universities, referees, tenure decisions, Department Chairs, and so forth, is a social institution. But this may not be quite true of its *content*. The Platonist could still argue that mathematicians "observe" a world of actually existing abstract objects, neither physical or mental such as numbers, groups, measure spaces, etc. Our concepts describe these objects but are not the objects themselves. Concepts are in individual peoples heads. How do we know that my concept of "two" is the same as your concept of "two" unless they have something in common, namely reference to the common abstract object "two"? Hersh is not really interested in such problems. He simply keeps asserting that mathematics is a certain kind of social construction because the alternative of Platonism is fantastic (and he thinks politically reactionary). This position has some awkward consequences. A mathematical truth cannot be said to be true before its discovery, for it did not exist then. To Martin Gardiner's claim that $2 + 2 = 4$ must be a timeless natural law independent of human thought because two dinosaurs meeting two other dinosaurs in the Jurassic era made four dinosaurs in the absence of any human being, Hersh replies by saying that in this context $2 + 2 = 4$ is a truth of elementary physics applied to dinosaurs (Hersh follows Mill here). In mathematics, on the other hand "2" is a *conceptual object* which human beings share and $2 + 2 = 4$ is a property of the number system.[44] Similarly, according to Hersh there is no reason to expect that our mathematical concepts to hold on Quasar X9 (in our absence) anymore than for our other social institutions to exist there.

---

[42] Hersh (1997), p. 16.
[43] These non-mental, non-physical parts of human culture amount to Popper's "World 3."
[44] *Ibid.*, p. 15. In the former case "two" is an adjective modifying dinosaurs in the latter it is a noun.

To interpose a brief philosophical digression here, the problem of the status of mathematical objects is a form of the ancient (still unsolved) problem of universals. How do we get from individuals to classes? We say that x_1 is in the same class as x_2 because both have something in common. This something is not found exclusively in x_1 or in x_2; since it is shared by both, it must consist of a certain relation between them. This relation is real, but not physical or simply an an idea in my head.[45] Of course we could say that the construction of classes is a social decision. Classes really do not exist independently of us. We group certain creatures together in the category "horse" for no intrinsic reason, but just because we decide to do so. But somehow this is not very convincing. It will not do to say: "We don't do this arbitrarily. We do it for a reason." But that reason in the end must be the result of some property shared by all the creatures we decide to call horses, and so we are back to the same problem. Nor is the mysterious faculty of abstraction posited by Aristotle a solution. For when we abstract the concept "Horse" from a group of creatures with four legs who like sugar, we are isolating a property which is either shared by all of them or simply in our minds. Both alternatives lead to Platonism because in the first case the property must be real, and in the second because the property in my mind (like the concept of "2" must have some real relation to the property in yours. But we must admit that this analysis in favor of Platonism has its own problems. It leads to a family of paradoxes collectively called the Third Man Argument.[46] On such issues the writer is neither a Platonist or a follower of Hersh; rather he is an agnostic.

Imre Lakatos also in our opinion over stresses fallibility, although to be fair he like Hersh locates fallibility in the "back" or in the process of conjectures and refutations by which mathematics develops rather than in the embalmed mathematical corpse resulting from the cosmetic attentions of the mathematical undertakers known as "deductivists" (or the competing funeral home of Formalism, inc.). A more serious objection is that Lakatos' effort to show that mathematics follows the same path as Popper envisaged for science is a bit rigid. To be sure, instances of this kind of development can be found in the history of mathematics. His examples of polyhedra and the Euler-Descartes formula or Cauchy's problems over convergence of con-

---

[45] This is the argument that Russell in Russell (1999) uses to conclude that relations are real. They exist independently of us and we perceive them.

[46] That is, suppose we group individuals $x_1, x_2, \ldots x_n, \ldots$ into a class C because the x_i are all pairwise related by a relation R. But what makes the *individual* relation (call it R_1 between x_1 and x_2 the same as the *individual* relation R_2 between x_2 and x_3. Plainly, this line of reasoning will lead to an infinite regress.

tinuous functions are certainly illustrations. But the scheme is too rigid. Lakatos tries to force mathematics into a bed of Procrustes. Certainly mathematicians make conjectures and try to find counter-examples. But they also generalize, pursue analogies, modify the assumptions of somebody else's paper to see if something more interesting or general can be proved, axiomatize, define, wipe a dirty mathematical plate clean in the "kitchen" and re-serve it with a new entrée,[47] and do all manner of odd things. Mathematical research either by one mathematician or in the development of a field seldom follows a dialectical path of a conjecture followed by counterexample, followed by a more subtle conjecture, and so on; nor can its disorderly course be "rationally reconstructed" in this form.

Let us bring this discussion to a close by noting that the questions SSK raises about the epistemological status mathematics such as to what degree it is a "social construction," its purported "fallibility," etc., however interesting to philosophers of mathematics or sociologists, are as irrelevant to the *practice* of mathematics as neo-Marxist academic criticism is to Microsoft. Mathematicians will shrug off the former as Bill Gates does the latter. They simply don't care about these issues. Also the ruminations of academic philosophers concerned with "foundations," or "alternative systems" of mathematics are far from the activities, interests, or literary competence of journeymen mathematicians. In particular, the so called "constructive mathematics," pioneered by Brouwer and Bishop is today a minor mathematical sect. Philosophical theses may still be churned out about it, and one may find a few true believers in good philosophy or second-rate mathematics departments but the question of nonconstructive existence proofs or the heinous sins committed with the axiom of choice arouses little interest in the average mathematician.[48] Like Ol' Man River, mathematics just keeps rolling along and produces at an accelerating rate "200,000 mathematical theorems of the traditional handcrafted variety ... annually."[49] Although sometimes proofs can be mistaken—sometimes spectacularly—and it is a matter of contention as to what exactly a "proof" is—there is absolutely no doubt that the bulk of this output is correct (though probably uninteresting) mathematics.

---

[47] As Orwell observed an elite Parisian restaurant doing in the early 1930s when he worked in the kitchen.

[48] Besides logic or set theory where such questions may arise is hardly the thing these days to study in order to get a job; in fact these disciplines may be nearing the professional oblivion of point-set topology at least in *mathematics* departments.

[49] Davis and Hersh (1981), p. 24.

The problem of the status of mathematical truth which seems to obsess some of the writers we have considered here is in our opinion not very interesting or productive; it leads to a straw man like "absolutism" or sterile philosophical debates, remote from the actual practice of mathematics. But there remain some possibilities for an SSK approach to mathematics which have room for additional development. What *is* a more interesting topic of investigation, we suggest, is not the "truth" (or why we believe the truth) of mathematical theorems, but their status as well as the status of the mathematicians who create them in both the world of mathematics and in society. Here there is a fertile field for the investigations of sociologists (and even anthropologists or psychiatrists). It is at this point that SSK can make a contribution which is potentially greater than in the case of physical science since mathematics is less tied to "reality" than the sciences. But, at the risk of repeating ourselves, we believe that SSK needs to quit trying to prove some kind of Wittgensteinian conventionality or "rule following" thesis concerning the sources of mathematical credibility and instead pursue different "Mertonian" directions. A starting point would be to investigate how—as is the case in science—the policy decisions of NSF or other government agencies can indirectly shape the content of mathematics (especially applied mathematics) by deciding which areas to fund.[50] Several writers, notably H. J. M. Bos and Herbert Mehrtens,[51] have begun studies in this direction. An even more intriguing possibility would be to look for "microsociological" explanations for mathematical growth and change. Paradoxically, the observations of a Bruno Latour may fit the "conference life" of mathematicians better than they do the "laboratory life" of scientists.[52] The effects and interplay of actual mathematical achievements, the academic status of rival mathematicians, the prestige of departments and journals, networking skills, grantsmanship, the methods by which recognition is obtained, the rise and fall of research areas, small scale group dominance, and other factors in the "agonistic field" of competitive mathematics would be a good opportunity to develop a "weak sociology" of mathematics focusing on development, change, and extinction rather than

---

[50] A case in point would be the growth of operations research in World War II or game theory during the Cold War.
[51] Bos and Mehrtens (1977) and Mehrtens (1996).
[52] Because in mathematics the epistemological confusions associated with Latour's "Third Rule of Method" simply vanish while "politics" and "negotiations" are still present.

foundations or truth claims.[53] Such a sociological description would imply that the development of mathematics is a contingent phenomenon, not a majestic unrolling of theorems subject to logical laws.[54] But to demonstrate this would be an entirely new project which we can only suggest is worth doing and cannot consider here.

---

[53] Our proposal is not completely original. An early paper by C. S. Fisher, "Death of a Mathematical Theory" (1966) on the sociology of the decline of Invariant Theory in Algebra was a pioneering effort in this direction which does not seem to have been followed up.

[54] For the contrary view that mathematics follows predictable laws of development, see Crowe (1975). To the this writer such a thesis is an example of naive historicism that has vanished elsewhere.

Chapter 13

Madison 1973

One beautiful spring day in 1973 I had just finished having a bratwurst and beer at a sandwich place on State Street, a few blocks from the Madison campus of the University of Wisconsin. As I walked out I noticed to my left a small mob (about fifteen to twenty individuals) running down State Street from the direction of the Capitol. Evidently the anti-MRC demonstrators were out again! They were headed for a Mathematics Research Center (MRC) Symposium in applied mathematics being held at a conference center on the edge of campus. I was attending the Symposium and was on my lunch break. Although at age 34 I was a bit old to be a recent Ph.D. in mathematics, I was in fairly good shape as I was in the habit of jogging daily a few miles along the shore of Lake Mendota, so it was a simple matter for me to lope at an easy pace in front of the approaching mob and take shelter behind the police lines guarding the entrance to the conference center.

The talks were just about to begin. They were the usual fare offered in most of the more than thirty mathematics conferences I would attend during my career: endless sequences of unintelligible over-detailed handwritten slides, speakers concentrating on the most technical and involved details of esoteric theorems rather than the main ideas or motivation behind them, and a bemused, sleepy, and totally bored audience with the possible exception of one or two members who were working in exactly the same field and actually understanding the speaker. Generally, the talks at such a conference represented a good occasion for a relaxing nap; but one could be thankful that MRC at least provided doughnuts of high quality during the coffee breaks. At this conference, however, more people than usual were paying attention. These were a half dozen or so delegates from "Science for the People" who were allowed to attend. Two or three were mathematics students with the others seeming to be nontechnical types—perhaps humanities majors or dropouts living in the Mifflin area next to the campus—who knew no mathematics at all. Judged from its magazine, the general level of mathematical knowledge of Science for the People was not high. A recent issue had featured an article on the crimes associated with spline functions. It defined a a spline to be a curve with "infinitely many points"; somehow this property was correlated (if my memory is correct) with the napalming of infinitely many Vietnamese villages. But in spite of their probable ignorance, the delegates were enthusiastic participants in the conference. They hoped to find evidence against the mathematical "war criminals" of MRC, and some were furiously taking notes. They were mostly dressed in standard 1970s hippy attire: dirty army jackets and jeans with greasy ponytails, beards, and sideburns. Their most prominent psychological characteristic was a complete lack of a sense of humor. In other eras they would have made good Puritans, witch hunters, followers of Teichmüller—when he drove the Jewish mathematician Landau from Göttingen—or the absolutely serious students I would meet at the University of Essen in 1989 who were passing out GDR propaganda in the university *mensa* a few months before the regime collapsed.

Finding the entire atmosphere depressing, I left the auditorium for the anteroom by the front entrance. Half a dozen police were there. Outside an orator was declaiming the evils of MRC to the demonstrators. This too was boring. But suddenly the boredom ceased! As if by a signal, the orator and his audience rushed in unison to the entrance and into the anteroom. Some were brandishing sticks or what seemed to be lengths of plumber's pipe that they had hidden in their jackets. The police retreated to the narrow hallway between the anteroom and the auditorium and started to whack away with their truncheons. Like the pass at Thermopylae it was an excellent defensive position. The outcome was more fortunate for the police than for the Spartans; no radical sympathizer showed the demonstrators a way to outflank the police line through a side door, and a bit bruised they eventually retreated from the building. But not before one screamed at a MRC official that they knew where he lived and that his wife had just had a baby.

I had arrived in Madison in the fall of 1972 to begin a two year postdoctoral appointment at MRC. This organization, founded in 1955, was funded by the Army and its formal purpose was to develop mathematical tools useful for defense purposes. The staff consisted of a few permanent members who were generally top ranked mathematicians and a greater number of visitors. The latter included a few postdoctoral appointments for young mathematicians such as myself who were just starting their careers, visitors from other universities on sabbatical leave, and Wisconsin faculty on half time appointments. MRC's directors were especially distinguished. The first was Rudolph Langer (1894–1968), a member of the talented Langer family.[1] He was succeeded by the logician J. Barkley Rosser (1907–1989),[2] and then by Robert C. Buck (1920–1998). The Center was well equipped and had excellent technical typists.[3] Its product consisted of "Technical Service Reports," many of which were later published in refereed journals. As far as I could tell, little if any classified work was done here. It was probable, however, that the permanent staff did some consulting for the military and gave talks at Army bases and defense laboratories. But as a whole the institution resembled a standard academic research center in mathematics, such as Berkeley's Mathematical Sciences Research Institute. The "flavor" of the mathematics MRC produced was mildly applied with no obvious application to military needs.[4] Evidently the Pentagon agreed, for a few years after my visit Madison lost the MRC contract and it was moved to Cornell. But this arrangement was still not satisfactory to the Army and MRC was eventually abolished. These developments saddened me. As originally conceived MRC was an enjoyable place to do research and especially valuable to young mathematicians, since at a critical time in their development they were free from the incubus of teaching.

The minor riot against MRC that day illustrated the nasty political atmosphere at American universities in the last few years of the Vietnam war.[5] Similar scenes of disorder and activism could be found on almost

---

[1] One of his brothers was the Harvard diplomatic historian William L. Langer; another was a psychiatrist who had written a psychological study of Hitler during the war. His sister was Suzanne Langer, a well known philosopher and critic of the arts.

[2] Rosser had been a student of the logician Alonzo Church at Princeton. He made many notable contributions especially in logic and number theory. These included improvements of Gödel's Incompleteness Theorems and a proof that the nth prime $p_n > n \ln n$. His wife Anetta was First Violin at the Madison Symphony Orchestra and an amateur ornithologist.

[3] The typing pool which consisted mainly of women with high school educations had typed so much mathematics that they had an uncanny ability to notice errors in the text. On several occasions my typist asked "Dr. Brown, are you sure that this is correct?" Usually this meant that I had made a major mathematical mistake!

[4] I spent my time thinking vaguely heretical—and from hindsight rather shallow thoughts about splines—and writing research papers which would win me a tenure track job in a difficult market when my appointment ended in 1974.

[5] However, by the time of the riot the war had ended. A truce with North Vietnam was signed in Paris on January 27, 1973.

every American campus from 1967 to 1973.⁶ What explained the unrest? Certainly the war was the primary cause followed by student perceptions of the racial situation in the US. Not only were both considered immoral, but the war involved the draft. In the 1960s this institution was a real threat, since students were being drafted out of college and some became casualties.⁷ The "baby-boom" generation of the 1960s was, moreover, psychologically different from that of their parents who had endured both the Depression and World War II. This generation was affluent, used to economic security, and brought up according to the precepts of Dr. Spock in affluent suburbs by loving and indulgent parents. They were unused to the common frustrations engendered by large impersonal bureaucracies encountered for the first time at the university. There were also enormous numbers of them; total college enrollment had increased by a factor of five or more between the early 1950s and 1970. This produced an atmosphere of alienation, particularly on large impersonal state university campuses which had experienced the most growth. All these factors created a combustible situation particularly on large state university campuses which would be ignited by the Vietnam war.⁸

Beyond these general causes, specific conditions in Wisconsin contributed to the situation. Had he been a contemporary, Hippocrates in his *On Airs, Waters, and Places* would have doubtlessly argued that the extreme weather bred excess. Whether for this reason or others, the state lacked political moderation. It was simultaneously the home of various progressives and socialists together with Joe McCarthy. By 1973 Madison, in particular, was extremely polarized. The tensions bred by the war expressed themselves in unusual hostility at the ordinary political level as well as on campus. The very conservative Republican ex-mayor of Madison Bill Dyke had just been replaced by the liberal Democrat Paul Soglin in a bitterly contested election in April 1973.⁹

In August 1970 MRC, then housed in the Physics building Sterling Hall, had been blown up by an oil and fertilizer bomb stuffed into a Ford Club Wagon, one of the worst acts of political terrorism in US history. Much

---

⁶Even the University of Alabama, certainly not noted for radicalism, had seen major disruption during the 1970 Cambodian crisis.

⁷For instance a fellow graduate student of the writer and his brother were both drafted. The brother was badly wounded, and the Ph.D. of the graduate student seriously delayed.

⁸For an excellent analysis of these and other underlying causes of student discontent, see Searle (1972), Chapter 5.

⁹The polarization was not only over Vietnam or electoral politics; in 1973 the city was also seriously divided over whether and how to honor its most famous son, the architect Frank Lloyd Wright.

to the disappointment of some Madison radicals (somewhat evident also in the attitude of the city's liberal newspaper) no MRC "war criminals" had been killed. The only fatality was a physics postdoctoral researcher Robert Fassnacht, father of three children, who was working late in his laboratory. Some sympathizers with the bombing expressed regret. But in war these things happen! Had not the army killed thousands of innocent Vietnamese. Besides, who could be sure that a *physicist*, funded by "Amerika," was innocent? Two of the four perpetrators Karl and Dwight Armstrong fled to Canada and took refuge in the Toronto antiwar underground. Karl was caught and extradited to Madison in 1972; Dwight suffered the same fate in the spring of 1976. Slightly earlier (January, 1976) David Fine was arrested in California.[10] When Karl Armstrong was returned to Madison the mayor Paul Soglin, who had only narrowly defeated Dyke, and others in the liberal establishment may have realized that a trial would be expensive, polarizing, and damaging to their future political prospects. For these or other reasons, Armstrong was persuaded to plea-bargain, even though given the political atmosphere of the time a trial would have probably resulted in either a hung verdict or jury nullification. His sentence for second degree murder and the other charges was 23 years; however, he was paroled in 1980. Dwight Armstrong and David Fine were each sentenced to seven years but were respectively released in 1980 and 1979. By 1992 Karl Armstrong was operating the "Loose Juice" fruit juice stand next on UW-Madison's Library Mall[11] and Dwight, having served a second three year sentence for drug manufacturing, was in the produce business.[12] Both regretted the accidental death of Robert Fassnacht, but continued to believe that MRC was an essentially criminal organization, and are reported to be still angry over their imprisonment which they feel was unjust.

In 1973 Wisconsin was still roiled by student unrest; but since the war had ended, the frightening intensity of New Left radicalism in the late 60s had diminished, doubtlessly because the students no longer feared the draft. Now besides a few anti-MRC demonstrations there were teach-ins celebrating the Chinese Cultural Revolution as a manifestation of "participatory democracy," far superior to the soulless corporate capitalism of the

---

[10] The gang had previously firebombed a university gymnasium used by ROTC and had attempted to air bomb an ordnance works in Baraboo. Its fourth member Leo Burt has never been found. One theory promulgated by sympathizers with the bombers is that he was a government *agent provocateur*.

[11] Later he established the "Radical Rye" sandwich shop.

[12] I am indebted to Bates (1992) for most of the details concerning the MRC bombing and subsequent events.

West. Conservative faculty were still occasionally harassed and denounced especially if they were "Cold War Liberals," a political species apparently more evil to radicals than outright "fascists." For security purposes after the bombing, MRC had been moved to an upper floor of the WARF building on the western edge of campus. There was a guard on the ground floor, and on occasion a jeep mounting a machine gun parked in front. From our office windows we could sometimes see and hear demonstrators waving placards and shouting anti-MRC slogans, but this was infrequent. One had to be careful, of course, and not brag of one's affiliation with MRC on State Street. But after the horror of the 1970 bombing the place was essentially safe, and I was too young to be frightened or worried. The Madison scene was dramatic and colorful;[13] the restaurants (especially the "Ovens of Brittany" and "Paisan's Pizza") were excellent. Lake Mendota was beautiful, and after taking common sense precautions I thoroughly enjoyed my time in Madison. Also, as an ex-history student I felt that the atmosphere of leftist radicalism at the University in 1967–73 was a valuable experience since it appeared to have shared much in common with, say, that of Heidelberg in 1932–33, as recorded by Golo Mann.[14] Making allowances for the trivial difference in ideology, the Madison New Left radicals in their celebration of the Vietcong and hunt for academic war criminals psychologically paralleled the idealistic and patriotic Nazi radicals of Heidelberg who denounced the shame of Versailles and hunted for Jewish Bolsheviks.

Yet student radicals although they may have been prominent on campus had no power in the nation as a whole. They and the Vietnam war may have helped to destroy the presidency of Lyndon Johnson; but Nixon, the *bete noir* of the Left since his days as a tormentor of Helen Gahagan Douglas and pursuer of Alger Hiss, had with the help of the "silent majority" won the election of 1968. When the 1960s counterculture succeeded in capturing the Democratic party their candidate George McGovern was wiped out in the election of 1972. The Watergate *coup d'etat* in 1973 represented only a temporary revenge of the Liberals. From 1968 to the present the country has slowly moved to the right. Even Democratic presidents like Clinton abandoned many liberal positions and indulged in the politics of "triangulation" to move to the center.

---

[13] Paul Feyerabend had similar feelings about the Eastern Front. He found it so colorful and the stuff of grand opera, that he wanted to observe the drama rather than taking shelter in a fox hole. In 1973 Madison was not the Eastern Front, but when I read Feyerabend's account in his autobiography *Killing Time* in 1994, I could sympathize and realized that I had similar feelings.
[14] In Mann (1990).

Even at its most intense the New Left only involved a minority of students. Fraternity-oriented people, majors in business, engineering, or members of professional schools like dentistry were hardly touched by it. The radicals tended to be verbal types, used to playing with ideas, and were concentrated in the liberal arts, such as philosophy, history, English, political science, and the like.[15] The students were encouraged and sometimes led by charismatic faculty in these areas.[16] It was not surprising that many of these people went to graduate school (often as a means to obtain a draft deferment) and together with their radicalism began a "long march" through academia. The consequence was rich with historical irony. The radicals in their student days despised the academic establishment; now in middle age they were academically influential tenured professors, and in some departments even dominant. The whole politically correct, multicultural, and feminist atmosphere on the modern campus would be largely their work. And just as they had tormented square or merely liberal faculty in the 1960s, they would be harassed by a new generation of conservative students and critics outside the academy, although the intensity would be much less than in the 1960s.[17] At the same time, they had almost no external political influence. Relative to the wider society and the universities, the now middle-aged New Left was in the curious position of representing an "idea whose time had passed and whose power had come."[18] Regardless of their prominence inside the academy, outside of it their only purpose seemed to serve as scary examples to fear-mongering authors of books like *Tenured Radicals* or the *The Fall of the Ivory Tower*[19] and politicians who could use their excesses as excuses to pursue agendas like abolishing tenure, turning the university into a clone of the corporation, or cutting state funding for public universities. In the Reagan-Bush era, however formidable on campus, these ex-New Left faculty had become political dinosaurs a mile or two outside of it. The local beer wholesaler or used car dealer who hap-

---

[15] The writer noticed that both at Berkeley and Madison pure mathematicians also often joined the radical current. For instance, MRC faced considerable opposition from some pure members of the Mathematics Department, while applied types sought contracts and visits there.

[16] In Madison William Appleman Williams and Harvey Goldberg of the History Department were among those who played such a role.

[17] There are now web sites sponsored by conservative organizations where students can complain about the liberal or leftist tendencies of their professors. This is a common type of student complaint to the higher administration.

[18] Diggens (1992), p. 298. I am indebted to Diggens for the general point of view of the next few paragraphs.

[19] Respectively by Roger Kimball and George C. Roche.

pened to be on the city council had more political power than the holder of the most prestigious Chair of Cultural Studies or expert in "repressive tolerance."

A satisfying explanation of profound intellectual changes in social groups is often problematic. One can correlate such changes with underlying social or historical change and then argue that the former are an ideological superstructure of the latter. But causal connections are hard to come by. The precise mechanism by which the Zeitgeist changes is poorly understood. The difficulty is similar to that faced by the Occasionalist philosophers of the seventeenth century in trying to explain the causal interaction of the radically different substances: mind and matter. Besides, the association generally does not give any real information. Have we really explained Newton by viewing him as Boris Hessen did as a product of the transition from feudalism to capitalism? Do we understand Darwin any better by claiming with Marx that Evolution "encodes" the values of the nineteenth century British bourgeoise. It is certainly true that social changes may supply values and metaphors that affect ideas; but there is also an element of contingency or chance involved. In its fine structure a transformation of the Zeitgeist usually involves something new, a mutation that is not predictable from previous conditions.

But the US adventure in Vietnam, which to quote Talleyrand was "worse than a crime, a mistake," seems up to a point an exception to the opacity of the connection between intellectual culture and society. Not only did it kill around two million Vietnamese together with nearly sixty thousand Americans, ignite the student unrest we have just described, and come close to destabilizing the American political order, but it plainly caused a permanent change in scholarship and in the intellectual atmosphere. This change—of which SSK/PIS is just one component—was profound in nearly all the humanities. It consisted of many confusing and tangled currents which are conventionally labeled "postmodern." It is true that most of its fundamental ideas had, as we have tried to show in Chapter 3, had been present in embryonic form decades earlier and had been perfected by much older orthodox scholars in the 1950s and early 1960s. For instance, Herbert Marcuse (1898–1979), a guru of the New Left with his theories of "repressive tolerance" and "democratic totalitarianism," was 70 in 1968. His three books *Reason and Revolution*, *One Dimensional Man*, and *Essay on Liberation*, which were sacred texts for many in the New Left, were written respectively in 1941, 1964 (a year still in the afterglow of the Kennedy era), and 1969. And Thomas Kuhn's *Structure of Scientific Revolutions*, which had enormous (but unintended) influence on the SSK/PIS project of deconstructing science and on the humanities in general, was published in 1962.[20] But without the enthusiasm of a young new generation of faculty who had drunk deep from the cup of late 1960s antiwar radicalism, such ideas might have either remained academic curiosities having little impact or at least infiltrated the academic disciplines much more slowly. Whether or not war is itself a creative force—in the words of Heraclitus "the father of all and the king of all"[21]—it is certainly an accelerant of underlying trends. The most enduring intellectual legacy of the Vietnam tragedy would be the shaping out of largely preexisting raw materials of a new and different kind of scholarship that mirrored its founders political activism.

Despite the challenges we have noted in Chapter 3 by historians such as Charles Beard, Crane Brinton, and others, probably the majority of scholars prior to the 1960s had attempted to model their disciplines on science in an effort to imitate what they conceived to be its cumulative and progressive success. This was also true of literary criticism, history, sociology, psychology, and anthropology. Some form of positivism was the dominant "paradigm" in the humanities and social sciences; in particular, the possibility of "objective" evaluation of evidence by dispassionate observers was accepted. Also, after the flirtations with Marxism (and to a lesser extent with fascism) in the 1930s the attraction of ideology had greatly diminished by the 1950s. It was felt that the "end of ideology"[22] was at hand; the future lay with the bureaucratically administered liberal capitalistic state, its rough edges smoothed by generous policies of social welfare. This whole approach was abandoned almost *in toto* by the late 1960s and early 1970s.

The form the replacement would take, of course, depended on the discipline; but there would be common structural features regarding questions of value and epistemology. As Peter Novick has observed:

> In one field after another distinctions between fact and value and between theory and observations were called into question. For many, postures of objectivity and neutrality increasingly appeared as outmoded and illusory. It ceased to be axiomatic that the scholar's or scientist's task was to represent accurately 'what was out there.'[23]

---

[20] Kuhn was younger than Marcuse; but by 1970 he was an established scholar nearing 50.
[21] Fragment 53. See Kirk and Raven (1963), p. 195.
[22] This was a title of a famous (and much criticized) book by Daniel Bell written in 1960.
[23] Novick (1988), p. 523.

"Objectivity," in fact now became an insult, essentially a code word for a desire to pursue some reactionary agenda. It was now cheerfully admitted that scholarly work did not discover "truth" whether about the foreign policy of Philip II or the novels of Jane Austin. Rather it was a "representation" reflecting the scholar's gender, class background, ideology, or historical situation. All scholarship was critically shaped by such factors. In fact, representations, might well be all that existed. What the French Revolution, for example, was "in itself" was "unknowable" in the same sense as Kant's *noumenon* or Kuhn's "stimuli." Since Thomas Kuhn had apparently dethroned science as a reservoir of objective truth, the same contextual dependence would be true for the theories of high-energy physics or the chemical bond. It seemed to follow that the both the humanities and the sciences amounted to sophisticated propaganda. But at least as far as the humanities were concerned this was not a bad thing if the propaganda could be used for good causes such as the overthrow of corporate capitalism and the Pentagon.

Politically, the new scholarship would be quite far to the Left and identifiable by some core themes:

- It would be anti-establishment and would especially oppose Cold War Liberalism in all its works which had dominated the academy for a generation after 1945.

- The West, its culture, history, form of government would be attacked and the Third World celebrated, its many problems and abuses either blamed on the West or ignored. In particular, the US would be painted as the *fons et origo* of contemporary international evil.

- There would be attempts to "gut" the "canon", i.e. literature traditionally recognized as classic. Instead attention would be directed away from it and the DWEMS (Dead White European Males) who had produced most of it to fairly obscure female and Third World authors.

- Emphasis would be placed on the most degrading historical aspects of race, class, and gender, with every color other than white celebrated.

- Matthew Arnold's advice that the humanities should teach "the best that has been thought and done in the world"[24] would be held in contempt. Instead,

---

[24] As quoted in Diggens (1992), p. 292.

> ... the authority of excellence, greatness, truth, and morality must give way to the realities of power, the experience of oppression, the lives of the common, average, and ordinary.[25]

- In all areas of the humanities class conflict, invidious economic factors, as well as wider themes of oppression and hegemony would be emphasized.
- Older concepts like "assimilation" or "melting pot" in reference to the diverse multi-ethnic population of the US would become dirty words. Instead, there would be emphasis on "identity politics." This emphasis would not only concern straightforward political questions, like fair representation for hitherto excluded groups; but it would be argued that each such group would and should develop its own "way of knowing" and consciousness which would be incommensurable with those of other groups.[26] In particular, women and Afro-Americans would have their own insights into science and mathematics, incapable of being appreciated by white males.

Although such theses would be loudly proclaimed, their true influence is difficult to evaluate. In mathematical jargon they represent the "extreme points" of post 1970 scholarship. Perhaps the "silent majority" of scholars, especially those trained before the 1960s continued to work in traditional modes; and in an age of Nixon and Reagan the above attitudes had little effect on public consciousness outside of academia, or even within academia outside the humanities building. But they were definitely "trendy," forming the cutting edge of many humanistic disciplines, and also "set the tone" for less radical merely liberal scholarship. They came to dominate some departments and disciplines. If they were under-represented in the average state university, small private college, or "red state" institutions generally, they were (and still are) over-represented in key departments at the flagship institutions—especially in the North—such as Berkeley, Wisconsin, Stanford, or Harvard.

At the same time there would be no return on the part of the New Left to the mechanical quasi-Stalinist Marxism of the Old Left. This had begun to be discredited after Krushchev's revelations of Stalin's crimes in a secret 1956 speech to the Twentieth Party Congress. After the brutal suppressions of the Hungarian revolt and the Prague Spring in 1956 and 1968 followed by the defeat in Afghanistan, the erosion of the moral authority of the USSR was complete by the time the system collapsed in 1989. The rejection was

---

[25] Ibid.
[26] A favorite motto printed on T-shirts worn by certain minority groups was "you just don't understand." The writer contemplated putting the same motto on a shirt printed just under the Poincaré inequality.

also stimulated by the fact that many of the new generation of radicals had been "red diaper babies". For them traditional Marxism was the ideology of the parents; therefore it was "square" and would be rejected along with all the other boring values of the previous generation. The primary ingredient of student revolutionary movements would be moral indignation rather than a well articulated ideology.[27] Yet, the Marxist frame of mind was fundamentally still present, it just assumed more exotic forms. Intense moral rather than than strictly ideological inspiration would be found in some of Marx's more humanistic early manuscripts of the 1840s, the Eurocommunism of Tito's Yugoslavia, Mao's China (especially his cultural revolution and little red book), the tragic adventures of Che Guevara in Latin America, Castro's Cuba, in various Third World liberation struggles, or in Franz Fanon's celebration of anti-colonial violence.[28]

Ironically, what enhanced the influence of the more radical strands of the new scholarship was the caution of its creators. Having long passed thirty, few were interested in continuing the activism of their own youth; they valued their own tenure and TIAA/CREF portfolios too much. It was sufficient to undermine the system from within while playing the academic game and enjoying the luxury of a tenured professorship. These newly empowered faculty would play a role similar to that of a much earlier generation of intellectuals, slowly eating away at the moral foundations of "Late Capitalism" in the same way that the *philosophes* of the 1750s had undermined the *ancien régime*. Revolution could be left to the future when the rotten structure would surely collapse. Consequently, faculty activism did not involve teach-ins or the leadership of turbulent student demonstrations, but mainly took the form of mildly subversive articles in obscure journals written in an opaque jargon and read at best by a few fellow believers. Within the University they often became allied with the administration or had even joined it, and (as we have noted in ironic contrast to their anti-adminstration violence of the 1960s) could safely and respectably push politically correct nostrums in the form of ethnic studies, compulsory ideological conditioning sessions for entering freshmen, and

---

[27] Searle (1972), p. 59.
[28] A similar phenomenon occurred in Europe in the 1970s. The writer recalls graffiti in the mid-70s at the Free University in Berlin: "Death to Carter and Brezhnev. Long live Mao [or Baader-Meinhof or Pol Pot]." A friend of his from the CSSR who was quite used to the staid and rather senile communism of the regime was genuinely alarmed by the Red Brigade types among the students he met while visiting a mathematical institute in Florence.

demands for minority hiring (or even affirmative action in grading[29]) with respect to race and gender. Unlike anti-war protests of the 1960s this was a risk free activity which by the 1990s was part of a new semi-official academic orthodoxy. Opponents could be safely and officially demonized as "reactionary" or "racist."

The precise form of this generation's contributions to specific disciplines, however, was not completely predictable from the intellectual background of the 1960s. Although it was characterized by the value system we have described, it exhibited some novel features as well as dependence on what pre-1960 intellectual ingredients were available to be rearranged or transmuted.

In the historical disciplines one got Hayden White who argued that the "stories" that either historians or novelists constructed in their work were equally "real." Both were "made", not "found"; hence there was little significant difference between them. Influenced by Thomas Kuhn, White also maintained that the work of great historians were effectively closed thought systems which were mutually incommensurable. The only reason for preferring one over the other was moral or aesthetic.[30] How commonplace a relativist epistemology could be even among lesser historians can be illustrated by remarks like "the poststructuralist effects of Jacques Derrida, Michel Foucault, and others further undermined claims for scientific truth and objective reality", found in articles having nothing obviously to do with the theory of knowledge;[31] Fortunately, these instances still remained less frequent than references to the Dialectic or to the Party which littered articles in every area of scholarship in the Soviet period.

Historical fashion also turned from traditional political, diplomatic, or military history, centering on "great men" and elite classes generally, to an emphasis on social history—especially in the form of "history from the bottom up" with a concentration on factors of race, class, and gender, and often accompanied by the thesis that one culturally distinct group such

---

[29] This was never made explicit for obvious political reasons. However, at certain universities a non-tenured instructor could get into real trouble if the failure rate of protected minorities was viewed as excessive. Such policies continued at the graduate level. The writer recalls seeing in the early 1970s the announcement at a major US mathematics department that the Ph.D. qualifying exams would be given separately to minoritiies. The situation mildly paralleled educational policies of communist regimes such as Czechoslovakia where the writer was informed after 1989 that the Ministry of Education secretly dictated what grades certain individuals and groups should receive regardless of their actual performance on examinations.
[30] For an excellent analysis of White, see Novick (1988), pp. 599–607.
[31] Our particular example is from the first page of Crawford (2006).

as white males could not write another group's history. All these new attitudes are evidence that a common denominator of much, if not all, of the cutting edge work by historians of the late 1960s and afterwards would be contemptuous rejection of Leopold von Ranke's dictum that one should write history *"wie es eigentlich gewesen ist"* together with the more general idea that history should be a disinterested and "objective" investigation of the past.[32]

Radically new topics of study which previous generations of historians would have found distasteful or unworthy of investigation would also be introduced. Two examples of the new research possibilities would be the subjects of Thomas Laqueur's *Solitary Sex: A Cultural History of Masterbation* (1992)[33] and Robert Darnton's *The Great Cat Massacre: And Other Episodes in French Cultural History* (1984).[34] We wish to make no judgment concerning the worth of such subjects and the new directions historical research pursued in the 1970s and beyond, but it is certain that most of them could not have been studied or written about in the Eisenhower era.

If we turn to literary criticism, the post-1960s saw the rise of "theory," principally manifested by deconstruction and French poststructuralism. To a new generation of literary scholars texts no longer had stable meanings which could be discovered by scientific techniques. Whatever the "meaning" of a text intended by the author might have been, it was in the opinion of Stanley Fish inaccessible. Instead, it was the creation of the critic and was a function of his membership in an "interpretive community." Different communities could and did produce different and incommensurable mean-

---

[32] For penetrating analyses of the changes we are attempting to describe see Handlin (1979), Himmelfarb (2004), and especially Novick (1988). Handlin, in particular, feels that the overproduction of historians in the 1960s led to a degradation of professional standards. Novick's analysis is especially valuable. We have been especially influenced by Chapter 15 of his book.

[33] Laqueur, whose nicknames I was informed on at least one internet site was "Professor Wank," is not a fringe author or pornographer. He is the Helen Fawcett Distinguished Professor of History at the Berkeley Campus of the University of California. He is also chronologically a member of the generation who came of age during Vietnam, graduating from Swarthmore in 1967 and earning the Ph.D. from Princeton in 1973. Laqueur is not alone in a preoccupation with historical aspects of sex. This has become a fashionable field, perhaps pursued more assiduously in leading universities than in adult bookshops. The writer's random search of the venerable *Journal of Modern History*, for instance, produced articles with the titles "The Manly Masquerade: Masculinity, Paternity, and Castration in the Italian Renaissance," "The Renaissance of Lesbianism in Early Modern England," "Sodomy in Early Modern Europe," etc., (See Volume 78, No 2 (2006).)

[34] Darnton, born in 1939, is a little older than the generation we are discussing. He was Shelby Cullen Davis Professor of History at Princeton, and in 2007 was appointed the Carl Pforzheimer University Professor of History at Harvard.

ings (here, as in the case of Hayden White, Fish was influenced by Kuhn). In the end the obligation of the critic was not to be "right" but only "interesting."

In philosophy, Nelson Goodman spoke of the actual existence of "many worlds." Contradictory propositions could be simultaneously "true" if one proposition was true in one world and the other in another world. To Goodman it is impossible to speak of a uniquely existing world. What exists is a plurality of co-equal interpretations that produce a plurality of worlds. Late in his career Hillary Putnam also mounted the bandwagon of anti-objectivism. Rejecting his earlier realism he argued in a way similar to Feyerabend or Fleck that "truth" made sense only within a prior belief system. It is according to Putnam a statement of coherence, not an assertion of correspondence with the world as it is since this would require a "God's eye view."

In Law we witness the rise of "Critical Legal Studies," a left wing movement, centered in the nation's most prestigious law schools such as Stanford and Harvard, which sought to delegitimize and relativize liberal jurisprudence, demonstrating that it was essentially an intellectually dishonest con game devoted to shoring up the existing unjust order.

In anthropology there was Clifford Geertz who became the "patron saint" of anthropology meetings and who argued for "Anti-Anti-Relativism,"[35] and so on and so on.

We could go on nearly forever describing these analogs of SSK/PIS in the humanities. It would be an interesting but massive project, for which we have no space here, to trace in detail the rise of such exotic fauna from their origins in the political climate of the 1960s, and to show how at the same time they built on earlier work and yet incorporated novel and unpredictable elements.

In the case of science, criticism *could* have been restricted to the same mode exhibited by Science for the People and other early radicals. The focus could have been on a critique of science's role as an adjunct and prop for the modern capitalist system. The new academics could have continued to argue that science is dehumanizing and alienating, responsible for environmental pollution, and (like MRC) a "culture of death" which had sold itself to big corporations and the military. Thinkers like Herbert Marcuse had pursued such themes, and they still are commonplace among critics of science. This type of criticism, however, accepted the fact that science produces *knowledge*, that it gives us understanding of the natural

---
[35]Novick (1988), pp. 551-552.

world in a way superior to any other "way of knowing," which was why it is so dangerous in the wrong hands. Few members of the New Left, unless stoned, would have put astrology on the same level as astronomy—"just another paradigm, man!" Their objection was that they did not like the knowledge that big science produced. Too much of it, they felt, had to do with horrific weaponry and not enough with human welfare. They would like to have seen science changed and redirected to serve what they regarded as human needs. But all this remained in the tradition of modernism. The contingent mutation that actually gave rise to PIS was a shift of the anti-objectivism we have already described in the humanities into science studies. It took the concrete form of a union of Mannheim's SK and Kuhn's deconstruction of scientific realism. This represented a much more powerful assault on science than mere left wing criticism of its applications. For what better way was there to knock science down from its pedestal than to attack it at its apparently strongest point—the generation of objective knowledge faithfully representing the world—and instead to try to show that it is just another culturally relative "discourse"!

At this point we need to make a distinction. So far we have mainly concentrated on the doctrines and origins of various forms of SSK. The proponents of SSK are mainly to be found in sociology and STS departments. They at least "study" science closely and make semi-rigorous arguments about its practice, epistemological foundations, role in society, social consequences of technological change, etc. In Chapter 9 we have looked at several scholars of this type. We have so far, however, (with the possible exception of Bruno Latour) only vaguely spoken of the "postmodern" or PIS component of the movement associated with some of the later propositions in our list. The representatives of PIS are found more in various humanities departments and in Education. They include philosophers such as Richard Rorty who are oriented towards continental rather than analytic philosophy. With some notable exceptions they are characterized more by an attitude of generalized hostility to the Enlightenment ideals of truth, reason, and objectivity rather by specific arguments.[36] They (more or less) accept the "findings" of SSK as an intellectual background. But they combine them with postmodern values and stances characteristic of their own disciplines, which have become fashionable since the early 1970s. For instance, liter-

---

[36] Andrew Ross' book *Strange Weather* (1991), for example, is full of brilliant aperçus, but the author seems incapable of a tightly woven rigorous argument. An exception, however, is Rorty's *Philosophy and the Mirror of Nature* published in 1979, a milestone in the anti-realist stance characteristic of both SSK and PIS.

ary types may regard science as a collection of "texts" and deconstruct it using the full resources of "theory" normally used on fiction (with which they often conflate science).[37] To such scholars science is a cultural production on the same epistemological plane as other cultural productions such as religion, politics, or art, and it is a mistake to separate it from or elevate its authority above the others as was done by nineteenth century positivism.[38] Its so-called knowledge is not universal but an artefact of Western androcentric and racist values. Furthermore, since all inquiry is from a "perspective," the "objectivity" that Western science pretends to value is a hypocritical illusion. Scientists are as biased as anyone else and their particular perspective is just one of many other equally valid possibilities. By pretending to divorce fact and value, scientists actually "encode" the political values of capitalist male elites, and delay the triumph of attempts to impose more socially just, antimilitarist, feminine, or Afrocentric values on science. The type of PIS proponent who believes in such doctrines also tends to be influenced by continental philosophers such as Hegel, Nietzsche, Heidegger, French postructuralists such as Foucault, Lyotard, and Derrida, or psychoanalysis. He or she also puts greater stress on cultural relativism, and pursues overt political implications and goals more than orthodox SSK. Most feminists and multiculturalists are closer to this camp than to SSK or the Strong Program. And as a veneer overlaying everything else, there is in PIS a certain elitist personal distaste for science and scientists who are viewed as culturally illiterate, socially irresponsible technophils. Andrew Ross, for example, brags of his ignorance of science, dedicating *Strange Weather* to "all the science teachers I did not have, it could only have been written without them." And Paul Forman stresses the moral irresponsibility of scientists who seek "transcendence," i.e., complete personal freedom to discover timeless scientific truths without thought of the social and political consequences.[39] Forman also proposes that scientists are so culturally limited as to be incompetent to judge what actually constitutes good science and scientific progress.[40] This role should be reserved for historians of science. Many more examples of such attitudes may be found in the collection of essays *Science Wars* (1996) written as a counterblast to *Higher Superstition*, the writings of Richard Rorty, articles

---

[37] Some interesting remarks on this methodology may be found in Levine (1996).

[38] Ross' approach in *Strange Weather* is to attack the presumed authority of orthodox scientists which he views as a "secular priesthood" over New Age "alternative science" movements with which he broadly sympathizes.

[39] Forman (1991).

[40] *Ibid.*

in *Social Text*, and in the many references given by Gross and Levitt and Sokal and Bricmont.

It should be clear from the above analysis that SSK/PIS is itself a group of socially conditioned ideologies, and in its radical forms certainly not a disinterested investigation into the nature of science. Rather, it reflects the political interests and psychological states of its believers. From our account of the antiwar, anti-MRC atmosphere in Madison which was mirrored on many other US campuses (and also in Europe) roughly between 1967 and 1973, it is obvious that more directly than for almost any other variety of academic postmodernism, SSK/PIS:

> ... was the child of the antiwar movement. Born from moral outrage against the Tabus of our science in their dirty war against Third World peasants, the radical science movement gradually learned that science was not outside of culture and history, to be used well or ill, but rather was integral to both. From rage against imperialist powers misusing science and technology, the movement slowly struggled into learning that science as a global system of the production of knowledge was intimately connected to political, economic, cultural, and military powers.[41]

While the initial anti-science feelings provoked by the Vietnam war may explain the origins of the movement, thirty years later they remain as a generalized hostility to the perceived exploitation of science and technology by the corporate elite and the military and industrial complex. This is frankly admitted by Barry Barnes as he tries to explain what he feels is an increasingly negative attitude towards science in SSK:

> One possibility [for the explanation] is that real changes in the institutional base of science lie at the root of the matter. Since its purchase by government, industry, and the military, it has become impossible to combine hostility to existing major institutions, or to capitalism as a whole with an entirely positive evaluation of science. A minority opposed to science has thereby almost automatically been generated ...[42]

The disdain of postmodernists in general for ordinary concepts of truth and rationality has an origin parallel to that of SSK/PIS. It is a natural continuation of the anti-intellectualism of the radical student movement of the Vietnam war era. From the Free Speech Movement onward "reason," "objectivity," and dispassionate academic analysis generally were viewed with suspicion by student radicals. They were code words for academic con-games by which the Establishment defended indefensible things like

---
[41] Rose (1996), p. 87.
[42] Barnes (1974), p. 136.

capitalism, the Vietnam war, or the bureaucratic structure of the university. This attitude survives, dressed up and "intellectually" defended by certain strands of continental poststructualist philosophy, among the humanities faculty these students have often become.

It is tempting to diagnose such permanent political feelings at least partially in psychiatric terms. Although touched off by the Vietnam war, the draft, and perceptions of racism, they grew, as we have already observed, in a psychological climate characterized by suburban affluence and a permissive upbringing. On the part of the baby boom generation there were no memories of war or the depression of the 1930s. Instead, there was a perception of entitlement to instant gratification mixed with uncritical idealism and an inability to tolerate frustration. Combined with a "moral" cause this was a recipe for an explosion.

More recently and although sincerely held, the continuing power and intensity of this generation's opinions reflect at a deep emotional level rage and jealously at the growing power and influence of technoscience and mathematics (often in the service of corporations and the military) at the expense of the humanities in modern societies. They are made more intense by the fact that academic intellectuals in the humanities continue to feel deprived of political power just as they had been as students by groups they regard as their intellectual inferiors (university administrators then, scientists and corporate types now). Even on the intellectual and cultural level their careers reflect an "increasing impotence as political-cultural 'legislators'."[43] Their activities are "increasingly unnecessary and irrelevant both to those who hold power and to those who produce our mass culture."[44]

More wounding than outright hostility is the feeling of being ignored by both the wider society and by one's peers within the university. The tremendous overproduction of contemporary research which according to Paul Forman is a critical characteristic of postmodernity[45] implies that one's scholarly work even if published in a quality journal is in the majority of cases read by almost no one other than a referee and consequently has near zero impact. For one who had dreamt in his youth of instigating revolutionary political change where real power would be wielded by intellectuals (under the camouflage of 'participatory democracy'), it is an unsatisfying life being a $50,000 Associate Professor grading hundreds of papers written or plagiarized by semi-literate and very juvenile students at

---

[43] Forman (1997b), p. 180.
[44] *Ibid.*
[45] *Ibid.*

Persepolis State—who frequently bitterly complain about his "unfairness" to the Dean, Provost, Vice-President for Academic Affairs, the local representative in the State Legislature, the ACLU, et al. Such an environment would seem fertile soil for postmodernism. In this context an important psychological characteristic of SSK and PIS is its ability to overcome latent inferiority complexes that humanists may feel towards scientists and mathematicians. Although several of the people we have discussed have had significant training in a scientific or mathematical discipline,[46] we conjecture that the majority of postmodernist critics of science either never had such training or had an unpleasant experience from their exposure to it.[47] Mathematics, in particular, seems a *bête noire* to them.[48] Fortunately, however, SSK and PIS are excellently designed to transform feelings of inadequacy *vis a vis* science into feelings of superiority. Their theses seem to reveal (1) the shaky epistemological foundations of a hated discipline and (2) the fact that scientists are slaves of external social and political forces which they do not understand (while, of course, SSK/PIS does). Moreover, these theses are quite simple and may be mastered without the need of learning the details of a science. The effect is similar to that of vulgar Marxism in the 1930s. Then one could dismiss (and ignore) the work of "bourgeois" historians and economists by invoking the necessary progression from feudalism through the bourgeoise to the triumph of the proletariat, supplemented by a few phrases concerning "the means of production" or "the ideological superstructure." In both cases the illusion of a complete and superior understanding of otherwise difficult subjects is created.

All this, of course, can only be rank speculation. But it may be significant that SSK/PIS *project* feelings similar to their own onto their opponents, the scientists. Scientists, they say, lash out at SSK or PIS in books such as *Higher Superstition* and *Intellectual Impostures* because they are disgruntled at their declining political power. Since the halcyon days of the immediate post-Sputnik era, the funding and influence of hard scientists, especially high energy physicists, has diminished. Although there is some truth in this observation, such projection is a well known psychological cop-

---

[46] Notably Shapin, Pickering, and Bloor who early in their careers had been trained respectively in biology, physics, and mathematics.

[47] We mean no criticism by this observation. Although the writer knows a professor of physics who as a student at Berkeley in 1962 took a graduate seminar on Joyce's *Ulysses* and got the only "A," most scientists would be as helpless in such a course as an English major would be in advanced calculus.

[48] Several otherwise brilliant people in the humanities have over the years complained to the writer of their difficulties with high school mathematics or college algebra.

ing mechanism. Actually, very few scientists "lash out" at SSK/PIS for the simple reason that most are unaware of the movement's existence even if it is present at their own universities.[49]

Another emotional source of PIS's disdain for science has been suggested by Stephen Hicks.[50] He feels that it is part of the *ressentiment*, that poisoned bitterness first described by Nietzsche, which postmodernists feel towards all aspects of Western society, caused at bottom by their disappointment at the failure of socialism. The fact that capitalism proved stronger is unbearable to them and calls forth a "slave morality" (i.e., postmodernism) designed to both lash out against and sow doubt and feelings of guilt among the victors. This thesis also has some validity, especially since 1989. However, the main ideas of PIS go back to the early 1970s and before when socialism (in some form) still had prestige among intellectuals despite the failures of the Soviet Union and was still confidently thought to be ultimately victorious, perhaps in the form of Yugoslav Eurocommunism or Maoism, a judgment reinforced by the perceived collapse of morale and stability in US society because of Vietnam. Hicks is partially correct. There is definitely *ressentiment* at work in postmodernism, but it reflects the curdled attitudes of academic intellectuals who have no power in a society that cares little for ideas and pays people in the liberal arts half the salary or less of accounting, marketing, or advertising professors.[51]

However fashionable SSK/PIS is in certain humanities departments, it is unlikely to achieve any of the global revolutionary aims that its proponents hoped for thirty years ago, at least if the US social order remains stable and the present conservative mood continues. Science and technology will continue to develop much as they have since World War II, supported by universities, corporations, NASA, the Pentagon, and other government agencies. For most scientists the various ideologies of SSK/PIS will remain

---

[49] In an informal survey of scientists and mathematicians at various universities, the writer found that almost none had either heard of the Science Wars or had read Gross and Levitt or Sokal.

[50] Hicks (2004), pp. 193–195.

[51] This may explain the sympathy many intellectuals had prior to 1989 for "actually existing socialism." If one was a US professor of a certain ideological stamp who was met at the Prague airport in the 1970s by a black Tatra limousine from the Ministry of Culture and wined and dined at, say, the "Opera Grill" by the editors of *Rude Pravo* and other members of the Party intellectual elite, then one could be excused for feeling that here was a society where ideas mattered and where intellectuals and poets were important enough to be either celebrated or persecuted. This experience must have been in exhilarating contrast to the stupefying boredom of teaching History 101 at Persepolis State.

invisible, and their status will only be obviously affected by the increasing penury of the public universities. At least for the foreseeable future those proponents of SSK/PIS hoping to effect a major transformation of the relations of science and society will be disappointed as they have been over the last three decades.

However, in the Hobbesian world of academia[52] the movement has rather successfully pursued more modest goals. As we have noted, its adherents have certainly increased their political power and prestige in various disciplines if not yet in the wider society. SSK/PIS theses are common or even dominant across a wide range of areas in the humanities and social sciences. Stephen Cole has observed in the particular case of sociology that:

> ... within the short time span of roughly one decade, this group [SSK] has come to completely dominate the sociology of science and the interdisciplinary called the social studies of science ... their control of all the major associations and specialty journals is clear to anyone participating in the field. ... [SSK] is an interest group that tries to monopolize rewards for its members and fellow travelers and exclude from any recognition those who question any of its dogma.[53]

Institutionally, STS has also enjoyed a period of rapid growth over the past fifteen to twenty years. As in the early post-Sputnik era in the case of science and mathematics this has enhanced the job prospects of newly minted Ph.D.s in the field and the job mobility of established scholars at a time when the job market in core scientific subjects like high-energy physics became worse than stagnant.

In educational theory SSK/PIS ideas have become very influential and seem especially to have pervaded K–12 education. Noretta Koertge has pointed out how textbooks at this level often contain excessive amounts of multiculturalist ideology and other forms of political correctness characteristic of PIS. Pictures of scientists emphasize people of color and women; few white males are shown. Distrust of science and technology is promoted. In a unit on scientific misconduct one textbook asks students to show how so-called "great" scientists like Galileo, Newton, Mendel, and Millikan (all of whom happen to be white males) faked their data and results. At the same time little emphasis is put on the actual content of science.[54] In el-

---

[52] Or should we say "academentia," a newly coined word that the writer discovered written on an elevator door at his university a few years ago.
[53] Cole (1996), p. 274.
[54] Koertge (2008), pp. 6–9. At the end of twelve years of such indoctrination one wonders why anyone would want to become a scientist!

ementary and high school mathematics education similar ideas have also furnished the intellectual foundations of one side of the "Math Wars," especially thanks to the work of Paul Ernest and the "personal constructivism" of Erich von Glasersfeld.[55] Trendy applications to education which are not limited to the high school level have also benefited SSK/PIS through an increased share of Federal funding. University mathematics instruction offers a good illustration of how this process works. In the first place, funding, particularly in pure mathematics, has almost collapsed since 1970.[56] This situation unfortunately coincides with an ever more pressing need of public universities for large grants since they have also suffered ever decreasing financial support from the State. Secondly, mathematics teaching at the university level is in crisis. For a variety of reasons, the average university student now seems incapable of learning college mathematics at the level at which it has been traditionally defined. Casualty rates among calculus students are very high, especially among some minority groups and to a lesser extent women.[57] On both political grounds and because it destroys the university's "market share" this situation is unacceptable to administrators. Consequently, as a way of solving both the financial and educational problem universities have backed the hiring of PIS oriented educational theorists by mathematics departments as these people have access to large HEW grants[58] and presumably "know how to teach." However, even if (as

---

[55] See Chapter 10. This is not the place to assess the effects of these ideas on educational policy, but the writer thinks they have been tremendously damaging.

[56] NSF research grants that remain available are mainly for summer salary support and usually involve only paltry sums.

[57] During his teaching career it was not uncommon for the writer to meet students who had spent half a decade or more trying to pass calculus at several universities. They were desperate since passing calculus was the necessary condition to gaining access to potentially lucrative careers. At most universities there is great hostility towards mathematics departments, since other departments and programs use mathematics as a filter to get rid of students they deem incompetent. This procedure is politically astute since blame for the resulting carnage can be deflected onto the "poor quality" of mathematics instruction. Machiavelli praised Cesare Borgia for similar tactics in disposing of political enemies. But fortunately mathematics departments have not yet the met the fate of Remirro de Orco. (See *The Prince*, Chapter 7.)

[58] In 1994 an ex-MRC colleague informed me of the pressure to do this exerted by the administration in his prestigious department (one of the top twenty in the US). Having hired one Mathematics Education specialist, it was found that his grants were equal in value to the rest of the Department combined. So the pressure to continue such hiring increased. My informant was not hostile to the idea; he was simply puzzled and a little envious. He had no idea what "constructivism" was or what the new hires in the Department actually *did* (he had difficulty understanding the jargon); he just wanted for himself a little support for his algebraic topology program which twenty years earlier he used to get from the Air Force.

is likely) the educational problem cannot be "solved" in this way, it can at least be defined away. Given the relativism in SSK/PIS that has penetrated into educational theory, it may be tempting to raise the question of "whose standards" and "whose knowledge" the contemporary university curriculum "encodes." Perhaps calculus and rigorous mathematical argument generally is just an cultural expression of "Western logocentrism," and is no longer politically acceptable in a democratic age of multiculturalism, and should be replaced by "other ways of knowing."

As STS has matured, however, there is some evidence that the kind of radical constructivism and relativism characteristic of classical SSK and the Science Wars of the 1990s has become a bit outmoded. While such positions continue to have influence, most academic practitioners of STS tend to be "moderate constructivists," which is to say that they argue that while political or social bias may enter into the construction of a theory while, at the same time, the theory may represent an "accurate map" of reality."[59] In other words, science is constrained by a real material world, but not completely; social factors can and do have their influence. Evidence may be theory-laden and require social negotiation for its interpretation, but it is still important. The precise nature of the mixture of sociological and empirical factors and their relative importance is, as we have seen, a matter of dispute among STS scholars.

Much STS research also continues in an earlier pre-SSK mode. There are investigations of scientific institutions, status, networking, and regulatory policy as well as hot issues such as stem-cells, abortion, nanotechnology, technology versus the environment, global warming, and the general relationship of science to the state and private corporations. This research would have been found recognizable and normal by Robert Merton. On the other hand, a large portion of STS research and theory—especially the wing which David Hess has called "cultural studies of science and technology"— still has a definite "progressive" tinge reminiscent of the radical origins of SSK/PIS in the 1960s. According to Hess:

> Additional features of cultural studies of science and technology include the tendencies to focus on questions of culture and power (particularly as theorized from feminist, postcolonial and antiracist standpoints), to problematize contemporary science and technology historically as part of the postmodern condition, to examine how nonexperts and historically excluded groups reconstruct science and technology, and to forge

---

[59] The term is Hess's. See Hess (1997a), pp. 35–36.

alliances between researchers and activist/interventionist social agendas.[60]

In this ideological universe ideas of the Frankfurt school, Antonio Gramsci, or Marx himself are frequently invoked. Much is made of the Gramsci's concept of "hegemony"—that process "whereby the ruling class(es) support the creation and diffusion of a general system of values and ideas that percolate through the major institutions of society"[61] and by which voluntary acceptance of the system is created. It is a safe conjecture that some mixture of political values consistent with the above quotations plus a certain distaste for "Late Capitalism" is standard for a large subgroup working in STS, even if it is not always explicit.[62]

So far as SKS/PIS is ensconced in STS departments, it is just another academic field (or shall we say "form of life" practicing a "language game"). It may not be to everyone's taste, but academic tolerance requires an attitude of *de gustibus non est disputandum*. After all, the American university is full of nontraditional subjects like hotel management, marketing, advertising, forensic science, orthodontics, etc.,[63] and there is generally enough conflict in one's own department to satisfy aggressive urges without having to pick a fight with another discipline. But this assumes, to use the jargon of point set topology, the equivalent of the $T4$ axiom of separation between different academic areas. They should be safely isolated from each other (with no shared limit points) unless their faculty should accidentally meet on the University Heating and Cooling Committee. This separation between STS and academic science has been the case since the late 1970s; a positive feature of it is that the harassment of scientists, as at MRC, during the

---

[60] *Ibid.*, p. 113.
[61] *Ibid.*, p. 115
[62] At least the writer has not yet found evidence for the presence of neoconservatism, hostility to feminist claims, love (in opposition to, e.g., Lewis Mumford) of the latest technological fix, and so on, in any of the published literature of STS. If it is exists its quantity and influence must be small.
[63] This is part of the genius of the US higher educational system as it has developed since World War II. It is pleasant to live in a university town, and sometimes even to work at a university. There are good restaurants, coffee bars, art cinemas, book stores, concerts, theatrical events, and an ample supply of nubile young things of the opposite or same sex. But to enjoy these opportunities, it is no longer necessary to be a specialist in Yang-Mills theory, the Corona problem, the genetic structure of fruit flies, Akkadian verb structure, Napoleon III's Italian policy, or the poetry of Pindar. Instead, one can now enjoy university life with a much less onerous intellectual investment, or even with no investment at all as an administrator.

Vietnam era by Science for the People types (probably the ancestors of STS) is no longer significant.[64]

But the very institutional success of STS means that it is eventually possible that this desirable situation may break down. Many STS Ph.D. graduates have followed the time honored and harmless academic practice of reproducing themselves in graduate programs at other universities. This is possible in an academic field which is expanding. However, the time may be approaching when the majority of new jobs will be in government agencies such as NSF, the National Institutes of Health, and HEW, or perhaps organizations like AAAS and university administrations. Since these agencies influence science policy, the old dream of SSK/PIS of controlling science and serving as a mediator between it and the wider public may be capable of realization. Should, however, the political values of SSK/PIS begin to seriously impact the practice of science via mandates of bureaucrats infused with the ideas we have been describing there will be conflict. Below we sketch some possible flash points which have not yet been completely actualized, but are certainly in the cards given the heated rhetoric of some forms of PIS.

One can imagine the discomfort of a physicist should "sexism" be officially discovered in his latest model for solving a problem in solid state physics. Or suppose one branch of mathematics was determined by NSF to be "biased" in some way against under represented minorities while another was not, and that this "discovery" affected NSF funding of grant proposals in these two fields.[65] The danger here is that once certain areas of science are found to incorporate "bad" ideological bias, as feminists have argued in the case of some biological fields, the pressure to hunt for these sins everywhere will be overwhelming—if only to gain some advantage for one's group in hiring or research funding. Of course, the impact on science will depend on whether or not such policies have real teeth. Scientists, like all successful faculty, are skilled at writing cant and boilerplate for grant proposals, mission statements, and five year plans for Academic Vice-Presidents. If verbal conformity is necessary to satisfy STS inspired PC demands of gov-

---

[64] But one exception is the sometimes violent action taken against biologists, by the *Animal Liberation Front* and similar organizations. Fortunately, however, their members are not commonly found in STS departments.

[65] Lest such examples be found implausible, the writer can testify that a Chinese graduate student, a middle aged survivor of the Cultural Revolution, asserted that in the early 1970s Topological Vector Spaces at his institute were declared "bourgeois" while point-set topology was "proletarian" (or perhaps the order was reversed).

ernment agencies, this will be done; it would be just one more piece of the mindless busy work that infests the modern university.[66]

Another development which is perhaps much more probable is far more vigorous and rigid affirmative action policy than now exists in the service of identity politics. If this happens, one can expect that the usual measures of merit now used in physics or mathematics for hiring and tenure will be found to be "sexist" or "racist" and universities compelled to hire, tenure, and promote according to a unacknowledged but *de facto* quota system. SSK/PIS arguments would, of course, be ideally suited to justify such a policy.

As long as proposals of this kind are limited to a leftist fringe within academia they will probably have limited impact. PC lip service will be given, but the present policy of mild affirmative action, usually confined to the construction of "diverse" short lists followed by the application of standard hiring criteria will not change. Unfortunately, there are signs that this benign situation may be ending. Quite radical ideas are infiltrating from the fringes into the center. For instance, at least one moderate Canadian philosopher of science, James Robert Brown, who otherwise opposes some of the excesses of the Strong Program and PIS, accepts arguments supporting an affirmative action policy potentially far more radical than anything existing now. He agrees with feminists and the partisans of other minority groups that a scientist's ethnic and social background, as well as gender does produce unavoidable but probably unconscious bias in the scientist's work. However, he does not believe that such bias is necessarily "bad" for it is compatible with good science. Brown, however, believes that this type of bias should not be restricted to that carried by white males. Since all groups have their biases, for the sake of both democracy and a "more objective science" he advocates that as many of these biases as possible be incorporated into the scientific work force, which can only happen via a state policy of iron fisted "affirmative action imposed on the scientific community."[67] Still another proposal which may be close to implementation in the US is to employ the Federal law Title IX. Title IX calls for gender equity in education, but its main application so far has been to increase

---

[66]The East Bloc scientists that the writer met in the 1970s and 80s were also masters of this art. Since evidence of one's "commitment to socialism" (functioning much like "commitment to teaching" or "cultural sensitivity" today) was necessary for promotion, they were good at fabricating the necessary paper evidence. It helped to put the Laws of the Dialectic or quotations of the General Secretary both into the introduction and concluding section of one's paper on algebraic number theory.

[67]Brown (2001), p. 187. See also pp. 186–187 and Chapter 9.

opportunities for women in sports by forcing universities to treat men's and women's sports as equally as possible (football, however, being exempted). Now it is seriously proposed to use Title IX to force "equality" at all levels in university science and mathematics between men and women. This would entail an affirmative action program that would dwarf all previous efforts; given the difference in numbers that now exist between male and female faculty in the sciences, such a policy would necessarily almost shut down male hiring.[68]

An area where PIS-type ideas may already be applied is the funding of research proposals. NSF has added a second criterion besides "intellectual merit" to the evaluation of grant applications. This is the "impact" of the proposal on US society. As Noretta Koertge has pointed out, this "second criterion" has usually in the past been used to promote research infrastructure and teaching,[69] But the potential for political manipulation here is obvious, since it can be argued that the funding of a weaker minority proposal would have far greater "impact" than that of a somewhat stronger proposal from a white male.

Given the number of aggrieved ethnic or social groups who would like to exploit such policies in either hiring or funding, it is not difficult to imagine the Pandora's box that their implementation would open up. What, for example, would be done with the hiring or funding over-represented groups such as Jews and Asians? Would a candidate for a position or a grant proposal be defined by a social-ethnic-gender "matrix"? What role, if any, would "merit" play? How could the "merit" of members of different categories be compared and by whom? *Quis custodiet ipsos custodes* of affirmative action? How could such a policy be limited?[70] Its practical (but unintended) social effect would be similar to throwing a truck load of red meat plus the driver to a pack of starving hyenas. No one objects to fair treatment of and reasonable attempts to improve the production of qualified people among under-represented groups. But all ultimately must be measured by common standards of excellence appropriate to the discipline. To claim an *a priori* merit for someone simply on the basis of who he or

---

[68]For a discussion of Title IX see Koertge (2008).

[69]*Ibid.*, p. 9. The policy also seems to have resulted in a bias towards "useful" applicable science as opposed to the ultra-pure (after all, what "impact on society" could a proposal in algebraic geometry possibly have?).

[70]Given the perspectivism favored by PIS, slight modifications of existing arguments would justify hiring or funding as many different "perspectives" as possible by universities. We would need therefore both fascists and communists in political science departments, holocaust revisionists in Jewish studies departments, etc., etc.

she *is* is to repeat the disastrous policies of both communist and fascist regimes who used criteria of class and race to confer advantages to some and deny them to others.[71]

Equally alarming, but fortunately less probable, would be large scale implementation of a proposal of David Hess. He imagines that those trained in STS could be employed to evaluate "the scientific merits of different positions in a [scientific] controversy rather than merely analyzing them."[72] Hess mainly focuses on controversies that seem apparently "settled" in the sense that there is a "consensus" theory that expresses scientific orthodoxy as well as a heterodox position, which has been perhaps long rejected. The evaluation by STS of the merits each side would depend on a more subtle analysis than provided by SSK and would involve four basic principles:

(1) The analysis is political; it explores the operations of power in a field that becomes constituted by a consensus and attendant heterodoxies ...

(2) The analysis is cultural in the sense that it develops a sophisticated noninstrumentalist explanation and explication of the dynamics of power that have been described in the first step. ...

(3) The analysis is evaluative; it draws on the philosophy of science to weigh the accuracy, consistency, pragmatic value, and potential social biases and knowledge claims of the consensus and alternative research traditions. ...

(4) The analysis is positional; it provides an evaluation of alternative policy and political goals that could result in beneficial institutional and research program changes. ...[73]

Hess has applied these ideas in an earlier book[74] to the theory that bacterial infections may be a major cause of cancer which had been mostly rejected by the cancer establishment by the early 1940s. Hess concludes that the theory may have some merit and was rejected more for sociological/political rather than for scientific reasons. He argues that a modest fraction of the cancer research budget ought to be devoted to the exploration of this theory. But it is difficult to see why scientists would voluntarily accept this role, suggested by Hess, of STS as a referee in scientific conflict. Even if scientists violently disagree with each other over a possible paradigm

---

[71] A mathematical friend of the author was almost denied entrance to the university in Czechoslovakia since an uncle owned a country store which made him "bourgeois."
[72] Hess (1997a), p. 153. Of course, as Hess admits, a necessary condition to do this successfully would be to learn the science in question, just as Marx in one of his theses on Feuerbach argued that one had "to understand how the world is represented if one is to attempt to change it and such a task would would take time and effort. (*Ibid.*)
[73] *Ibid.*, pp. 153–54.
[74] Hess (1997b).

change, they *would* agree that *they* should be the ones to settle the issue, not somebody "outside the field" who isn't even a scientist; even more perverse to them is the idea that the referee should be a sociologist, a field—like Rodney Dangerfield—which does not get much respect. If alternatively the conflict has been "settled" and the heterodox opinion, such as the bacterial aetiology of cancer, long since rejected, scientists simply would not accept a new and possibly revisionist look at the discarded theory. The only way that Hess's idea could work would be if it was enforced by some agency that controlled funding. But this would validate the worst nightmares of STS critics such as Gross and Levitt or Sokal and Bricmont that STS is really just a cover for an effort to gain power over science. If this should turn out to be the case and radical currents of STS should try to leave the sandbox of their own departments, where almost all scientists are happily unaware of them, and influence science policy in any major way, then the Science Wars of the 1990s would in hindsight be only a preliminary skirmish to a far more destructive conflict.

It is delightfully ironic, considering its origins in the radicalism of the 1960s, that SSK/PIS should now be criticized from the Left. In fact, leftist criticism has done more damage to its claims and drawn more blood than anything put out by the Right.[75] Although they have been accused by their victims of pandering to reactionaries, the authors of both *Higher Superstition* and *Intellectual Impostures* represent criticism from an older pre-postmodern liberal/leftist perspective. Norman Levitt, for instance, is a self-described "socialist in economics, a liberal in politics, and a conservative in culture,"[76] and Alan Sokal was sympathetic with the Sandinistas and taught mathematics in Nicaragua. Both feel that politics should be a continuation of the Enlightenment project to improve society, an effort grounded in reason and having a symbiotic relation to scientific progress. In their opinion the PIS tendencies of the "Academic Left" are both intellectual nonsense and damaging to progressive politics. Another more radical critique of STS/PIS has been given by Steve Fuller. Fuller's *bête noire* is Thomas Kuhn. He sees him first of all as being coopted by the Cold War militarism of his mentor James Conant and secondly thinks of Kuhn's concept of "paradigm" as a device by which science can be isolated from social criticism and control. Even worse, inspired by Kuhn STS has developed their own "paradigms" (such as the Strong Program, SSK, etc.).

---

[75] Conservative critics like Roger Kimball have been too interested in the "tenured radical" phenomenon in the humanities generally to focus much on science.
[76] Levitt (1999), p. 29.

This process Fuller calls "Kuhnification" and causes STS to become a network of minor academic cults safely isolated from politics and having no real impact. Any apparent leftism becomes merely a matter of empty talk and journal articles, and actually helps to protect existing technoscience rather than helping to bring it under democratic political control.

The characteristic relativism and deconstruction of scientific claims of SSK/PIS has also been employed by causes that would be anathema to many of its creators. The religious right has, for instance, borrowed the constructivist arguments of SSK to criticize Darwinism. Phillip Johnson, a Berkeley law professor and a leader in the intelligent design (ID) movement identifies Darwinism as a Kuhnian paradigm. The core of this paradigm is "metaphysical naturalism," the attitude (based Johnson feels on no compelling evidence) that the natural world and its laws are all there is and whose development reflects no purpose or intelligence. Like other scientific paradigms naturalism dictates what answers are acceptable and what questions are forbidden. It shapes the entire culture of the biological sciences, and like other paradigms it has nothing to do with truth. It's chief puzzle, an anomaly in Kuhn's sense, already noted by Darwin, is the apparent failure of the fossil record to support Natural Selection's model of continuous change by small variations.[77] To Johnson this anomaly has never been successfully solved, but the cultural power of the paradigm allows Darwinian believers to continually invent ad hoc solutions or explain it away.[78] Johnson's arguments are not a distortion of SSK/PIS doctrine. His point of view is partially shared by Steve Fuller, a leading sociologist and philosopher in the movement. Fuller's belief that all scientific theories are socially constructed and reflect wider non-rational cultural values leads him to assert that ID and the neo-Darwinian view of evolution have equal legitimacy as scientific theories and should be accorded equal time in high school biology classes.[79] Like Andrew Ross, Fuller seems to regard scientists as a privileged elite who without any justification except power impose their beliefs on society.

---

[77] That is, species usually appear and disappear suddenly in the geological record. Before their extinction they enjoy long periods of stability with little or no change. This seems to contradict Darwin's view that one species changes to its successor via many small changes due to natural selection.

[78] Johnson (1981), pp. 118–21. Elsewhere Johnson begins to sound a bit like Bruno Latour in his analysis of science. See e.g., Johnson (2000), pp. 173–4. For a reply to Johnson see Pennock (1999).

[79] He has maintained this position as an expert witness for the ID side in a well-known trial *Kitzmiller v. Dover* involving the Dover, Delaware school board.

But anti-Darwinism is not the only cause where the tenets of SSK/PIS may be exploited. To take a different example, Sheila Jasanoff, a well known partisan of SSK mentioned in Chapter 11, relates that on various occasions corporations in legal or regulatory cases involving dioxin, asbestos, or alcohol have requested that she testify on their behalf:

> For each of these actors, moreover, the objective was not to win my support for one or another specific scientific claim (for example, that alcohol consumption in pregnancy does not cause foetal alcohol syndrome, or that asbestos exposure in buildings creates no risk of cancer), but rather to resist an adversary's claims to scientific credibility. The sociology of knowledge was for them an instrument to undercut their opponent's expert status, and thus to deprive them of 'science' itself as a political resource.[80]

The irony in these examples puts one in mind of Hegel's comments on the "cunning of reason" in his *Lectures on the Philosophy of History* by which historical actors accomplish purposes which they do not foresee and which in some cases are contrary to their intentions. One feels, however, that the creators of SSK who for the most part are not apologists for the religious right or corporations ought to have anticipated such uses for their work.

---

[80] Jasanoff (1996), p. 399. Elsewhere in her article, Jasanoff relates how constructivist SSK notions were incorporated into opinions at the Supreme Court level in a case where the petitioners consisted of parents and their two children claiming claiming that consumption of drug manufactured by Merrill Dow had caused birth defects (*Ibid.*, pp. 403–409).

Chapter 14

Kto Kogo?

The central aim of this essay has been to critically analyze the doctrines of SSK/PIS as they relate to science and mathematics. We reach conclusions hostile to the more extreme members of the school, and in the particular case of mathematics regard the issue of "fallibility" as a straw man whose significance Paul Ernest and other social constructivists have grossly exaggerated. Many of the individuals discussed in Chapters 3—9 have helped cause at least two fundamental changes in the interpretation of science. In the first place, they have almost destroyed the internalist "conceptual analysis" approach to the history of science pioneered by Alexandre Koyré in such works as *Études Galiléennes*, so influential at his death in 1964,[1] and replaced it, as we have seen, by the constructivism and contextualism so characteristic of SSK. Secondly, they have ignored traditional philosophy of science and its perennial project of analyzing "scientific method" in order to find prescriptive rules that science *should* follow and substituted a descriptive historically oriented analysis via case studies of the way science "warts and all" actually has developed. This approach has persuaded them that there are no epistemological differences between science and other historically contingent cultural "discourses." Both of these revolutionary changes have had similar consequences: Concern over the "truth" of a theory is replaced by an analysis of the sociology of its acceptance. The hagiography accorded "great scientists" is replaced by the kind of treatment reserved for other historical actors, such as politicians. And the view that the history of science shows progression towards truth (i.e., Whig history) is replaced by a view of it as a random process having no particular direction.

---

[1] For evidence of Koyré's influence in the 1960s see Cohen and Clagett (1966).

An obvious corollary of these new approaches is a quasi-political debunking of the claims of science as a means to produce reliable knowledge. Inspired by Karl Mannheim SSK/PIS wishes to "disintegrate" such claims rather than to take them seriously. The technique is one of "unmasking" by focusing on the contextual nonrational factors influencing scientific research. For Barry Barnes, David Bloor, and Steve Shapin these factors, as they attempt to show by historical case studies, are often essentially political and reflect the scientist's position in the social order. The scientist's actual arguments, however sincerely held, amount to a kind of intellectual superstructure which "encodes" these more fundamental determinants of belief. For Bruno Latour and Steve Woolgar, on the other hand, global political factors (such as, for example, Newton's Whiggishness) are not so important as the micropolitics and "trials of strength" in the "agonistic field" which is the modern research laboratory. Science produced there is the work of the locally dominant group using age old techniques of rhetoric and political warfare. Once victory has been secured, the resulting science is made through effective propaganda to seem like it was inevitably "caused" by nature. For Andrew Pickering victorious theories of particle physics are not direct products of sophisticated experiments, but owe their victory to the institutional interests of powerful research groups. For Harry Collins and Trevor Pinch the "experimenter's regress" make it impossible to settle a scientific controversy (like e.g., the existence of gravity waves) by experiment. Hence, any settlement is sociological having little, perhaps nothing, to do with nature. The above strategies are combined with a philosophical anti-realism. By recycling skeptical arguments going back to Descartes, Berkeley, Hume, Kant, and even Sextus Epiricus, SSK/PIS attempts to show that the existence of a mind-independent world is either impossible or unknowable. More specifically, this belief is often linked to a neo-Kantian view that the "world in itself" is inaccessible to us. Only our "representations" of the phenomenal world are available. They may be compared with each other, but never with the world as it actually is. It follows that that everything we can say about our experience is a human construction, which for SSK originates in the conventions of the social group. As Ian Hacking[2] has pointed out this viewpoint entails a form of modernized nominalism. The terms we use to describe and classify objects in the world reflect our needs and values and not inherent properties in the

---

[2] Hacking (1998).

objects. It is not difficult to show such doctrines collapse, especially in the case of the Strong Program, into a vacuous social solipsism.

Over and over again, extreme relativists in the SSK or Strong Program camp have argued that science is no more rationally grounded than the "knowledge" produced by other cultures such as the Azande (perhaps their favorite example). Since "reality" (whatever it is) cannot determine the content of scientific theories, this task is reserved, just as in the case of any ideology, for a variety of contextual, social, and above all *political* factors which SSK in particular spares no effort and ingenuity in "unmasking." By such arguments the privileged position of science with regard to its access to "truth" as well as its social autonomy can be deconstructed. Since science is just one belief system among many with no particular epistemological superiority, it should not be valued (especially by the State) over other enterprises such as, say, literary criticism, spoon bending, New Age philosophy, or alternative medicine. Frequently accompanying this position is a vigorous "trashing" of famous scientists. Their moral failures, political biases, and possible scientific mistakes are exposed in loving detail. Any celebration of their actual achievements is dismissed as hagiography or as an exercise in HSTS: "Hooray for Science, Technology, and Society."[3] Newton's neurotic tendencies, his possibly latent homosexuality, and Freudian fixation on his mother, for instance, completely trumps the *Principia* which itself according to critics like Patricia Fara is deeply flawed.

It follows that at best, science is only a tool for achieving prediction, control, and better technology (which SSK as a whole accepts, but seems to consider a rather trivial task).[4] However, some (such as Michael Mulkay), as we have seen, even doubt the usefulness of basic scientific knowledge for technology. Other critics such as Richard Rorty claim that science is *nothing but* technology, and hence can make no claims to knowledge which

---

[3] The acronym comes from Winner (1996), p. 106.

[4] Sarah Franklin, for instance, commenting on Richard Dawkins's comment "show me a relativist at 30,000 feet and I will show you a hypocrite." says

> But what is so self-evident about the fact that planes can fly? This feat could as easily be described as sophisticated tool use instead of as an indicator of epistemological certainty. And as certain as the fact that planes can fly is the fact that ... experiments such as those of the Wright brothers have as much to do with desire as with established scientific principles, that some airlines show prayer films during take-off to invoke the aid of Allah in order to remain safely airborne, and so on (Franklin (1996), p. 153.)

I leave it to the aeronautical engineers of Boeing to evaluate this passage.

is superior and different than that produced in the humanities. Physics, according to him, is merely the social construct of a group "with a shared interest in pushing objects around."[5])

But as we have noted in Chapters 11 and 13 the same "unmasking" strategy employed by SSK/PIS can be used to uncover its real motivations by applying its own methods to itself (as suicidally advocated by the "reflexivity" requirement (iv) of the Strong Program). When this is done, it is apparent that very little of the movement counts as "knowledge as such." Much of it is pure ideology packaged under a veneer of argument and saturated with politics. We can understand it perfectly by investigating its origins. We have seen that the majority of SSK/PIS advocates are men of the Left who in their youth were strongly influenced by the anti-war movement of the 1960s. They regard science and mathematics as central pillars of the modern bureaucratic capitalist state and wish for this reason to attack their prestige. Also plainly visible is simple jealousy of the hard sciences which still have a higher status and are better funded than the social sciences and humanities,[6] as well as the alienation of intellectuals with little power or influence in society. The result has been that the PIS bouillabaisse that the writer began to taste in 1996 has been simmering in its noisome tank of *ressentiment* for decades.

Although PIS fellow travelers of SSK like Richard Rorty dream of replacing science by some mixture of poetry, literary criticism, and the revolutionary aspirations of youth as the dominant contemporary cultural enterprise, most proponents of SSK/PIS realize that—like many deplorable aspects of our civilization—science in the modern West is here to stay, and that—even if they manage to discredit it in the minds of literary intellectuals—its hegemony will be hardly dented. But perhaps they can exercise some control over it. From the beginning of the movement there has been an unconcealed yearning to be the mediator between science and the public and to have the power to reshape science by requiring it to conform to what SSK/PIS believe are socially just goals. This may yet happen as we have argued in Chapter 13 if SSK/PIS begins to spread its orthodoxy among government/funding agency bureaucrats concerned with science policy or in departments at elite universities from which these bureaucrats are recruited. Even if this does not prove to be fully possible, at the very least

---

[5] Forman (1998), p. 4.

[6] Although there has been some improvement in the last decade or so, financial support of subjects like sociology or literature has been tenuous compared to the resources available to science, mathematics, and particularly engineering.

SSK/PIS hope to enhance the political power and prestige of its adherents and to secure a bigger piece of the Federal research funding pie. Given the institutional growth of academic STS over the past generation these ambitions may be in the process of being successfully realized. SSK/PIS ideology in particular, it must be admitted, is exquisitely designed for such purposes, for it permits an apparently sophisticated "take" on science and mathematics suitable for the cocktail party conversation of science policy bureaucrats, without the painful and time consuming necessity of actually mastering or even being familiar with the details of any science. It is much easier to discourse lightly on Gödel's theorems as symptoms of the bankruptcy of Western logocentrism than understanding either what these theorems really say or mastering their proof.

By contrast we have argued that there is value still in preserving the barrier erected by Mannheim at least between the formal *content* and justification of either the hard sciences science or mathematics and sociological analysis. Looking constantly for external political or social factors to explain the genesis or "cause of belief" of particular scientific theories or explaining them by a process of negotiation or local political manoeuvering among groups of investigators does not do justice to the science. A nonscientific reader would have no idea from this approach what the science was *about* or what exactly were the arguments held by defenders or opponents of a theory. Furthermore, neither this aspect of SSK or its epistemological relativism can explain why science works and why it is nearly universal. If there is no correspondence to reality, its prediction of unknown phenomena would be inexplicable, and no explanation can be given of how it differs (as it plainly does) from nonscientific ideologies. The same would be true of the incredible technology which is based on fundamental scientific discoveries. Without positrons and gamma rays (to say nothing of some extremely advanced mathematics) it is impossible to imagine, for instance, how PET or CAT scanners could even be imagined, let alone constructed. In our opinion such problems have never been successfully solved by either SSK/PIS or more philosophically inclined scientific anti-realists, despite numerous attempts to do so.

In the case of mathematics, we leave undecided its ontological status and the question whether mathematics is discovered, invented or is a mixture of both. None of the conventional answers are satisfactory. Platonism is burdened by a fantastic ontology consisting of objects which somehow exist outside of space and time.[7] Formalism cannot explain *why* mathematicians come up with the "meaningless" marks on paper that they do:

> Why do we give *this* particular definition, and not some other one? Why *these* axioms and not some others? Such questions, to the formalist, are premathematical. If they are admitted at all to his text or his course, it will be in parentheses, and in brief.[8]

Neither formalism or the reduction by Russell and Whitehead in *Principia Mathematica* of mathematics to logic can do justice to the actual practice and behavior of mathematicians.[9] On the other hand, the approach of SSK which oscillates between conventionalism and empiricism can—despite strenuous efforts—explain neither the universality or the necessity of mathematical conclusions. No convincing case, in particular, has been made for either the historical or present existence of "incommensurable" mathematical systems. The most common candidate for this role, intuitionism, is not satisfactory. Its differences with conventional mathematics are a matter more of disputes over the philosophy of mathematics than of actual mathematical content. That content, moreover, is merely a restriction of the content of classical mathematics and does not contain theorems that contradict those provable in the non-intuitionist framework.[10] Lastly, the search for political elements in the content of mathematical theory seems especially futile and wrongheaded.

Where does this leave us? The purpose of this essay has been only to show that SSK/PIS has serious deficiencies, as does every other known approach to the epistemology of science and mathematics. We have nei-

---

[7] Another perhaps more telling objection to the theory of Forms is contained in Plato's dialogue *Parmenides*. The dialogue is between the aged Parmenides, his disciple Zeno, and the young Socrates. It is one of the few dialogues where Plato has Socrates mention the theory of Forms, and he is refuted by Parmenides. Besides the usual objections, including a version of the Third Man Argument, Parmenides says that if Forms exist there must be Forms corresponding to "hair, mud, and dirt, or anything else which is vile and paltry" (130c, Jowett translation). Applied to mathematics, this would say that the ugly and boring mathematical objects in a bad ten minute contributed talk at a regional AMS meeting at Persepolis State exist eternally in the world of Forms. One perhaps might reply that that these objects are constructed out better quality but more elementary Forms; but the construction is poorly done—like East Bloc public housing, for example.

[8] Davis and Hersh (1981), pp. 341–342.

[9] For technical criticism of both formalism and logicism see Körner (1960), Chapter V. Körner claims that both approaches cannot explain the empirical aspects of mathematics.

[10] Besides it seems today a dying cult which is ignored by most mathematicians (although perhaps not by philosophers).

ther the ability or the intention to supply a "correct" approach. Perhaps the truth of the matter is that none is available. This does not entail a form of Mannheim-like relationism, however. Instead, we have some sympathy with the position of Colin McGinn who has argued that the problems of "knowledge" and "truth" as well as most other nontrivial philosophical problems are simply too difficult for humans to solve; our minds are just not constructed in a way to do this successfully. McGinn[11] suggests that "solutions" to philosophic problems oscillate endlessly around a series of positions which he has designated by the acronym DIME. What do these letters stand for? Consider a philosophic concept which we will call C. Then D stands for an effort to have C "domesticated, defanged, demoted, dessicated". I maintains that C is "irreducible, indefinable, and inexplicable," i.e., C just *is* and we should get used to it. Next "M stands for a magical, miraculous, mystical ... mad" explanation while:

> E is for elimination, ejection, extrusion. The E proponent despairs of domestication, balks at irreducibility, and scoffs at magic. His position is that C facts look impossible because that is what they are ... [12]

How do these alternatives match some of the positions discussed in this essay? M clearly corresponds to mathematical Platonism. The common sense realism of both scientists and G. E. Moore seems a D-type assertion, while the latter's ethical theory falls under case I.[13] If C is the standard textbook assertion of scientific realism and method, E corresponds to the denial of absolute truth and the relativism of SSK—the foundation of this position being Humean skepticism. The writer's own position towards mathematics inclines to I. Mathematics simply *exists*; where it comes from and its ontological implications are mysteries which we do not believe it is possible to solve. Formalism would fall under case D, and the "humanism" of Reuben Hersh in *What is Mathematics Really?* (1997) or the social constructivism of Paul Ernest appears to be an unstable mixture of D and E.

However we classify various philosophical positions under the DIME scheme. The realism anti-realism dispute as it pertains to science like most major philosophic disputes is probably unresolvable. What can even be meant by a mind-independent reality (and certainly what properties it can have independent of human perceptions) is unclear. Yet viewing it solely

---
[11] McGinn (1993), McGinn (1999).
[12] McGinn (1999), pp. 15–16.
[13] Moore in the *Principia Ethica* argued that the Good is a non-natural property which is indefinable.

as a collection of perceptions conditioned by the nature of the human mind leads to solipsism. Attempted solutions such as Kantianism seem an unstable compromise. But in any event metaphysical anti-realism does not imply the characteristic SSK/PIS thesis that the scientist's ideology, class, gender, politics, or race can or should determine his science, and that consequently we can freely change what we feel to be an undesirable scientific "reality" by changing the group that "produces" it. Anti-realism is also perfectly compatible with the normal practice of science if we view scientific theories as purely instrumental, i.e., a means of organizing our perceptual experience, "save the appearances," or to make predictions. But we do feel that a mild scientific realism, or the view that there is a way the world is which is independent of human cognitive structures or wishes, which science can *discover*, has some advantages. Scientists certainly *for the most part*[14] believe this, and it motivates their work. They want to find out what the structure of the world *is* instead of just constructing some fictional model to account for observations. Without a belief in realism there would be no need for "explanations" at all. Science need not amount to much more than a vast extension of the activities of Babylonian temple priests who could predict the exact position of the planets on the basis of exact records with no explanatory model of their motion at all. Further, as we and many others have argued realism seems the best explanation for the success of science, particularly its ability to predict previously unknown qualitative or quantitative phenomena. The existence of a mind-independent reality also seems to be the most believable explanation of the failures of a scientific theory. How else should we explain anomalies and the resistance or "push-back" our theories experience from nature?[15]

There is certainly a broad role for sociology in the study of science. But this would be more in the tradition pioneered by Merton than one centering on the SSK focus on the socially constructed status of scientific truth. Clearly society and politics can influence science in many ways, including the research direction that science pursues. A certain kind of social structure is needed for the development of science as are universities and laboratories. These are not to be found in the madrassas of rural Afghanistan. Government policy and funding decisions decide what kind of science is to be done. At the level closest to working scientists this function is performed

---

[14]The Copenhagen interpretation of quantum mechanics being an exception
[15]Interesting and accessible discussions of these issues and social constructivism together with references may be found in Klee (1997).

by NSF. More indirectly, as Paul Forman[16] has argued, the needs of the post World War II national security state has had a decisive impact in the United States on the development of several areas of physics including quantum electronics and especially in the invention of gadgets like the maser. Similarly, military needs were critical to the growth of space science and technology from 1946 onward;[17] and the efflorescence of mathematics in the 1960s was, as we have noted, driven by a panicked reaction to Sputnik. But these factors cannot explain the details of particular theories; nor can the theories be criticized on the basis of a military origin (although the use made of them certainly can be). Their substance is simply a consequence of the direction taken by the research. Of course, this direction and why one project or area is funded and another rejected can be a function of government policy and as such is open to criticism. But this has no bearing on the validity of the results, at least in the physical sciences. Finally, we do not deny that scientists are human beings. They engage in behaviors common to both human and primate groups and are subject to the same cultural and social influences that affect ordinary people. The vicious infighting that Bruno Latour discerns in the laboratory *does* on occasion occur and can, at least temporarily, be a key factor in the victory of both correct and incorrect theories. Also, not only the direction but also the conclusions of scientific research can (especially in some examples we consider below) be affected by social, cultural, and political factors. At the very least, the cultural environment can supply metaphors and concepts which the scientist can apply in other contexts. Neo-Platonism, Malthus, and the fashionable denunciation of causality in post-World War I Germany probably played such roles for Copernicus, Darwin, and the Weimar physicists creation of quantum mechanics.[18] In these respects science does not differ from the humanities. Yet, at the same time science *is* different. At least some parts of it (especially the hard sciences) are anchored to a mind-independent reality. The history of science may be a messy affair which does not conform to the canons of scientific methodology promulgated by philosophers. But no matter how shady or ambiguous the origins of a scientific theory—even in physics—may be, it still must in the long run *work*. It must be compatible with that mind-independent reality we call nature. Mere spin and propa-

---

[16] Forman (1996).
[17] Devorkin (1996).
[18] However, we feel that Forman's thesis on the connection of the Weimar cultural climate and its physicists is far too extreme. There is little evidence that the abolition of causality in nuclear physics represented a cynical or conscious ploy by Weimar physicists to regain their lost prestige.

ganda by scientists can merely postpone the inevitable fate of an incorrect theory. In this feature science does not resemble other areas of cultural production such as political ideology, philosophy, and religion.[19]

Despite our many criticisms of the SSK approach as it has been applied to the hard sciences and to mathematics, there is a realm which we have not yet discussed where it may have some application. This includes the so-called "soft sciences" such as psychiatry, anthropology, parts of biology and medicine, and of course sociology itself (whether of the SSK persuasion or not), or more generally any scientific discipline whose questions and results affect significant social interests (e.g. nuclear engineering, climatology). These are often obvious and unstable mixtures of "objective" and ideological elements. The latter especially can determine what questions *can* be asked and the form and content of the answers it is legitimate to consider. In Kuhnian language the paradigm may have a conscious or unconscious political content which may change with the historical Zeitgeist. For example, anthropology in the age of "scientific" racism and social Darwinism is very different in values and "results" from the anthropology of the heirs of Franz Boas in the late twentieth century. Some of the questions which concerned the former can not even be legitimately *asked*[20] by the latter, and it is not clear that the difference reflects the growth of objective knowledge rather than a change in the political climate. Similarly, in psychiatry theories on the origin of serious mental illness have oscillated over the last century from a frankly "eugenic" hereditarian point of view (c. 1900) to purely environmental (e.g., the uncaring remote "refrigerator" mother or a poorly functioning family) causation (c. 1950), and then at the present time to an emphasis in discovering chemical imbalances in the brain (e.g., low serotonin levels). Here too it would be a fascinating task to distinguish purely "objective" from contemporary "political" factors. We could ask, for instance, if the postwar disgust at Nazi excesses made it politically unfashionable to seek innate or biological causes of mental illness or personality traits or to seek other than environmental/cultural explanations

---

[19] We do not deny that what are felt to be incorrect or obsolescent theories in these areas are also rejected. But the process by which this is done is essentially "sociological"; the reasons for change often have to do with changes in intellectual fashion–analogous to changes in the length of women's dresses—or changes in society. But what can change for these reasons can also return. We often find essentially the same positions oscillating between domination and rejection over the centuries. PIS itself is a revival in part of the attitudes of ancient Greek sophists, while their realist opponents owe much to Plato. One rarely finds this kind of oscillation in the development of science.

[20] So doing would mark the questioner as the worst kind of reactionary or racist.

for characteristics such as apparent "IQ" or behavioral differences between human groups. In many of these situations one set of global assumptions is temporarily dominant until the underlying political Zeitgeist changes. The institutional role of the science incorporating these assumption is to ensure "correct" and suppress unwanted, indeed politically forbidden answers. No experiment or data is *allowed* to challenge the dominant paradigm. Any attempt to do so will call forth countless papers demonstrating its defects and the stupidity or evil intent of its authors. If the contending factions are more evenly matched but reflect different and strongly held political interests, then one can obtain a scientific analog of the dispute between Salafist Muslims and Zionists sketched in Chapter 11. Each side simply will *not* accept (and will find rational reasons not to do so!) the other's results no matter how convincing they may be to an neutral observer.[21] In still other cases the "science" in question is at a pre-paradigm stage in the Kuhnian sense. Different schools differ in methodology and in the philosophy of what the science should be. One can think, for example, of sociobiology versus strictly environmental approaches and the various competing sects of psychoanalysis (e.g., Jung versus Freud), or at a more fundamental level psychoanalysis versus orthodox psychiatry. Often these examples are difficult to separate from the previous ones since they may also involve explicit and enduring political disagreements.

Such situations tend to be rich in "social constructions," i.e., concepts which both bear some relation to "reality" and at the same time are products of complicated social interests. They often are either vast extensions of real phenomena or concepts for whose validity experimental evidence is ambiguous.[22] Most also have "extra-theoretical" functions in the Mannheim sense. In the most obvious sense they reinforce the social or economic power of their proponents. Psychiatry, for example, by diagnosing more and more clusters of deviant or merely eccentric behavior as specific named mental illnesses[23] has increased the the need for more psychiatric experts

---

[21]We leave it to philosophers to decide whether or not the correct application of "scientific method" can settle such "incommensurable" disputes. Even if it is theoretically possible, as a practical matter it often is not when huge political issues are at stake. Conversely, of course, accepted methods of scientific decision may work if the only extra-scientific issue at stake is one's rank in the professional food chain.

[22]"Child abuse" and and "ritualized Satanic sexual abuse" would be respectively examples of the former and latter. See Hacking (1998), Chapter 4.

[23]Recorded in *The Diagnostic and Statistical Manual of Mental Disorders*, successive editions of which in 1980, 87, and 94 name an ever increasing number of previously unknown conditions. They also remove a few earlier diagnoses of "disease," such as homosexuality.

who in turn "construct" more psychiatric conditions and "find" an increasing prevalence of them, thus creating a positive (and lucrative) feedback effect. In his fascinating book *The Social Construction of What?* (1999) Ian Hacking has given many more examples of such constructions.

But all this is very obvious. The reader doubtlessly may supply many examples of his own. Paradoxically, however, SSK seems not to be very interested in this kind of analysis. Perhaps it is too easily made. Perhaps also, as we have argued above, SSK has a political dislike of the "hard" sciences and therefore prefers to find ideology at work in areas like physics, astronomy, or paleontology.[24] Properly done, however, the application of sociological analysis to the sciences where it is appropriate might produce a sophisticated and nuanced theory of social construction along the lines of Ian Hacking's work.

So far despite its dramatic institutional growth over the last twenty or thirty years SSK/PIS seems to have had little influence on other disciplines except indirectly on Education. In spite of exceptions like Paul Gross and Norman Levitt or Alan Sokal, few scientists or mathematicians know anything about the movement and it certainly has not affected their work. The so-called Science Wars may have been launched by the 1994 publication of *Higher Superstition* but, leaving aside a few high profile scientists, it has essentially been a conflict *within* the humanities rather than between scientists and postmodern humanists. If this situation continues SSK or PIS will remain harmless academic fiefdoms found in STS or in some humanities departments, generating ideas no more peculiar than what are commonly found on campus and which may be safely tolerated by scientists. More literate and aware scientific types may find the writings of SSK/PIS a source of amusement, and some—concentrating on policy issues—may (if sane) even be useful. If, however, the movement should impact science through absorption of its ideas by activist groups, government agencies, or PC university administrators, then we may have a new outbreak of the Science Wars, this time involving more scientists.[25]

The issue is a potentially important one. Paradoxically science may be worse off suffering from the "good intentions" of a liberal democratic state than under a frankly oppressive system such as the old Soviet Union. Since Marxist ideology valued science, the USSR supported "big science" with

---

[24] For a PIS approach to dinosaurs see Mitchell (1998).
[25] This possibility would probably be accompanied by a general radicalization of a significant portion of society as we had in the Vietnam era; one could imagine this developing as the result of some prolonged foreign policy crisis such as a second Vietnam-like situation.

few questions asked. In the eyes of the Party excellence in high energy physics and mathematics were a symbol of the superiority of socialism. Consequently, considerable resources were committed to these and other hard sciences, and scientists were granted material perks unavailable to ordinary citizens. Even better, scientists and their theories were more or less left alone provided they accepted the authority of the Party. Elementary particle theory, for example, was seldom examined by official ideologists for signs of "bourgeois formalism." The only serious exceptions were Stalin's interference in a few politically sensitive areas like biology and genetics, resulting in a harsh twenty year domination of the pseudoscience of Lysenkoism and some periodic hostility to Relativity. However tragic these episodes were for individuals and specific disciplines, they did not affect science as a whole. But with STS influencing policy, scientists may face a far more choking bureaucracy than they ever did before either in the West or in the Soviet Union. We predict an increasing ideological interference in the name of buzzwords like "feminism", "democratic social control," "social justice," "equity," or "diversity." We can already see an analogy of this phenomenon in educational policy where the teaching of rigorous mathematics (like axiomatic geometry) at the high school level and the tension between "excellence" and "equity" is felt by educational theorists to be as replete with ideological dangers as proposals for larger private plots on collective farms were in Moscow in 1950.[26] Especially ruinous would be the bypassing of traditional standards of excellence in the interests of hiring or tenuring previously under represented groups. SSK, especially in its feminist form, could justify this by arguing that:

> ... the goals methods, theories, and even the actual data of the natural sciences are not written in nature; all are subject to the play of social forces. ... Social, psychological, and political norms are inescapable, and they too influence the questions we ask, the methods we choose, the explanations we find satisfying, and even the data we deem worthy of recording.[27]

Since according to SSK/PIS science encodes dominant political values just as any other cultural activity does, feminists and other advocates of "de-

---

[26] See Schoenfeld (2004) and the many papers on social justice and mathematics education published on Paul Ernest's website: http://www.people.ex.ac.uk/PErnest/. One feels that the demands of "social justice" make it very difficult to teach any mathematical subject that is not accessible to the majority of students. That would be "elitist" and violate "equity."

[27] Keller (1990), quoted in Gross and Levitt (1994), p. 139.

prived" groups argue that contemporary science is, perhaps unconsciously, molded by the values of white males in a capitalist society. Other groups such as women, people of color, the poor would have values that would mold science in a different way. They would bring insights that are unavailable to science as it is currently practiced. Nor should objections to the new groups on the grounds of "merit" be taken seriously, contaminated as they are by androcentric and racial bias. It is a case of "whose science, whose knowledge."[28]

It should be obvious that this entire stance is simply a political weapon designed to produce jobs and entitlements to one group and to justify discrimination against another—white males. The political power of this position is not a function of the quality of its arguments which as we have seen are structurally similar to Aryan arguments in the 1930s against "Jewish Science," but is likely to be be enhanced both by the changing demography of the nation as reflected on its campuses and by the weakening of the idea of "assimilation" into a common American public culture. Policies based on such arguments (ignoring the fact that there is not the slightest evidence that they are true) could lead to the balkanizing of science (and academia generally) at least in the US. For administrators this might be a small price to pay for political peace on campus. Superficially business would continue as usual. Departments would continue to function. Grants would be awarded—but perhaps on unstated political grounds—as before. Only very slowly would it be noticed that the quality (whatever that means)[29] would decline and that mainly political opportunists would choose scientific careers.[30]

We hope that this situation never happens both because of its harmful effect on science and because academic peace and toleration are values worth preserving. If it does, however, scientists would need to emulate the tactics of their enemies. Because of their own academic background this may be difficult. Scientists want to concentrate on their research.

---

[28] See the book by Sandra Harding of the same title.
[29] Accusations of lack of quality, however, would very probably be explained away as the disgruntlement of the previously dominant who had lost their privileges. Even if the new order produced very eccentric science, this would be interpreted as a sign of previously suppressed originality.
[30] In the USSR the best and the brightest went into the hard sciences because of the opportunities and relative freedom. Conversely, the humanities—especially philosophy—were filled with political hacks who churned out ideology rather than scholarship. It would be ironic if aggressive affirmative action and ideological monitoring using SSK arguments produced the same effect in the hard sciences in the US.

They are inclined to ignore academic policy and leave such matters to administrators from whom they expect very little. They also tend to have an automatic respect for other academic fields even if they don't understand them. Moreover, a scientist, especially in the hard sciences, is trained to take his opponent's arguments seriously. He will honestly attempt to comprehend these arguments and present them clearly. Only then will he try to find to find mistakes or gaps, and is prepared to change his own mind if the facts require it. The same is true for most analytic philosophers or historians of an old-fashioned type who take seriously a distinction between analysis and propaganda.

We have tried to do the same in this essay regarding SSK/PIS. But if there is a recurrence of the Science Wars, this time over substantial material issues, such a policy may be both unnecessary and excessively charitable. We have argued that SSK/PIS is simply an ideology directed against the prestige of science and designed to change the balance of power within academia; its assertions are mostly propaganda rather than factual discoveries, of which it has made very few. Having understood this, scientists are in a position in true Mannheim fashion to "unmask" or "disintegrate" it. Serious argument should be reserved against positions where a modicum of communication and shared respect is present; it is useless here. It would be more effective for scientists simply to follow the tactics of the "oldest politics of all" recommended by Bruno Latour for success in the laboratory.[31] That is, the methods of Cleon, Gorgias, and other ancient Sophists are to be preferred. One should mobilize propaganda, rhetoric, and persuasion, seek allies, isolate the opposition, and sow discord among them. These tactics are necessary since the question is really one of academic power to determine—in Lenin's phrase *"Kto Kogo?"*— roughly translated as "who dominates whom?"—rather than a disinterested quest for academic "truth," a word which itself according to PIS is merely a propaganda weapon. If all else fails, calls for institutional action can be employed. In an era of tightened budgets there is no reason for engineering or physics to be sacrificed for women's studies or sociology. Financial stringency requires quality control, the need for which can be clearly illustrated even for the most obtuse politician by the writings of SSK/PIS itself. In 1803 Napoleon closed down Destutt de Tracy's "School of Ideology"; in the future for similar reasons it may be necessary for politicians close down obstreperous STS

---

[31] Quoted in Chapter 9.

or sociology departments should they intrude on scientific turf.[32] However, just as feminists admit that their desires for a radical change in the practice of science would have to accompanied by "large scale social changes,"[33] the kind of counteroffensive against the radical forms of STS we are proposing would probably require a further rightward shift in the social order, perhaps catalyzed by the Christian Right, together with a political purge of PIS sympathizers in Federal agencies such as NSF and HEW. But this would bring new and perhaps even worse problems of its own to academia, since there is no evidence that the Right has much appreciation for science unless it makes money or weapons. Again, let us hope that this situation never arises, and that due to "Kuhnification" the various strands of STS remain harmless academic disciplines, cultivating their paradigms—like Candide his garden—without having any political impact on science policy. It would be better for all concerned if SSK/PIS continued as a form of intellectual Ghost Dancing, although in the long term it will have no greater benefits for those fearing science than the original form did for the Lakota Sioux fearing settlers. In this happy case the Machiavellian tactics we have described will need never to be employed and there would be no necessity for an academic version of the Wounded Knee Massacre.

We are aware that though out this essay we have been (to quote Paul Forman) "assailing the seasons"[34] since some form of PIS together with a strong dose of perspectivism and relativism is probably the dominant contemporary interpretation of both science and its history among humanists at the present time. But whatever the current situation, a long-term view of STS/PIS gives some comfort. Since, like everything else in the humanities, it is a historically contingent phenomena, arising at a certain time under certain conditions, it too will eventually disappear—as most other dogmas in the humanities have—once the conditions which stimulated its birth change. While Maxwell's Laws, Gödel's Theorems, and even an occasional theorem produced at Persepolis State will survive, we would hazard a prediction that in a few academic generations[35] SSK/PIS will probably

---

[32] Very valuable for this purpose are "tenured radicals." They do not have to be in STS. English or Ethnic Studies will suffice. To the average politician the differences between different species of left wing academics is vague and they can be easily conflated.

[33] See Harding (1991), pp. 299–302.

[34] See Forman's extremely hostile review of the same title of a 1995 anti-PIS conference organized in 1995 by Gross and Levitt in *Science*.

[35] A period of about 12 to 15 years which is the sum of the average time it takes to get a Ph.D., the duration of a few temporary appointments, and the time until tenure is awarded or denied.

be known only to specialists in the intellectual pathologies of the post-Vietnam era; for them the only important question may be how thinkers such as Bruno Latour or Steve Shapin were possible. The doctrine may still be marginally more important in the wider intellectual setting then than Jansenism or fourteenth century Franciscan quarrels over the Poverty of Christ are for us. But we are willing to bet that it will rank no higher than Comte's Positivism or the thought of Herbert Spencer do now. But even if this comes to pass, we should not mourn too much the demise of SSK/PIS. When it was fashionable it satisfied real needs: it provided a potent psychological therapy for the the frustrations and *ressentiment* of its practitioners. On a more mundane level it helped to secure jobs, tenure, grants, administrative positions, and some influence in the scientific bureaucracy. One can ask for little more in the humanities.

Bibliography

Abbott, E. A. (1884). *Flatland: A Romance of Many Dimensions* (Seely&Co., London).
Adams, H. H. (1990). *African-American Baseline Essays—Science Baseline Essay: African and African-American Contributions to Science and Technology* (Portland Public Schools).
Aronowitz, S. (1988). *Science as Power: Discourse and Ideology in Modern Society* (University of Minnesota Press, Minneapolis).
Aronowitz, S. (1996). The politics of the science wars, *Social Text* **46/47**, pp. 177–197.
Barash, D. P. (2005). C. P. Snow: Bridging the two-culture divide, *The Chronicle of Higher Education*, Vol. 52, Issue 14, p. B10.
Barnes, B. (1974). *Scientific Knowledge and Sociological Theory* (Routledge & Kegan Paul, Boston, London and Henley).
Barnes, B. (1982). *T. S. Kuhn and Social Science* (Columbia University Press, New York).
Barnes, B. and Bloor, D. (1982). Relativism, rationalism, and the sociology of knowledge, in M. Hollis and S. Lukes (eds.), *Rationality and Relativism* (Blackwell, Oxford), pp. 21–47.
Barnes, B., Bloor, D, and Henry, J. (1996). *Scientific Knowledge: A Sociological Analysis* (University of Chicago Press, Chicago).
Bates, T. (1992). *Rads: The 1970 Bombing of the Army Math Research Center and Its Aftermath* (Harper Collins Publishers, Inc., New York).
Bernal, M. (1987). *Black Athena: The Afro-Asian Roots of Classical Civilization* (Free Association Press, London).
Berger, P., and Luckmann, T. (1966). *The Social Construction of Reality* (Doubleday, New York).
Beyerchen, A. D. (1978). *Scientists under Hitler: Politics and the Physics Community in the Third Reich* (Yale University Press, New Haven).
Bird, A. (2000). *Thomas Kuhn* (Princeton University Press, Princeton).
Bloor, D. (1976). *Knowledge and Social Imagery* (Routledge & Kegan Paul, Boston, London and Henley).
Bloor, D. (1981a). Hamilton and Peacock on the essence of algebra, in H.

Mehrtens, H. Bos, and I. Schneider (eds.), *Social History of Nineteenth Century Mathematics* (Boston), pp. 202–232.

Bloor, D. (1981b). The strengths of the Strong Programme, *Phil. Soc. Sci.* **11**, pp. 199–213.

Bloor, D. (1982a). Wittgenstein and Mannheim on the sociology of mathematics, in H. M. Collins (ed.), *Sociology of Scientific Knowledge* (Bath University Press, Bath), pp. 39–57.

Bloor, D. (1982b). Durkheim and Mauss revisited: Classification and the sociology of knowledge, *Studies in History and Philosophy of Science* **23**, pp. 267–298.

Bloor, D. (1983). *Wittgenstein: A Social Theory of Knowledge* (Macmillan, London).

Bloor, D. (1997). *Wittgenstein: Rules and Institutions* (Routledge and Keegan Paul, Boston, London, and Henley).

Bloor, D. (2004). Sociology of scientific knowledge, in I. Niiniluoto, M. Sintonen, and J. Woleński (eds.), *Handbook of Epistemology* (Kluwer Academic Publishers, Dordrecht/Boston/London), pp. 919–962.

Bos, H. J. M., and Mehrtens, H. (1977). The interactions of mathematics and society, in History: Some Exploratory Remarks, *Historia Mathematica* **4**, pp. 7–30.

Brown, J. R. (1989). *The Rational and the Social* (Routledge, London and New York).

Brown, J. R. (1989). *Smoke and Mirrors: How Science Reflects Reality* (Routledge, London and New York).

Brown, J. R. (2001). *Who Rules in Science: An Opinionated Guide to the Science Wars* (Harvard University Press, Cambridge).

Brown, J. R. (ed.) (1984). *Scientific Rationality: The Sociological Turn* (Reidel, Dordrecht).

Callon, M. (1992). Some elements of a sociology of translation: Domestication of the scallops and the fishermen of St. Brieuc bay, in A. Pickering (ed.), *Science as Practice and Culture* (University of Chicago Press, Chicago), pp. 196–233.

Callon, M. and Latour, B. (1992). Don't throw the baby out with the Bath school: A reply to Collins and Yearly, in A. Pickering (ed.), *Science as Practice and Culture* (University of Chicago Press, Chicago), pp. 343–367.

Campbell, M., and Campbell-Wright, R. K. (1995). Towards a feminist algebra, in Sue Rosser (ed.) *Teaching the Majority: Breaking the Gender Barrier in Science, Mathematics and Engineering* (Teachers College Press, New York).

Castiglioni, A. (1947). *A History of Medicine* (Alfred A. Knopf, New York).

Cohen, H. F. (1994). *The Scientific Revolution: A Bibliographical Inquiry* (The University of Chicago Press, Chicago).

Cohen, I. B. and Clagett, M. (1966) Alexandre Koyré (1892–1964): Commemoration, *Isis* **57**, pp. 157–166.

Cole, S. (1992). *Making Science: Between Nature and Society* (Harvard University Press, Cambridge).

Cole, S. (1996). Voodoo sociology, in P. Gross, N. Levitt, and M. L. Lewis (eds.), *The Flight from Science and Reason*. Annals of the New York Academy of Science, Vol. 775 (New York Academy of Sciences, New York), pp. 274–287.

Collins, H. M. (1981a). Stages in the empirical programme of relativism, *Social Studies of Science* **11**, pp. 3–10.

Collins, H. M. (1981b). What is TRASP?: The radical programme as a methodological imperative, *Philosophy of the Social Sciences* **11**, pp. 215–224.

Collins, H. M. (1982). Special relativism: The natural attitude, *Social Studies of Science* **12**, pp. 139–143.

Collins, H. M. (1983a). An empirical relativist programme in the sociology of scientific knowledge, in *Science Observed: Perspectives on the Social Study of Science* (Sage Publications, London and Beverly Hills), pp. 85–143.

Collins, H. M. (1983b). The sociology of scientific knowledge: Studies of contemporary science, *Annual Review of Sociology* **9**, pp. 265–285.

Collins, H. M. (1985). *Changing Order: Replication and Induction in Scientific Practice* (Sage Publications, London and Beverly Hills).

Collins, H. M. (1996). In praise of futile gestures: How scientific is the sociology of scientific knowledge, *Social Studies of Science* **26**, pp. 229–244.

Collins, H. M. and Pinch, T. (1982). *Frames of Meaning: The Social Construction of Extraordinary Science* (Routledge and Keegan Paul, Boston, London, and Henley).

Collins, H. M. and Pinch, T. (1988). *The Golem: What Everyone Should Know About Science* (Cambridge University Press, Cambridge).

Collins, H. M., Pinch, T., and Yearly, S. (1982). Epistemological chicken, in A. Pickering (ed.), *Science as Practice and Culture* (University of Chicago Press, Chicago), pp. 310–325.

Crawford, K. (2006). Privilege, possibility, and perversion: Rethinking the study of early modern sexuality, *Journal of Modern History* **78**, pp. 412–423.

Crowe, M. J. (1975). Ten 'laws' concerning patterns of change in the history of Mathematics, *Historia Mathematica* **2**, pp. 161–166.

Davis, P. J. and Hersh, R. (1981). *The Mathematical Experience* (Houghton Mifflin, Boston).

de Kruif, P. (1926). *Microbe Hunters* (Harcourt Brace and Company, Boston and New York).

Devitt, M. (1984). *Realism and Truth* (Princeton University Press, Princeton).

Devitt, M. (2001). Incommensurability and the priority of metaphysics, in Hoyningen-Huene and H. Sankey (eds.), *Incommensurability and Related Matters* (Kluwer Academic Publishers, Dordrecht/Boston/London), pp. 143–157.

Devorkin, D. H (1996). The Military origins of the space sciences in the American V–2 era, in P. Forman and J. M. Sánchez-Ron (eds.), *National Military Establishments and the Advancement of Science and Technology* (Kluwer Academic Publishers, Dordrecht/Boston/London), pp. 233–260.

Diggens, J. P. (1992). *The Rise and Fall of the American Left* (W. W. Norton, New York and London).

Earman, J. and Glymour, C. (1981). Relativity and eclipses: The British eclipse

expeditions and their predecessors, *Historical Studies in the Physical Sciences* **11**, pp. 49–85.
Editors of *Lingua Franca* (eds.) (2000). *The Sokal Hoax: The Sham That Shook the Academy* (University of Nebraska Press, Lincoln and London).
Eldredge, N. and Gould, S. J. (1972). Punctuated equilibria: An alternative to phyletic gradualism, in T. J. M. Schopf (ed.), *Models in Paleobiolgy* (Freeman Cooper & Co., San Francisco), pp. 82–115.
Ernest, P. (1991). *The Philosophy of Mathmatics Education* (The Falmer Press, London and New York).
Ernest, P. (1998). *Social Constructivism as a Philosophy of Mathematics* (State University of New York Press, Albany).
Fara P. (2002). *Newton: The Making of Genius* (Columbia University Press, New York).
Farley, J., and Geison, G. L. (1982). Science, politics and spontaneous generation in nineteenth-century France: the Pasteur-Pouchet debate, in H. M. Collins (ed.), *Sociology of Scientific Knowledge* (Bath University Press, Bath), pp. 1–38.
Feyerabend, P. (1967). Consolations for the specialist, in *Pilosophical Papers*, Vol. 2 (Cambridge University Press, Cambridge), pp. 131–161.
Feyerabend, P. (1969). Explanation, reduction, and empiricism, in H. Feigl and H. Maxwell (eds.), *Minnesota studies in the Philosophy of Science*, Vol. III (University of Minnesota Press, Minneapolis), pp. 28–97.
Feyerabend, P. (1965). Problems of empiricism, in R. Colodny (ed.), *Beyond the Edge of Certainty* (Prentice Hall, Englewood Cliffs), pp. 145–260.
Feyerabend, P. (1975). *Against Method* (Humanities Press, New York and London).
Feyerabend, P. (1978). *Science in a Free Society* (Lowe & Brydone Ltd., Thetford, Norfolk).
Feyerabend, P. (1981). *Philosophical Papers*, 2 vol. (Cambridge University Press, Cambridge).
Feyerabend, P. (1987). *Farewell to Reason* (Verso, London).
Feyerabend, P. (1991). *Three Dialogues on Knowledge* (Basil Blackwell, Oxford).
Feyerabend, P. (1992). Atoms and consciousness, *Common Knowledge* **1**, pp. 28–32.
Feyerabend, P. (1995a). B. Terpstra (ed.), *Conquest of Abundance* (The University of Chicago Press, Chicago).
Feyerabend, P. (1995b). *Killing Time* (The University of Chicago Press, Chicago).
Feyerabend, P. (1999). How to defend society against science, in J. Preston (ed.), *Knowledge, Science and Relativism* (Cambridge University Press, Cambridge), pp. 181–191.
Fine, A. (1984). The natural ontological attitude, in J. Leplin (ed.), *Scientific Realism* (University of California Press, Berkeley), pp. 83–107.
Fine, A. (1996). Science made up: Constructivist sociology of scientific knowledge, in P. Galison and D. J. Stump (eds.), *The Disunity of Science: Boundaries, Contexts and Power* (Stanford University Press, Palo Alto), pp. 231–254.

Fisher, C. S. (1966). The death of a mathematical theory: A study in the sociology of knowledge, *Archive for History of Exact Sciences* **iii**, pp. 137–139.

Fleck, L. (1979). T. J. Trenn and R. K. Merton (eds.), *The Genesis and Development of a Scientific Fact* (The University of Chicago Press, Chicago).

Forman, P. (1971). Weimar culture, causality, and quantum theory, 1918- 1927: Adaption by German physicists and mathematicians to a hostile intellectual milieu, *Historical Studies in the Physical Sciences* **3**, pp. 1–115.

Forman, P. (1985). Behind quantum electronics: National security as basis for physical research in the United States, 1940–1960, *Historical Studies in the Physical Sciences* **18**, pp. 149–229.

Forman, P. (1991). Independence, not transcendence, for the historian of science, *Isis* **82**, pp. 71–86.

Forman, P. (1996). Into quantum electronics: The maser as a 'gadget' of cold-war America, in P. Forman and J. M. Sánchez-Ron (eds.), *National Military Establishments and the Advancement of Science and Technology* (Kluwer Academic Publishers, Dordrecht/Boston/London), pp. 261–326.

Forman, P. (1997a). Assailing the seasons, *Science* **276**, pp. 750–752.

Forman, P. (1997b). Recent science: Late-modern and post-modern, in Thomas Söderquist (ed.), *The Historiography of Contemporary Science and Technology* (Harwood Academic Publishers, Newark), pp. 179–215.

Forman, P. (1998). In postmodernity the two cultures are one and many, Lecture given at Stanford University, January 26, 1998, http://www.stanford.edu/dept/HPS/FormanThinkPiece.html.

Fowler, D. (1987). *The Mathematics of Plato's Academy: A New Reconstruction* (Clarendon Press, Oxford).

Franklin, A. (1981). Millikan's published and unpublished data on oil drops, *Historical Studies in the Physical Sciences* **11**, pp. 185–202.

Franklin, S. (1996). Making transparencies: Seeing through the science wars, in A. Ross (ed.), *Science Wars* (Duke University Press, Durham and London), pp. 151–167.

Franzén, T. (2005). *Gödel's Theorem: An Incomplete Guide to its Use and Abuse* (A. K. Peters, Wellesley).

Frege, G. (1959). *The Foundations of Arithmetic*, translated by J. L. Austin (Blackwell, Oxford).

Fuller, S. (2000). *Thomas Kuhn: A Philosophical History for Our Times* (The University of Chicago Press, Chicago).

Fuller, S. (2006). A step towards the legalization of science studies, *Social Studies of Science* **36**, pp. 827–834.

Geison, G. (1995). *The Private Science of Louis Pasteur* (Princeton University Press, Princeton).

von Glasersfeld, E. (1995). *Radical Constructivism* (Falmer Press, London and Washington, D.C.).

Golinski, J. (1998). *Making Natural Knowledge: Constructivism and the History of Science* (Cambridge University Press, Cambridge).

Goodman, N. (1978). *Ways of Worldmaking* (Hackett Publishing Company, Indianapolis and Cambridge).

Gould, S. J. (2007). *Punctuated Equilibrium* (Harvard University Press, Cambridge).
Grenz, S. J. (1996). *A primer of Postmodernism* (William B. Eerdmans Publishing Company, Grand Rapids and Cambridge, UK).
Griffiths, H. B. and Hilton, P. J. (1970). *A Comprehensive Textbook of Classical Mathematics: A Contemporary Interpretation* (Springer Verlag, New York, Heidelberg, Berlin).
Grinell, G. (1973). Newton's *Principia* as whig propaganda, in P. Fritz and D. Williams (eds.), *City and Society in the 18th century* (Toronto), pp. 181–192.
Gross, A. (1990). *The Rhetoric of Science* (Harvard University Press, Cambridge).
Gross, P. and Levitt, N. (1994). *Higher Superstition: The Academic Left and Its Quarrels with Science* (The John Hopkins University Press, Baltimore and London).
Guillemin, R. (1977). Peptides in the Brain. The New Endocrinology of the Neuron, Nobel Lecture, December 8, 1977, http://nobelprize.org/nobel_Prize/medicine/laureates/1977/guillemin-lecture.html.
Haack, S. (1974). *Deviant Logic* (Cambridge University Press, Cambridge).
Haack, S. (1978). *Philosophy of Logics* (Cambridge University Press, Cambridge).
Haack, S. (1998). *Manifesto of a Passionate Moderate* (The University of Chicago Press, Chicago).
Haack, S. (2007). *Defending Science—Within Reason: Between Science nd Cynicism* (Prometheus Books, Amherst and New York).
Hacking, I. (1998). *The Social Construction of What?* (Harvard University Press, Cambridge).
Handlin, O. (1979). *Truth in History* (The Belknap Press of Harvard University Press, Cambridge, Mass. and London, UK).
Hanson, N. R. (1958). *Patterns of Discovery* (Cambridge University Press, Cambridge).
Harding, S. (1986). *The Science Question in Feminism* (Cornell University Press, Ithaca).
Harding, S. (1991). *Whose Science? Whose Knowledge? Thinking from Women's Lives* (Cornell University Press, Ithaca).
Harding, S. (ed.) (1993). *The "Racial" Economy of Science: Toward a Democratic Future* (Indiana University Press, Bloomington).
Harding, S. (ed.) (1998). *Is Science Multicultural: Postcolonialisms, Feminisms, and Epistemologies* (Indiana University Press, Bloomington).
Harding, S. (ed.) (2004). *Feminist Standpoint Theory Reader: Intellectual and Political Controversies* (Routledge, London and New York).
Hardy, G. H. (1992). *A Mathematician's Apology* (Cambridge University Press, Cambridge).
Hayes, N. K. (1990). *Chaos Bound: Orderly Disorder in Contemporary Literature and Science* (Cornell University Press, Ithaca).
Hegel, G. W. F. (1953). *Reason in History: A General Introduction to the Philosophy of History* (The Liberal Arts Press, New York).

Heilbron, J. L. (1996). Thomas Samuel Kuhn, 18 July 1922—17 June 1996, *Isis* **89**, pp. 505–515.
Heilbron, J. L. (2005). *Leviathan and the Hot Air Machine*, unpublished manuscript.
Heilbron, J. L. and Kuhn, T. S. (1969). The genesis of the Bohr atom, *Historical Studies in the Physical Sciences* **1**, pp. 211-290.
Hersh, R. (1997). *What is Mathematics Really* (Oxford University Press, New York and London).
Hess, D. J. (1997a). *Science Studies: An Advanced Introduction* (New York University Press, New York).
Hess, D. J. (1997b). *Can Bacteria Cause Cancer? Alternative Medicine Confronts Big Science* (New York University Press, New York).
Hesse, M. (1980). *Revolutions & Reconstructions in the Philosophy of Sciences* (Indiana University Press, Bloomington).
Hessen, B. (1931). The social and economic roots of Newton's Principia, in *Science at the Crossroads* (Kniga, London).
Hicks, S. (2004). *Explaining Postmodernism: Skepticism and Socialism from Rousseau to Foucault* (Scholargy Publishing, Tempe, New Berlin, and Milwaukee).
Himmelfarb, G. (2004). *The New History and the Old: Critical Essays and Reappraisals.* (The Belknap Press of Harvard University Press, Cambridge, Mass. and London, UK).
Horkheimer, M. (1941). The end of reason, *Philosophy and Social Sciences* **IX**, pp. 316–389.
Horkheimer, M. (1947). *The Eclipse of Reason* (Oxford University Press, New York and London).
Hoyningen-Huene, P. (1993). *Reconstructing Scientific Revolutions: Thomas S. Kuhn's Philosophy of Science*, translated by A. L. Levine (The University of Chicago Press, Chicago).
Hoyningen-Huene, P. (1999). Paul Feyerabend: An obituary, in J. Preston, D. Lamb, and G. Munevar (eds.), *Worst Enemy of Science?: Essays in Memory of Paul Feyerabend* (Oxford University Press, London and New York), pp. 3–15.
Isaacson, W. (2007. *Einstein His Life and Universe* (Simon&Shuster Paperbacks, New York, London, Toronto, Sydney).
Jackson, A. (1997). Chinese acrobatics, an old-time brewery, and the "much needed gap": The life of the *Mathematical Reviews*, *Notices of the AMS* **44**(3), pp. 330–337.
Jasanoff, J. (1996). Beyond epistemology: relativism and engagement in the politics of science, *Social Studies of Science* **26**, pp. 393–418.
Jasanoff, J., Markle, G. E., Petersen, J. C., and Pinch, T. (eds.) (1995). *Handbook of Science and Technology Studies* (Sage Publications, Thousand Oaks, London, New Delhi).
Jay, M. (1973). *The Dialectical Imagination: A History of the Frankfurt School and the Institute for Social Research* (Little Brown and Company, Boston, and Toronto).

Jencks, C. (ed.) (1991). *The Post-Modern Reader* (Academy Editions, St Martins Press, New York, and London).
Johnson, J. E. (1981). *Darwin on Trial* (Regnery Gateway, Washington, DC).
Johnson, J. E. 2000). *The Wedge of Truth: Splitting the Foundations of Naturalism* (Intervarsity Press, Downers Grove).
Keller, E. F. (1984). *Reflections on Gender and Science* (Yale University Press, New Haven).
Keller, E. F. (1990). Long live the difference between men and women scientists, *The Scientist* **4**, (October 15, 1990).
Kennedy, E. S. (1966). Late medieval planetary theory, *Isis* **57**, pp. 365–378.
Kirk, G. S. and Raven, J. E. (1963). *The Presocratic Philosophers* (Cambridge University Press, Cambridge).
Klee, R. (1997). *Introduction to the Philosophy of Science: Cutting Nature at its Seams* (Oxford University Press, New York and London).
Kline, M. (1950). *Mathematics The Loss of Certainty* (Oxford University Press, New York and London).
Knorr-Cetina, K. D. (1981). *Manufacturing Knowledge: An Essay on the Constructivist and Contextual Nature of Science* (Pergamon Press, Oxford and New York).
Knorr-Cetina, K. D. (1999). *Epistemic Cultures: How the Sciences Make Knowledge* (Harvard University Press, Cambridge).
Körner, S. K. (1960). *The Philosophy of Mathematics: An Introductory Essay* (Hutchinson University Library, London).
Koertge, N. (ed.) (1998). *A House Built on Sand: Exposing Postmodernist Myths about Science* (Oxford University Press, New York and London).
Koertge, N. (2008). Political correctness in the science classroom, National Association of Scholars, http:$//$www.nas.org$/$polArticles.cfm?Doc_Id= 193.
Koyré, A. (1994). *From the Closed World to the Infinite Universe* (Johns Hopkins University Press, Baltimore and London).
Kraft, P. and Kroes P. (1984). Adaption of scientific knowledge to an intellectual environment. Paul Forman's 'Weimar Culture, Causality, and Quantum Theory, 1918–1927: Analysis and criticism, *Centaurus* **27**, pp. 76–99.
Kragh, H. (1987). *An Introduction to the Historiography of Science* (Cambridge University Press, Cambridge).
Krige, J. (1980). *Science, Revolution, & Discontinuity* (Harvester Press, Sussex).
Kuhn, T. S. (1957). *The Copernican Revolution* (Harvard University Press, Cambridge).
Kuhn, T. S. (1970a). *The Structure of Scientific Revolutions* (University of Chicago Press, Chicago).
Kuhn, T. S. (1970b). *The Essential Tension: Selected Studies in Scientific Tradition and Change* (University of Chicago Press, Chicago).
Kuhn, T. S. (1970). Logic of discovery or psychology of research, in I. Lakatos and A. Musgrave (eds.), *Criticism and the Growth of Knowledge* (Cambridge University Press, Cambridge), pp. 1–23.

Kuhn, T. S. (1974). Second thoughts on paradigms, in F. Suppe (ed.), *The Structure of Scientific Theories* (The University of Illinois Press), pp. 500–517.
Kuhn, T. S. (1978). *Black-body Theory and the Quantum Discontinuity, 1894–1912* (Oxford University Press, London and New York).
Kuhn, T. S. (2000a). Reflections on my critics, in J. Conant and J. Haugeland (eds.), *The Road since Structure* (The University of Chicago Press, Chicago), pp. 123–175.
Kuhn, T. S. (2000b). J. Conant and J. Haugeland (eds.), *The Road since Structure* (University of Chicago Press, Chicago).
Kuhn, T. S. (2000c). The natural and human sciences, in J. Conant and J. Haugeland (eds.), *The Road since Structure* (The University of Chicago Press, Chicago), pp. 216–223.
Kuhn, T. S. (2000d). Possible worlds in the history of science, in J. Conant and J. Haugeland (eds.), *The Road since Structure* (The University of Chicago Press, Chicago), pp. 58–84.
Kuhn, T. S. (1995). Afterwords, in Paul Horwich (ed.), *World Changes* (The MIT Press, Cambridge), pp. 311–341.
Lakatos, I. (1976). *Proofs and Refutations: The Logic of Mathematical Discovery* (Cambridge University Press, Cambridge).
Lakatos, I. (1978a). Infinite regress and the foundations of mathematics, in *Mathematics, Science, and Epistemology*, Philosophical Papers, Vol. 2 (Cambridge University Press, Cambridge), pp. 3–23.
Lakatos, I. (1978b). A renaissance of empiricism in the recent philosophy of mathematics, in *Mathematics, Science, and Epistemology*, Philosophical Papers, Vol. 2 (Cambridge University Press, Cambridge), pp. 24–41.
Latour B. and Woolgar, S. (1979). *Laboratory Life: The Social Construction of Scientific Facts* (Sage Publications, London and Beverly Hills).
Latour, B. (1987). *Science in Action* (Open University Press, Milton Keynes).
Latour B. (1988a). *The Pasteurization of France*, translated by A. Sheridan and J. Law (Harvard University Press, Cambridge and London).
Latour B. (1988b). A relativistic account of Einstein's relativity, *Social Studies of Science* **18**, pp. 3–44.
Latour B. (1999). *Pandora's Hope: Essays on the Reality of Science Studies* (Harvard University Press, Cambridge and London).
Laudan, L. (1981). The pseudo-science of science, *Phil. Soc. Sci.* **11**, pp. 173–198.
Laudan, L. (1984). A confutation of convergent realism, in Jarett Leplin (ed.), *Scientific Realism* (University of California Press, Berkeley), pp. 218–249.
Leavis, F. R. (1963). *Two Cultures? The Significance of C. P. Snow: And an Essay on Sir Charles Snow's Rede Lecture By Michael Yudkin* (Pantheon Books, New York).
Lefkowitz, M. R. (1996). *Not Out of Africa: How Afrocentrism Became an Excuse to Teach Myth as History* (Basic Books, New York).
Levine, G. (1996). What is science studies for and who cares, in A. Ross (ed.), *Science Wars* (Duke University Press, Durham and London), pp. 123–138.
Levins, R. (1996). Ten propositions on science and antiscience, *Social Text* **46/47**, pp. 101–111.

Levitt, N. (1999). *Prometheus Bedeviled: Science and the Contradictions of Contemporary Culture* (Rutgers University Press, New Brunswick and London).
Loader, C. (1985). *The Intellectual Development of Karl Mannheim: Culture, Politics, and Planning* (Cambridge University Press, Cambridge).
Longino, H. (1990). *Science as Social Knowledge: Values and Objectivity in Scientific Inquiry* (Princeton University Press, Princeton).
Mann, G. (1990). *Reminiscences and Reflections: A Youth in Germany*, translated by K. Winston (W. W. Norton & Company, New York and London).
Mannheim, K. (1936). *Ideology and Utopia*, Harcourt Brace & World, London).
Mannheim, K. (1936). *Essays on the Sociology of Knowledge* (Routledge & Kegan Paul, Boston, Massachusetts, London, and Henley, UK).
Manuel, F. E. (1968). *A Portrait of Isaac Newton* (The Belknap Press of Harvard University Press, Cambridge, Mass. and London, UK).
Martin, B. (1988). Mathematics and social interests, *Search* **19**, pp. 209–214.
Masterman, M. (1970). The nature of a paradigm, in I. Lakatos and A. Musgrave (eds.), *Criticism and the Growth of Knowledge* (Cambridge University Press, Cambridge), pp. 59–90.
McGinn, C. (1993). *Problems in Philosophy: The Limits of Inquiry*. (Blackwell, Oxford, UK & Cambridge, USA).
McGinn, C. (1999). *Knowledge and Reality: Selected Essays* (Clarendon Press, Oxford).
Mehrtens, H. (1996). Mathematics and war, in P. Forman and J. M. Sánchez-Ron (eds.), *National Military Establishments and the Advancement of Science and Technology* (Kluwer, Dordrecht/Boston/London), pp. 87–134.
Merton, R. K. (1937). The sociology of knowledge, *Isis* **27**, pp. 493–503.
Merton, R. K. (1980). Science, technology, and society in seventeenth century England, in *Osiris: Studies on the History and Philosophy of Science and on the History of Learning and Culture* (Harper & Row, New York).
Miller, D. (1974). Popper's qualitative theory of verisimilitude, *British Journal for the Philosophy of Science* **25**, pp. 178–88.
Mitchell, W. J. T. (1998). *The Last Dinosaur Book: The Life and Times of a Cultural Icon* (The University of Chicago Press, Chicago).
Mulkay, M. (1979a). *Science and the Sociology of Knowledge* (Allen & Unwin, London).
Mulkay, M. (1979b). Knowledge and utility: Implications for the Sociology of Knowledge, *Social Studies of Science* **9** No 1, pp. 63–80.
Nanda, M. (2000). The science wars in India, in *The Sokal Hoax: The Sham That Shook the Academy*, edited by the editors of *Lingua Franca* (The University of Nebraska Press, Lincoln and London), pp. 205–213.
Newcomb, S. (1911). Mercury, *Encyclopaedia Britannica* Eleventh Edition, Vol. 18, pp. 1895–1896.
Newton-Smith, W. H. (1981). *The Rationality of Science*. (Routledge & Keegan Paul, Boston, London, and Henley).
Nietzsche F. (1968). *The Will to Power*, translated by W. Kaufmann (Random House, New York).

Novick, P. (1988). *That Noble Dream: The "Objectivity" Question and the American Historical Profession* (Cambridge University Press, Cambridge).

Nye, M. (1980). N-rays: An episode in the history and psychology of science, *Historical Studies in the Physical Sciences.* **11**, pp. 126–156.

Parsons, K. M. (2001). *Drawing Out Leviathan: Dinosaurs and the Science Wars* (Indiana University Press, Bloomington).

Parsons, K. M. (2005). *Copernican Questions: A Concise Invitation to the Philosophy of Science* (McGraw Hill, New York).

Pennock, R. (1999). *Tower of Babel.* (Bradford, Cambridge).

Pickering, A. (1984). *Constructing Quarks: A Sociological History of Particle Physics* (Edinburgh University Press, Edinburgh).

Popper, K. (1965). *Conjectures and Refutations: The Growth of Scientific Knowledge* (Harper & Row, New York and Evanston).

Popper, K. (1968). *The Logic of Scientific Discovery* (Harper & Row, New York and Evanston).

Popper, K. (1970). Normal science and its dangers, in I. Lakatos and A. Musgrave (eds.), *Criticism and the Growth of Knowledge* (Cambridge University Press, Cambridge), pp. 50–58.

Popper, K. (1972). *Objective Knowledge: An Evolutionary Approach* (Clarendon Press, Oxford).

Psillos, S. (1999). *Scientific Realism: How Science Tracks Truth* (Routledge, London and New York).

Putnam, H. (1981). *Reason, Truth, and History* (Cambridge University Press, Cambridge).

Quine, W. (1970). On the reasons for indeterminacy of translation, *Journal of Philosophy* **67**, pp. 178–183.

Restivo, S. (1982). Mathematics and the sociology of knowledge, *Knowledge: Creation, Diffusion, Utilization* **4**, pp. 127–144.

Restivo, S. (1988). The social construction of mathematics, *Zentralblatt fur Didaktik der Mathematik* **20**, pp. 15–19.

Restivo, S. (1992). *Mathematics in Society and History* (Kluwer, Dordrecht).

Restivo, S. (1993). Social life of mathematics, in S. Restivo, J. P. Van Bendegem, and R. Fischer (eds.), *Math Worlds* (State University of New York Press, Albany).

Ringer, F. (1992). The origins of Mannheim's sociology of knowledge, in Ernan McMullin (ed.), *The Social Dimensions of Science* (University of Notre Dame Press, Notre Dame), pp. 47–67.

Roll-Hansen N. (1979). Experimental method and spontaneous generation: the controversy between Pasteur and Pouchet, 1859–64, *Journal of the History of Medicine and Allied Sciences* **34**, pp. 273–292.

Rorty, R. (1979). *Philosophy and the Mirror of Nature* (Princeton University Press, Princeton).

Rorty, R. (1991). Is natural science a natural kind?, in *Objectivity, Relativism, and Truth* Vol. 1 (Cambridge University Press, Cambridge), pp. 46–62.

Rorty, R. (1995). Response to Charles Hartshorne, in H. J. Saatkamp, Jr. (ed.), *Rorty & Pragmatism: The Philosopher Responds to His Critics* (Vanderbilt University Press, Nashville), pp. 29–36.

Rose, H. (1996). My enemy's enemy is—only perhaps my friend, in A. Ross (ed.), *Science Wars* (Duke University Press, Durham and London), pp. 80–101.

Ross, A. W. (1991). *Strange Weather: Culture, Science, and Technology in the Age of Limits* (Verso, London and New York).

Russell B. (1999). *The Problems of Philosophy* (Oxford University Press, London and New York).

Salmon, W. (ed.) (2001). *Zeno's Paradoxes* (Hackett Publishing Company, Inc., Indianapolis).

Santillana, G. de (1958). *The Crime of Galileo* (University of Chicago Press, Chicago).

Scheler, M. (1980). *Problems of a Sociology of Knowledge*, translated by M. S. Frings; edited and with an introduction by K. W. Stikkers. (Routledge & Keegan Paul, Boston, London, and Henley).

Schiebinger, L. (ed.) (2008). *Gendered Innovations in Science and Engineering* (Stanford University Press, Palo Alto).

Schoenfeld, A. H. (2004). The math wars, *Educational Policy* **18** No. 1, pp. 253–286.

Searle, J. (1972). *The Campus War* (Penguin Books Ltd., Harmondsworth, Middlesex).

Searle, J. (1995). *The Social Construction of Reality* (The Free Press, New York).

Shanker, S. G. (1987). *Wittgenstein and the Turning-Point in the Philosophy of Mathematics* (Croom Helm, London and Sydney).

Shapin, S. (1982). History of science and its sociological reconstruction, *History of Science* **20**, pp. 157–211.

Shapin, S. (1984). Talking history: Reflections on discourse analysis, *Isis* **75**, pp. 125–130.

Shapin, S. (1995). Here and everywhere: Sociology of scientific knowledge, *Annual Review of Sociology* **21**, pp. 289–321.

Shapin, S. and Schaffer, S. (1985). *Leviathan and the Air Pump* (Princeton University Press, Princeton, New Jersey).

Snow C. P. (1959). *The Two Cultures and The Scientific Revolution* [First American Edition] (Cambridge University Press, New York).

Sokal, A. (1996). Transgressing the boundaries: Toward a transformative hermeneutics of quantum gravity, *Social Text* **46/47**, pp. 217–252.

Sokal, A. (2008). *Beyond the Hoax: Science, Philosophy and Culture* (Oxford University Press, London and New York).

Sokal, A. and Bricmont, J. (1998). *Intellectual Impostures* (Profile Books Ltd., London).

Stove, D. (1982) *Popper and After: Four Modern Irrationalists* (Pergamon Press, Oxford, New York, Toronto, Sydney, Paris, Frankfurt).

Stove, D. (2000.) *Against the Idols of the Age*, edited with an introduction by Roger Kimball. (Transaction Publishers, New Brunswick and London).

Stove, D. (2001a). *Against All Idols*, R. Kimball (ed.), (Transaction Publishers, New Brunswick and London).
Stove, D. (2001). *Scientific Irrationalism: Origins of a Postmodern Cult* (Transaction Publishers, New Brunswick and London).
Thomas, D. W. (1995). Gödel's theorem and postmodern theory, *Proceedings of the Modern Language Association* **110**, pp. 248–261.
Tichý, P. (1974). On Popper's definition of verisimilitude, *British Journal for the Philosophy of Science* **25**, pp. 155–160.
Van Frassen, Bas C. (1990). *The Scientific Image* (The Clarendon Press, Oxford).
Watkins, J. (1999). Feyerabend among the Popperians, in D. Lamb, and G. Munevar (eds.), *Worst Enemy of Science?: Essays in Memory of Paul Feyerabend* (Oxford University Press, London and New York), pp. 47–57.
Watson, T. (1848). *Lectures on the Principles and Practice of Physic; Delivered at King's College, London* (Lea and Blanchard, Philadelphia).
Weinberg, S. (1998). The revolution that didn't happen, *New York Review of Books* Vol. XLV, pp. 48–52.
Wilder, R. L. (1981). *Mathematics as a Cultural System* (Pergamon Press, Oxford, New York, Toronto, Sydney, Paris, Frankfurt).
Wilson, A. L. (1999). *God's Funeral* (W. W. Norton, New York and London).
Winner, L. (1996). The gloves come off: shattered alliances in science and technology studies, in A. Ross (ed.), *Science Wars* (Duke University Press, Durham and London), pp. 102–113.
Wittgenstein, L. (1956). *Remarks on the Foundations of Mathematics*, G. H. von Wright, R. Rhees, G. E. M. Anscomb (eds.), Translated by G. E. M. Anscomb (B. Blackwell, Oxford).
Wittgenstein, L. (1976). *Wittgenstein's Lectures on the Foundations of Mathematics: Cambridge 1939*, C. Diamond (ed.), (The University of Chicago Press, Chicago).
Woldring, H. E. S. (1987). *Karl Mannheim. The Development of His Thought: Philosophy, Sociology and Social Ethics, With a Detailed Biography* (St. Marin's Press, New York).
Woolgar, S. (1986). On the alleged distinction between discourse and praxis, *Social Studies of Science* **16**, pp. 109–117.
Woolgar, S. (1988a). *Science: The Very Idea* (Ellis Horwood, London and New York).
Woolgar, S. (ed.) (1988b). *Knowledge and Reflexivity: New Frontiers in the Sociology of Knowledge* (Sage Publications, London and Beverly Hills).
Woolgar, S. (1992). Some remarks about positionalism: A reply to Collins and Yearly, in A. Pickering (ed.), *Science as Pracice and Culture* (University of Chicago Press, Chicago), pp. 327–341.
Zammito, J. H. (2004). *A Nice Derangement of Epistemes: Post-positivism in the study of science from Quine to Latour* (University of Chigago Press, Chicago).

Index

Ars Characteristica, 178
Demos, 126, 184
Principia, 180, 182, 219
The Mysteries of Udolpho, 224
Theaetetus, 169
modus ponens, 155, 162
ressentiment, 263, 278, 291
1919 solar eclipse expedition, 131

Aristotelian cosmology, 179

Abbé Sieyès, 83
absolutism in mathematics, 157, 161
Acheson, Dean, 18
Acquinas, Thomas, 45, 164
actants, 193
actor-actant network theory, 16, 193
Adorno, Theodore, 49
aether, 179
aether wind, 132
Afrocentrists, 148
Alexander I, 204
Alexander II
  and liberation of the serfs, 43
Animal Liberation Front, 268
Apollonius, 184, 219
Archimedes, 23, 184
Aristotle, 220
Armstrong, Dwight, 247
Armstrong, Karl, 247
Arnold, Mathew, 253
Aronowitz, Stanley, 10, 12, 144

astrology, 40, 96
Atwater, Lee, 178
Augustine of Hippo, 164
axiom of infinity, 160
Azande, 115, 116, 135, 155, 208, 277
Aztecs, 10

Baader-Meinhof, 254
Bacon, Roger, 45
Balmer, Johann J., 67
Banach Fixed Point Theorem, 229
Barnes, Barry, xiv, 82, 107, 112, 113,
  115, 135, 188, 209, 260, 276
Baudrillard, Jean, 10
Beard, Charles, 33, 251
Becker, Carl, 33
Beethoven, Ludwig van, 184
Bell, Daniel, 251
Bellarmine, Cardinal, 94
Benjamin, Walter, 49
Berger, Peter and Luckmann, T.
  Social Construction of Reality, 63
  similarity to Mannheim and
    Merton, 63
Bergson, Henri, 47
Berkeley Campus
  situation of in the 1960s, 91
Berkeley, George, 28, 109, 110, 164,
  229, 276
Bernal, Martin, 148
Big Bang, 66
Bishop, Errett., 239

black box, 119
Blake, William, 22, 29
Blondot, René Prosper, 121
Bloor, David, xi, xiv, 107, 112–116, 136, 139, 141, 142, 180, 183, 188, 209, 236, 262, 276
   and mathematics as inductive reasoning, 152
Boas, Franz, 33
Bohr, Niels, 32, 66, 67
Boole, George, 151
Borgia, Cesare, 265
Bossuet, Bishop, 225
Bourbaki, 21
Boyle, Robert, 117, 139
   and politics, 140
Brezhnev, Leonid, 254
Bricmont, Jean, x, 13, 17, 108, 148, 260, 272
Bridgeman, P. W., 33
Brinton, Crane, 33, 251
Brouwer, Luitzsen E., 164, 239
   and intuitionism, 159
Brown, James Robert
   and affirmative action, 269
Buck, Robert C., 245
Burali-Forti, Cesare, 157
Burt, Leo, 247
Bush, George W., 5, 105

Calculus Liberation Front, 225, 234
Callicles, 27, 126
Callon, Michel, 107, 144, 193
Cantor, Georg, 157
Cardan, Jerome, 40
Carnap, Rudolf, 24
Carter, Jimmy, 254
Cartesian vortices, 179
Cassirer, Ernst, 47
Castigioni, Arturo, 8
CAT scan, 206
CAT scanners, 279
Catiline, 127
Cauchy, Augustin, 157, 166, 219
Cavalieri, Bonaventura, 156
Ceauşescu, Nicolae, 220

Chadwick, James, 8, 200, 206
Charles I, 225
Charles II, 7
Chomsky, Noam, 180
Cicero, 127
Cleon, 127
Cohen, I. Bernard, 18
Cole, Stephen, 136
Collins, Harry, xi, xiv, 18, 107, 111, 128, 201, 202, 209, 276
   and 1919 test of Relativity, 132
   and experimental regress, 131
   and science as a golem, 129
Comte, Auguste, 45, 291
conic sections, 219
Constructivism, 171
   and anti-realism, 171
Copernicanism, 87, 94, 100, 135, 177, 179, 200
correspondence theory of truth, 11
Cosmas of Alexandria, 195
Coulter, Ann, 171
Crawford, Katherine, 255
Critical Legal Studies, 257

d'Alembert, Jean le Rond, 164
Darton, Robert, 256
Darwin, Charles, 13, 250
   and Malthus, 137
Darwinism, 138
de Orco, Remirro, 265
Debray, Régis, 159
Derrida, Jacques, x, 255, 259
Descartes, René, 122, 137, 164, 276
   and vortices, 184
Devitt, Michael, 170
Dewey, John, 31
Diaz, Bernal, 10
Dieudonné, Jean, 21
Dijsterhuis, E. J., xi
Dillinger, John, 101
Dilthey, Wilhelm, 46
Diophantus, 156, 220
Dirac, Paul, 67
discourse analysis, 192
Divine Right of Kings, 225

DNA molecule
  double helix structure of as
    rhetoric, 127
Douglas, Helen Gahagan, 248
Douglas, Mary, 14
Dr. Spock, 246
Duhem, Pierre, 111, 275
Duhem-Quine thesis, 11, 30, 40, 111, 131
Duns Scotus, 226
Durkheim, Emile, 47
Dyke, Bill, 246, 247

Eagleton, Terry, 5
Earnest, Paul, 233
Eddington, Arthur, 131, 200
Edward VI, 40
Einstein, Albert, 9, 128, 132
Elagabalus, 179
Eleatics, 164
Elkana, Yehuda, 14
Engels, Friedrich, 46
epicycles, 203
epistemological chicken, 192
Ernest, Paul, xiv, 157, 161, 164, 230, 234, 236, 265, 275
  and Gödel's Incompleteness Theorems, 158
  and elitism, 159
  and fallibility of mathematics, 157, 228
  and set theoretic paradoxes, 157
Euclid, 23, 184
Eudoxus, 219
Euler-Descartes formula for polyhedra, 166
Eurocommunism, 254
Evans-Prichard, Edward Evan, 116
Evolution, 77, 97, 138, 189
  as encoding bourgeois values, 13, 250

fallibility
  and mathematics, 234
false consciousness, 46
Fanon, Franz, 254

Fara, Patricia, 142, 184
Faraday, Michael, 206
Farley, J., 112, 138
Fassnacht, Robert, 247
Feigl, Herbert, 94
Fermat's Last Theorem, 161, 216
Feyerabend, Paul, xi, xiv, 1, 107, 109, 118, 128, 135, 177, 189, 207, 248
  and incommensurability of traditions, 100
  and atoms, 95
  and dadaism, 101
  and epistemological equivalence of traditions, 100
  and Galileo, 89
  and incommensurability, 90
  and intellectual anarchism, 101
  and Karl Popper, 89, 101
  and multiculturalism, 103
  and nonexistence of tradition-independent beliefs, 99
  and normal science, 101
  and objectivity, 95
  and Olympian gods, 95
  and Protagoras, 98
  and relativism, 98
  and science as dogma, 96
  and scientific method, 94
  and scientific realism, 89
  and separation of science from the state, 98
  and teaching, 91
  and the Vienna Circle, 89
  and theory choice, 91
  and theory-ladeness of observational statements, 90
  and traditions, 95
  and witch-doctors vs modern medicine, 97
  Kuhn, influence of, 92
  personality of, 88
  Popper, rebellion against, 93
  successful career of, 102
Fine, David, 247

Fish, Stanley, 28, 257
Fisher, Charles S., 241
flat earth cosmology, 235
flatlanders, 231
Fleck, Ludwik, xi, xii, xiv, 69, 91, 170, 189, 206
  and active associations, 61
  and facts, 61
  and passive associations, 61
  and thought collectives, 58–60
  and thought styles, 58–60
  and truth, 60
  as a precursor to postmodernism, 63
  life of, 57
formalism, 160, 280
Forman, Paul, xi, 158, 259, 261, 290
  and Weimar science, 87, 138
Forms
  criticism of, 280
Foucault, Michel, x, 31, 255, 259
foundationalism, 161
Four Color Theorem, 161, 163
Free Speech Movement, 260
Frege, Gottlob, 153, 164, 215
Freud, Sigmund, 46
Fromm, Erich, 49
Fuller, Steve
  and criticisms of Darwinism, 274
  and James Conant, 272
  and Kuhn, 272
  and Kuhnification, 273

Gödel's Incompleteness Theorems, 158, 160, 165, 233, 290
  postmodern consequences of, 158
Galen, 205–207
Galileo, 200, 264
Gardiner, Martin, 229, 237
Gauss, Karl Friedrich, 23, 184
Geertz, Clifford, 257
Geison, Gerald, 112, 138, 141
German historicism, 46
Gilbert, William, 94
Gillespie, C. Coulton, 18
Glasersfeld, Erich von, 225, 265

Goering, Hermann, 3
Goethe, Johann W. von, 22, 29
  and hostility to Newtonian science, 211
Goldberg, Harvey, 249
golem, 129
Goodman, Nelson, 30, 257
  and many worlds, 191
Gorgias, 28, 126
Gould, Stephen J., 77
Gramsci, Antonio, 267
gravitational waves, 130
Grenz, Stanley, J., 29
Grinell, George, 142, 183, 184
Gross, Alan, 126
  and science as rhetoric, 127
Gross, Paul, x, 10, 17, 108, 148, 188, 194, 260, 272, 286
Guevara, Che, 254
Guillemin, Roger, 117
Gumplowicz, Ludwig, 47

Hacking, Ian, 135, 286
Haggard, H. Rider, 222
Hahn, Otto, 200, 206
Halmos, Paul, xii
Hamas, 175
Hamilton, William Rowan, 141
Hanson, Norwood Russell, 1, 69, 109, 111
Harding, Sandra, 13, 145, 211, 212
Hardy, Godfrey H., 149, 223, 231, 236
Heaviside, Oliver, 206
Hegel, Georg Wilhelm, 29, 46, 143, 259
  and the World Spirit, 29
Heidegger, Martin, 22, 259
Heilbron, J. L., xi
Heisenberg Uncertainty Principle, 33
Heisenberg, Werner, 67
Helms, Jesse, 22
Hempel, Carl, 24
Henry VIII, 40
Henry, John, 107, 112
Heraclitus, 251
Hermeticism, 235

Heroditus, 28, 103
Hersh, Reuben, xiv, 160, 163, 164, 217, 234, 236, 281
   and absolutism, 161
   and elitism, 162
   and Humanism, 161
   and mathematical fallibility, 163
   mathematics and politics, 164
Hertz, Heinrich, 206
Hess, David, 266
   and bacteria as a cause of cancer, 271
   and the deciding of scientific disputes, 271
Hesse, Mary, 188
Hessen, Boris, 53, 250
Hilbert, David
   and formalism, 160
Hiss, Alger, 248
Hobbes, Thomas, 139, 221
Horkheimer, Max, 45, 46
Horowitz, David, 105
House Unamerican Activities Committee (HUAC), 92
Hufbauer, Karl, xi
Hume, David, 10, 28, 41, 164, 276
   and induction, 10, 37, 44
   and skepticism, 45

idealism, 111, 136
incommensurable, 1, 12, 16, 51, 68, 90, 100–102, 171, 172, 174, 175, 189, 207, 221, 253, 255, 256, 280, 285
inscription device, 118, 119, 126
intuitionism, 159, 217
Ionesco, Eugene, 20
Irigaray, Luce, 10, 146
Iron Guard, 3

Jacobins, 204
James, William, 31
Jameson, Frederick, 10, 171
Jansenism, 291
Jasanoff, Sheila, 190, 274
Jassy, 3

Jenkin, Fleeming, 177
John of Leyden, 100
Johnson, Dr., 229
Johnson, Lyndon, 248
Johnson, Phillip, 273
Jourdain, Monsieur, 9

Kant, Immanuel, 28, 109, 164, 228, 231, 252, 276
Kantianism, 137
Keller, Evelyn Fox, 147, 210–212
Kennewick man, 189
Kerensky, 143
Kimball, Roger, 249
Kipling, Rudyard, 34
Knorr-Cetina, Karin, 107
   and facts as social constructions, 111
Koch, Robert, 121, 179
Koertge, Noretta, x, 10, 17, 108, 148, 264, 270
Koyré, Alexandre, 18, 65, 85, 275
Kristeva, Julia, 159
Kruif, Paul de, 8
Krushchev, Nikita, 253
Kuhn, Thomas, ix, xi, xii, 1, 18, 40, 107, 109, 111, 128, 133, 135, 136, 170, 201, 203, 250, 255
   and anomalies, 66
   and analogies between scientific and political revolutions, 69
   and analogies to Darwinian evolution, 76
   and anti-realism, 75
   and Aristotle, 64
   and Gestalt shifts, 72, 79
   and incommensurability of, 68
   and James B. Conant, 64
   and John H. van Vleck, 64
   and Ludwik Fleck, 85
   and many worlds, 80
   and neo-Kantianism, 77
   and nominalism, 79
   and normal science, 65, 66
   and paradigms, 65
   and paradigms shifts, 68

and Quine, 70
and relationism, 75, 78
and relativism, 74, 80
and Sapir-Whorf thesis, 85
and scientific progress, 73
and semantic incommensurability, 70, 175
and sociology, 81
and stimuli, 79
and the actual world, 79
and the phenomenal world, 78
and truth, 73, 76, 116, 207
Kun, Béla, 47

Lévy-Bruhl, Lucien, 47
Lacan, Jaques, 10
Lagrange Four Square Theorem, 229
Lakatos, Imre, xiv, 83, 155, 164, 229, 233, 234, 236
   and conjectures and refutation in mathematical development, 167
   and deductivism, 166
   and mathematical fallibility, 166
   and mathematics, 165
   life of, 165
Landau, Edmund, 244
Langer, Rudolf, 245
Langer, Suzanne, 245
Langer, William L., 245
Laqueur, Thomas, 256
Latour, Bruno, xi, xiv, 3, 18, 107, 111, 117, 131, 133, 136, 144, 188, 191, 193, 225, 258, 276, 289, 291
   and actor-network theory, 16
   and anthropological approach to science, 118
   and fabrication of microorganisms, 121
   and historicity of things, 195
   and inscription devices, 118
   and Kant, 123
   and laboratory warfare, 119
   and N-rays, 121
   and nonexistence of TB before Koch, 121
   and Pasteur, 124
   and Rhetoric, 126
   and subject-object dicotomy, 122
   and Third Rule of Method, 11, 122
   and Whig history, 122
Latour, Hilbert Archimède, 196
Laudan, Larry, 208
Lavosier, Antoine, 66
Leavis, Frank R., 22
Leibniz, Gottfried Wilhelm, 73, 157, 177
Lenard, Phillipe, 211
Lenin, Vladimir, 143
Lentricchia, Frank, 5
Levitt, Norman, x, 10, 17, 108, 148, 188, 194, 260, 272, 286
Lewis, M. L., 108
Lewis, Wyndham, 19
Liebig, Justus, 124
Linderholm, Carl, 21
Locke, John, 164
logic
   as a social convention, 155
   nature of, 218
logicism, 160, 280
Longino, Helen, 147, 210–212
Lord Kelvin, 177, 206
Louis XIV, 225
Lovejoy, A. O., 85
Lowell, Percival, 198
Lukáks, Georg, 47
Luther, Martin, 83, 100
Lyotard, Jean-Francois, 31, 259
Lysenko, Trofim., 204

Münzer, Thomas, 100
Machiavelli, 265
Maier, Anneliese, 65, 85
Mainstream philosophy of mathematics, 161
Malthus, 181
Mann, Golo, 248
Mannheim, Karl, xi, xii, xiv, xv, 15, 76, 109, 113, 144, 276
   and *Institut für Sozizlforschung*, 49
   and false consciousness, 53

and free floating intellectuals, 51
and ideology
    particular theory of, 50
    total theory of, 50
and mathematics, 54
and reflexivity, 51
and relationism, 52, 210
and science, 47
and sociology of knowledge (SK), 49
and unmasking, 50
life of, 47, 49
Manuel, Frank, 142
Marconi, Guglielmo, 206
Marcuse, Herbert, 49, 250, 257
Martian canals, 198
Marx, Karl, 29, 46, 267
and ideology, 50
Marxism, 39, 40
Masterman, Margaret, 65
Math Wars, 265
mathematical necessity as moral compulsion, 154
mathematics
    as a social convention, 155
Mathematics Research Center (MRC), 243, 244, 268
    and 1970 bombing of, 246
    and riot against, 244
Maxwell's Laws, 290
Maxwell, James Clerk, 206
McCarthy, Joe, 246
McGinn, Colin, xiv, 281
McGovern, George, 248
Meister, Joseph, 141
Mendel, Gregor, 264
Mendeleev, Dmitri, 67
Mercury
    perihelion of, 112
Merton, Robert K., 54, 114, 266
Mesmerism, 235
metaphysical naturalism, 13, 273
Metzger, Hélène, 85
Meyerson, Emile, 65, 85
Michelson-Morley experiment, 132
Mill, John Stuart, 164

and mathematics, 153
Millikan, Robert H., 112, 199–201, 264
modus ponens, 233
monads, 191
Moore, Michael, 171
Morland, Catherine, 224
Mulkay, Michael, xiv, 107, 108, 110, 192
    and critique of technology, 205
Mumford, Lewis, 267

N-rays, 121, 122
Nanda, Meera, 13
Napoleon III, 125
Natural Selection
    anomaly of, 273
Nechayev, 37
neo-Kantianism, 172, 276
neo-Platonism, 87
neoconservatism, 235
Neuman, Alfred E., 196, 224
neutrons, 7, 8
Newton, Isaac, xiv, 9, 23, 117, 132, 142, 219, 250, 264
    and *Principia*, 182
    and Cartesian vortices, 184, 185
    and mathematics, 183
    and politics, 140
Newton-Smith, William H., 39, 41, 43
Nietzsche, Friedrich W., 23, 29, 31, 259
Nim Chimpsky, 180
Nixon, Richard, 248
noumenal world, 170
Novick, Peter, 251

O'Brian, 220
Oberwolfach, 3, 4
Occasionalism, 250
Orwell, George, 239
Osiander, Andreas, 89

Palestine, 175
Paracelsus, 205
    and feminine values, 147

paradigms
  incommensurability of, 12, 68–70, 85, 171, 175, 176
Pareto, Vilfredo, 46
Parmenides, 280
Pasteur, Louis, xiv, 9, 112, 117, 181, 196
  and political beliefs, 138
  and rabies vaccine, 141, 142
  and spontaneous generation, 138
Pasteur-Pouchet controversy, 138, 141, 182, 209
Peacock, George, 141
Persepolis State University, xvi, 2, 23, 198, 215, 222, 223, 225, 227, 234, 235, 262, 263, 280, 290
PET scan, 206
PET scanners, 279
phenomenal world, 170
phlogiston, 66
Piaget, Jean, 65, 85
Pickering, Andrew, 180, 262, 276
  and quarks, 134
Pinch, Trevor, 107, 129, 276
Planck, Max, 67
Plato, 45, 126, 164, 215, 231, 235
  and geometry, 184
Platonism
  and mathematics, 160
Poincaré Conjecture, 161
Pol Pot, 254
Polanyi, Michael, 1
Ponzi scheme, 24
Popper, Karl, xiv, 2, 93, 111, 164, 165, 237
  and falsifiability, 38
  and Jazz Age, 37
  and relativism, 42
  and scientific knowledge, 38
  and demarcation between science and non-science, 39, 40, 42
  and falsifiability, 66, 67
  and induction, 37
  and verisimilitude, 41
  and Vienna Circle, 43
  and Worlds 1, 2, and 3, 42

Porter, Cole, 37
positrons, 6, 206, 279
postmodern interpretation(s) of science (PIS)
  and anti-intellectualism, 261
  and criticisms of Darwinism, 273
  and criticisms of science, 259
  and education, 264
  and hostility to objectivity, 258
  and hostility to the Enlightenment, 258
  and hostility to truth, 258
  and politics, 260
  and SSK, 15, 258
  as a socially constructed ideology, 260
Pouchet, Felix, 207
  and spontaneous generation, 112
Pound, Ezra, 19
Poverty of Christ, 291
primes, 196
  infinity of, 185
prosthaphaeresis, 219
psychoanalysis, 39, 40
Ptolemaic astronomy, 135, 177, 202
Ptolemy, 220
Putnam, Hillary, 30, 201, 257
Pyrron, 28
Pythagoras, 184
Pythagorean Theorem, 185
Pythagoreans, 164

quantum mechanics, 117
Quine, Willard van Orman, xi, xv, 30, 111, 164
  and indeterminacy of translation, 112

Röntgen, Wilhelm C., 206
Radcliffe, Mrs., 224
Radon transform, 206
Radon-Nikodym Theorem, 225, 229
Rameses II, 121, 194
Ranke, Leopold von, 256
Red Brigades, 254
Reichenbach, Hans, 24

Relativity, 132
Restivo, Sal, xiv, 150, 236
   and George Boole, 151
   and logic as an inductive inference, 151
   and Marxism, 152
   and mathematics as a social construction, 151
   and pure mathematics as a political strategy, 151
   and relativism, 152
Rhazes, 209
Rhind papyrus, 219
Richard, Jules, 157
Robespierre, 2
Robinson, Abraham, 161
Roche, George C., 249
Rorty, Richard, xi, 14, 30, 143, 201, 258, 259
Ross, Andrew, 10, 258, 259
Rosser, Anetta, 245
Rosser, J. Barkley, 245
Roux, Emile, 141, 142
Rove, Karl, 178
Rubin, Jerry, 92
Russell, Bertrand, 37, 39, 157, 160, 164
Rutherford, Ernest, 8, 67

Sapir-Whorf thesis, 70, 175
Sarton, George, 18, 65
Satan, 13
Savio, Mario, 92
Schaffer, Simon, xiv, 12, 107, 113, 137, 139
Scheler, Max, 47
Schwartz, Laurent, 161
science
   a tool of the ruling elites, 13
   and contextual factors, 12
   and ideology, x
   and induction, 11
   and magic, x
   and oppression, x, 14
   nonobjectivity of, 13
   undetermined by nature, 11

Science for the People, 243, 257, 268
Science Wars, x, 10, 16, 27, 35, 108, 148, 263, 266, 272, 286, 289
Science, Technology, and Society (STS), 266
   and affirmative action, 269
   and politcal control of science, 268
   and science policy, 268
   recent growth of, 16
Searle, John, 92, 174
Serres, Michel, 159
Sextus Empiricus, 28, 276
Seymour, Jane, 40
Shapin, Steve, xi, xiv, 12, 18, 107, 113, 137, 139, 262, 276, 291
Siger of Brabant, 45
Simmel, Georg, 47
Snow, Charles P., 5, 19, 22
social construction, 133
   and science, 285
sociology of knowledge (SK), 46, 47
   and unmasking, 54, 289
sociology of scientific knowledge (SSK), 107, 264
   and affirmative action, 288
   and anti-realism, 109
   and correspondence theory of truth, 109, 190
   and difficulty with facts, 196
   and falsification, 111
   and mathematics, 216
   and neo-Kantian Constructivism, 197
   and non-uniqueness of theories, 112
   and relativism, 135
   and science, 284
   and science policy, 287
   and scientific prediction, 199
   and social construction of science, 113
   and social roots of linguistic meaning, 197
   and technology, 205
   and theory-laden obsevations, 109
   and unanimity of scientific belief, 203

main theses of, ix, 108
Socrates, 126, 225, 280
Soglin, Paul, 246, 247
Sokal hoax, xii, 9
Sokal, Alan, x, 9, 12, 17, 108, 148, 188, 194, 260, 272, 286
solipsism, 173
sophist, 27
sorites paradox, 155
Spencer, Herbert, 291
Spengler, Oswald, 218
   and mathematics, 156
Spinoza, Baruch, 164
Sproul, Robert Gordon, 92
SSK/PIS
   and science policy, 279
   political goals of, 279
Stark, Johannes, 211
Stevenson, Adlai, 92
Stove, David, 37, 43
Strong Program (SP), xiv, 113, 180, 227
   and causality, 114
   and causes of belief, 114, 117
   and impartiality, 114
   and reflexivity, 114
   and relativism, 135, 188
   and self-negation, 187
   and symmetry, 114
   and truth, 116, 188
   criticisms of, 178, 186, 187
student unrest
   causes of, 246
subject-object distinction, 137

Talleyrand, Charles Maurice, 250
Tarski, Alfred, 21
TEA laser, 130
Teichmüller, Oswald, 216, 244
Tenochtitlan, 10
Thackray, Arnold, x
theory-laden observations, 111
Third Man Argument, 238, 280
Thomas, D. W., 159
thought style, 91
Thrasymachus, 27, 31

Thurmond, Strom, 22
Thyrotropin Releasing Factor (TRF), 117
Title IX
   and science policy, 270
Torquemada, Tomás, 189
Tracy, Destutt de, 289
Troelsch, Ernst, 47
two world ontology, 170
Tycho de Brahe, 200

Ulam, Stan, 200
universal characteristic, 73
Uranus, 112
USSR
   erosion of moral authority of, 253

Verkhovensky, Stepan, 37, 41
Vespucci, Amerigo, 195
Vietnam era
   and emergence of radical faculty, 255
   and intellectual changes, 250
   and radical faculty, 249
   and radical scholarship, 251
   and student unrest, 246
Virilo, Paul, 159

Wahhabism, 204
Wallis, John, 156, 221
Weber, Joseph, 130, 131
Weierstrass, Karl, 157, 219
Weimar physicists
   and quantum mechanics, 181
Weinberg, Stephen, 126
Whewell, William, 65
White, Hayden, 255, 257
Wiles, Andrew, 216
William of Ockham, 28, 45, 226
Williams, William Appleman, 249
Wilson, Andrew N., 207
Wilson, Harold, 107
Wittgenstein, Ludwig, xiv, 149, 154, 164, 217, 222, 230
   and contradiction in mathematics, 233

and language games, 153
and mathematics as a form of life, 153
Wood, Robert, 121
Woolgar, Steve, 107, 109, 111, 118, 128, 133, 136, 150, 169, 191, 193, 202, 276
  and idealism, 110
  and pulsars, 110
  and representations, 110

World 3, 237
Wright, Frank Lloyd, 246

Yeats, Francis Butler, 19
Yudkin, Michael, 20

Zammito, John H., xv
Zeno, 280
Zermelo-Fraenkel axioms, 158, 160